Dietary Phenylalanine and Brain Function

Dietary Phenylalanine and Brain Function

Edited by
Richard J. Wurtman
Eva Ritter-Walker

Birkhäuser
Boston · Basel

Richard J. Wurtman, M.D.
Eva Ritter-Walker
Department of Brain and Cognitive Sciences
Massachusetts Institute of Technology
Cambridge, MA 02139
U.S.A.

Library of Congress Cataloging-in-Publication Data
Dietary phenylalanine and brain function/edited by Richard J. Wurtman and Eva Ritter-Walker.
p. cm.
Based on papers presented at a conference held in Washington, D.C. on May 8–10, 1987; sponsored by the Center for Brain Sciences and Metabolism Charitable Trust.
Includes bibliographies.

1. Phenylalanine—Physiological effect—Congresses. 2. Aspartame–
–Physiological effect—Congresses. 3. Brain—Effect of drugs on–
–Congresses. 4. Phenylketonuria—Nutritional aspects—Congresses.
I. Wurtman, Richard J., 1936– . II. Ritter-Walker, Eva.
III. Center for Brain Sciences and Metabolism Charitable Trust.
[DNLM: 1. Brain—physiology—congresses. 2. Phenylalanine–
–congresses. 3. Phenylketonuria—diet therapy—congresses. WD
205.5.A5 D565 1987]
QP562.P5D54 1988
612′.822—dc19
DNLM/DLC
for Library of Congress 87-27580

CIP-Kurztitelaufnahme der Deutschen Bibliothek
Dietary phenylalanine and brain function / ed. by
Richard J. Wurtman and Eva Ritter-Walker.—
Boston; Basel: Birkhäuser, 1988.

NE: Wurtman, Richard J. [Hrsg.]

Softcover reprint of the hardcover 1st edition 1988

ISBN-13: 978-1-4615-9823-7 e-ISBN-13: 978-1-4615-9821-3
DOI: 10.1007/978-1-4615-9821-3

Typeset by Asco Trade Typesetting Ltd., Hong Kong.

9 8 7 6 5 4 3 2 1

Preface

This volume contains the manuscripts of the full papers and posters presented at the conference "Dietary Phenylalanine and Brain Function," which took place at the Park Hyatt Hotel, Washington, D.C., on May 8–10, 1987. The conference was organized by a committee that included Drs. Louis Elsas (Emory University, Atlanta), William Pardridge (UCLA), Timothy Maher (Massachusetts College of Pharmacy), Donald Schomer (Harvard), and Richard Wurtman (MIT). It was sponsored by the Center for Brain Sciences and Metabolism Charitable Trust, a foundation which, during the past few years, had also organized seven other conferences related to interactions between circulating compounds (drugs, nutrients, hormones, toxins) and brain function. The Center's most recent other conferences were on "Melatonin in Humans" (Vienna, Austria; November 1985) and "The Pharmacology of Memory Disorders Associated with Aging" (Zurich, Switzerland; January 1987).

The decision to organize this conference was based on the perception that major changes had recently occurred in society's uses of phenylalanine and phenylalanine-containing products, and on the belief that a meeting of scientists and physicians who work on the amino acid's neurological effects could both catalyze additional research on these effects and assist regulatory bodies in formulating appropriate public policies relating to the use of these products: phenylalanine, in both its L- and D-forms, has apparently become a popular sales item at "health-food" stores, and thus is now being consumed by a fairly large number of people, in the absence of the other neutral amino acids which invariably accompany it in protein and which suppress its entry into the brain. Moreover, a very much larger number of people are now consuming phenylalanine, without these other amino acids, as the major constituent of the artificial sweetener aspartame. There is increasing recognition that an inborn impairment in the metabolism of phenylalanine [i.e., heterozygosity for the phenylketonuria (PKU) gene] is fairly common in the general population, as well as increasing evidence that phenylalanine, besides being neurotoxic in high concentrations, can, in concentrations much lower than those associated with clinical PKU, affect the production of neurotransmitters (and might thereby modify any of the numerous brain functions that depend on these compounds). Some years ago, the expanding public use of another pure amino acid, L-tryptophan, had spawned an organization—ISTRY—which has since held periodic meetings to discuss that amino acid's effects and its safety; apparently, no similar organization exists for phenylalanine, even though very many more people now consume it (as a constituent of aspartame) than have ever taken tryptophan, and even though a distinct neurological syndrome (i.e., PKU) is known to be associated with an excess in its brain levels.

The conference included papers concerned with four main fields: the effects of oral phenylalanine (or aspartame) on plasma (human) and brain (rat) phenylalanine levels, and on those of various breakdown products of phenylalanine; effects of the amino acid on brain composition and on particular brain functions in experimental animals; the behavioral and neurological effects of phenylalanine in humans; and the processes through which the sales of phenylalanine and aspartame to the public are currently regulated. A partial list of some of the individual topics that were covered includes:

1. Factors affecting the plasma phenylalanine level, e.g., its changes after consumption of dietary proteins, carbohydrates, aspartame, and phenylalanine itself; its variations in metabolic diseases (hepatic; renal; endocrine) and in genetic disorders affecting phenylalanine metabolism.
2. Factors affecting the "plasma phenylalanine ratio"—the ratio of the phenylalanine concentration to the summed concentrations of the other large neutral amino acids that compete with phenylalanine for brain uptake—which presumably determines brain phenylalanine levels.
3. Brain phenylalanine levels: evidence that they vary with the plasma phenylalanine ratio; their regional differences; effects on them of dietary proteins, carbohydrates, aspartame, phenylalanine, or other neutral amino acids; β-adrenergic mechanisms.
4. Systems mediating the uptake of phenylalanine into nerve terminals, neuronal cell bodies, and brain cells other than neurons.
5. Plasma and brain tyrosine levels in rats and humans after various phenylalanine or aspartame doses; the extent to which tyrosine protects the brain from damaging effects of excess phenylalanine (or, conceivably, enhances those effects); ways of calculating the phenylalanine dose for experimental rodents that would be equivalent (in terms of its relative effects on brain phenylalanine and tyrosine) to a particular dose for humans.
6. Mechanisms of brain cell death in PKU: Is the toxin phenylalanine, or a phenylalanine metabolite, or is the condition caused by the scarcity

of tyrosine? What is the threshold plasma phenylalanine level (or ratio) associated with cell death—and how might this be experimentally determined? Are particular groups of brain cells especially vulnerable?

7. The epidemiology and natural history of homozygous and heterozygous PKU, and of other forms of hyperphenylalaninemia in the general population; the genetic basis of PKU.
8. Neurochemical effects of elevated plasma phenylalanine levels and ratios: in vivo effects on the synthesis and release of monoamine (and other) neurotransmitters; other in vivo effects (e.g., on energy metabolism; on other enzyme systems; on protein and nucleotide synthesis); in vitro effects of phenylalanine—alone or in the presence of other amino acids—on cultured cells and brain slices.
9. Behavioral and neurochemical effects of phenylalanine or aspartame in normal people (e.g., on plasma DOPA; on cognitive functions; on appetite); in people with PKU or other types of hyperphenylalaninemia; effects in experimental animals; correlations with changes in brain composition.
10. Physiological and pharmacologic effects of phenylalanine: the expectations of people who obtain it at "health-food" stores (e.g., alleged antidepressant or analgesic effects) and the evidence, if any, supporting these expectations; its central and peripheral actions; the effects of phenylalanine or aspartame in people taking drugs which modify phenylalanine's metabolism (e.g., monoamine oxidase inhibitors) or which compete with it for brain uptake (e.g., L-DOPA; alpha-methyldopa).
11. Phenylalanine, aspartame, and seizure susceptibility in experimental animals and humans.
12. Phenylalanine, aspartame, and headaches and other central nervous system effects.
13. Alleged side effects of phenylalanine or aspartame, as reported by consumers or physicians to the Centers for Disease Control (CDC), the Food and Drug Administration (FDA), and individual investigators; the likelihood that the effects attributed to aspartame are or are not related to its phenylalanine content; estimates of the frequency of side effects in the general population and of the amounts of pure phenylalanine or aspartame that actually are consumed; and attempts to identify at-risk populations.
14. Regulatory aspects of the sale of phenylalanine in "health-food" stores, and of aspartame to the general public: What can be learned from the history of aspartame's introduction into the general food supply (i.e., recognizing that this sweetener is the first widely used new food additive in decades to have been admitted to the United States food supply, and that the laws relating to the safety assessment of new food additives predate the "Thalidomide Revolution" in drug regula-

tion)? What are the actual and ideal roles for the FDA in regulating the use of these compounds? How does the FDA assess possible pharmacologic and, specifically, neurochemical effects of new food additives, and do such procedures examine the changes they might cause in *thresholds* for undesired responses (like seizures) or only their ability to produce such responses? On what types of tests is the allowable daily intake (ADI) for a food additive like aspartame based, and do these tests adequately consider its possible brain effects? How should foods containing supplemental phenylalanine (as aspartame) be labeled? What systems do and should exist for monitoring possible adverse reactions to phenylalanine or aspartame, and for continuously updating the FDA's assessment of their safety?

Besides formal lectures on such topics, the conference also included a poster session, at which investigators currently carrying out related research had the opportunity to summarize that research. A few of the posters were also presented orally, in summary form, as part of a plenary session; extended abstracts of all of the posters are published in this volume.

The organization under whose aegis this conference was organized, the Center for Brain Sciences and Metabolism Charitable Trust, is a nonprofit, tax-exempt, United States foundation chartered in the public interest to promote the growth of knowledge in the scientific fields encompassed within its name. All of the funds for the Center's activities are derived from contributions from industry and from other foundations; no funds are ever solicited or accepted from any governmental source. A major goal of the Center's meetings is to facilitate interactions between academic scientists and their counterparts in industry; this latter group has the skills, the resources, and the responsibility to convert new laboratory observations into products that benefit the public. We hope the reader will share the organizing committee's belief that the present conference has been useful, both for its discussions of science and for its focus on the public health aspects of new food additives and individual amino acids.

RICHARD J. WURTMAN

Contents

X. Summary

Part I Factors Affecting Blood Phenylalanine Levels

CHAPTER 1

Control of Plasma Phenylalanine Levels

Benjamin Caballero and Richard J. Wurtman*

Virtually all of the molecules of the essential amino acids that enter the body derive from dietary proteins; high-quality protein sources such as eggs or milk provide around 5 g of phenylalanine (Phe) per 100 g of protein. A large fraction of dietary Phe is hydroxylated to tyrosine in the liver, a process catalyzed by the enzyme Phe-hydroxylase. Plasma Phe levels, like those of the other amino acids, are not regulated, and at any given time they reflect the proportions of protein and carbohydrates in the meal most recently consumed. Since people eat during the daytime, there is an apparent circadian rhythmicity in plasma amino acid concentrations. Protein consumption tends to decrease brain Phe concentrations because although it raises plasma Phe, it increases much more the level of the other neutral amino acids (e.g., valine, leucine, isoleucine, tyrosine, and tryptophan), which compete with Phe for brain uptake. In contrast, consumption of pure Phe or of aspartame increases brain Phe because of the lack of these competing amino acids. In rodents, which have a very active Phe-hydroxylating system, administration of pure Phe or aspartame causes larger increases in plasma and brain tyrosine than in Phe, unless very high doses are given. This does not occur in humans, because Phe hydroxylation is much slower.

The plasma phenylalanine (Phe) pool, like that of other amino acids, is very small compared with the total amount of Phe in the tissues or with the daily Phe turnover. Plasma Phe concentrations are subject to wide variations subsequent to protein intake. However, as discussed below, these variations normally have little or no effect on brain Phe because they are buffered by parallel changes of the other plasma amino acids that compete with Phe for brain uptake (Fernstrom and Faller 1978).

Phe is an essential amino acid, and its requirement, among healthy adults, is estimated at 1 to 2 g per day, equivalent to about 50 g of high-quality protein. When dietary protein is digested and its constituent amino acids absorbed, a large portion of Phe is taken up from the portal blood into the liver; a portion of this Phe is utilized for hepatic protein synthesis,

*Department of Brain and Cognitive Sciences and Clinical Research Center, Massachusetts Institute of Technology, Cambridge, MA 02139, USA.

and over 95% of the rest is hydroxylated to tyrosine (Tyr). Endogenously released Phe (from tissue proteins, catabolized during periods of fasting) is disposed of by the same biochemical pathway.

Phe is one of the large neutral amino acids (LNAA), a group that also includes the aromatic amino acids tyrosine and tryptophan; the branched-chain amino acids valine, leucine, and isoleucine; and such other amino acids as threonine and methionine. All of the neutral amino acids are transported into the brain by a common facilitated diffusion system present in the capillary endothelial cells comprising the blood-brain barrier, and Phe must compete with these other LNAA to be affected by this transport system. Hence, even when plasma Phe levels remain constant, an increase or decrease in the plasma concentration of the other LNAA will decrease or increase brain Phe levels. Moreover, since the affinity of Phe for the LNAA transport binding sites is somewhat greater than that of the other LNAA, ingestion of pure Phe can have a disproportionately greater effect in reducing brain levels of these other LNAA. The aromatic amino acids relate directly to neurotransmitter synthesis; tyrosine and tryptophan as precursors for the monoamines, and Phe as an inhibitor of the monoamine-synthesizing hydroxylase enzymes. The other LNAA also have an important effect on brain neurotransmitter synthesis by determining the rates at which the aromatic amino acids are allowed to pass from the blood into the brain.

The flux of Phe into and out of peripheral tissues is largely hormone dependent. Insulin secreted in response to food intake facilitates the uptake of phenylalanine into skeletal muscle, with a consequent reduction in plasma Phe levels (Schauder et al. 1983, Felig and Wahren 1971). Hormones that enhance protein breakdown, such as some of the steroid hormones, increase the output of amino acids from peripheral tissues into the plasma. Urinary excretion is a minor route of Phe disposal, accounting for only 120 mg of Phe per day in adults consuming a balanced diet and with average level of physical activity (Tewksbury and Lohrenz 1970).

Phenylalanine Metabolism

The liver disposes of most excess Phe by converting it to Tyr. Although products of other pathways of Phe catabolism, such as phenylpyruvic acid and acetyl-phenylalanine, can be detected in the plasma or urine under conditions when the conversion of Phe to Tyr is impaired (e.g., in phenylketonuria), Phe levels rapidly increase to the toxic range, causing tissue damage.

Phe-hydroxylase is a mixed-function oxidase present only in the liver. There are marked species differences in the basal activity of this enzyme, which is highly substrate inducible (Kaufman 1986). Isotopic studies in rats show that the rate of conversion of Phe to Tyr in the fasting state is about 75 μmol/kg per hour; this process contributes about 20% of the tyrosine entering the circulation (Moldawer et al. 1983). When fasting plasma Phe levels are increased eight fold, the conversion of Phe to Tyr increases, contributing over 70% of the tyrosine appearing into the plasma (Moldawer et al. 1983). Clarke and Bier (1982), using [$^2H_5$]Phe and [^{13}C]Tyr infusions, reported a hydroxylation rate of only 6 μmol/kg per hour in humans in the postabsorptive state. This difference is in agreement with the finding that after a Phe load, rats exhibit a major plasma Tyr peak (larger, in fact, than the Phe peak itself) whereas Phe loading in man produces the largest increase in Phe itself (Yokogoshi et al. 1984). This species difference in the ability to dispose of Phe is a major obstacle in using the rat to study the potential adverse effects of excess Phe consumption in humans.

Normal Fasting Plasma Phenylalanine Levels

Like those of most other amino acids, plasma Phe levels are very high in the fetus and fall rapidly after birth (McIntosh et al. 1984), remaining elevated relative to adult values, however, during the first few days of life. After four months of age, the age-related changes in plasma Phe are minor, and studies in a wide variety of populations have consistently reported mean fasting Phe concentrations of around 60 μM (Table 1.1).

Levels of branched-chain amino acids are normally higher in men than in women; hence, men usually have lower plasma Phe/LNAA ratios. However, the plasma branched-chain amino acid levels rise progressively with aging, and the sex difference tends to disappear in the elderly. Absolute plasma Phe levels are slightly lower in women than in men (Koh and Cha 1983).

TABLE 1.1. Normal fasting plasma phenylalanine levels at different ages.

Age	*n*	Phe (μM)	Reference
Fetus	12	102 ± 51	McIntosh et al. 1984
Maternal blood		54 ± 17	
3 days	11	70 ± 8	Volz et al. 1983
2 wks		77 ± 15	
2 mos		64 ± 13	
4 mos		56 ± 5	
8 yrs	52	62 ± 18	Gregory et al. 1986
16 yrs	80	60 ± 13	
18–24 yrs	7	55	Fernstrom et al. 1979
25–35 yrs	15 (F)	52 ± 5	Bjerkenstedt et al. 1985
	50 (M)	60 ± 7	
24–54 yrs	10	58 ± 14	Scriver et al. 1985
17–94 yrs	11 (F)	57 ± 6	Milsom et al. 1979
	13 (M)	57 ± 8	
20–82 yrs	245 (F)	58 ± 2	Koh and Cha. 1983
	184 (M)	61 ± 2	

Effects of Disease States

Liver diseases cause complex and severe changes in amino acid metabolism, directly affecting the rates at which phenylalanine, tyrosine, and tryptophan are hydroxylated, and causing secondary insulin-dependent changes in amino acid metabolism (by decreasing the hepatic clearance of this hormone). Patients with cirrhosis exhibited elevated plasma Phe levels and ratios (Fernstrom et al. 1979b): mean fasting levels were 60 to 80 μM, and rose to 100 μM when subjects consumed 75 g of protein per day; in contrast, mean fasting values were 55 μM in normal subjects and rose to 63 μM after the same protein intake (Fernstrom et al. 1979a). The plasma Phe/LNAA ratio in patients with cirrhosis was 0.160 and fell to 0.145 when they consumed a 75-g protein diet; the corresponding values in normal controls were 0.120 and 0.080. These observations indicate that while in the fasting state high plasma Phe/LNAA ratios may cause corresponding elevations in brain Phe levels among cirrhotics, ingestion of protein meal fails to increase the brain Phe further in these patients (as evidenced by the decrease in plasma Phe/LNAA ratios), just as dietary protein fails to elevate brain Phe in normal subjects.

Obesity is also associated with significant changes in the plasma amino acid profile. Obese subjects had significantly higher fasting plasma Phe levels than controls (Caballero et al. 1987) (Figure 1.1). Consumption of a 30-g carbohydrate snack two hours after lunch produced a fall in plasma Phe that was less marked in obese than in lean control subjects. The high fasting plasma Phe and blunted response to carbohydrate intake are manifestations of resistance to insulin's effect on plasma amino acid levels, evidenced also by high plasma concentrations of branched-chain amino acids and Tyr in the obese (Felig et al. 1969).

Effects of Food Intake on Plasma Phenylalanine Levels

Immediately after protein intake, during the absorptive period, plasma Phe concentrations rise. The Phe/LNAA ratio, however, is *lower* at higher levels of protein intake, because of the larger rise that proteins produce in plasma levels of the branched-chain amino acids (Fernstrom et al. 1979a, Maher et al. 1984). In one study, oral administration of 25 g of pure albumin produced peak Phe levels of 100 μM, one hour after its ingestion (Moller 1985). High-carbohydrate meals cause an insulin-mediated fall in branched-chain amino acid levels, thus increasing the plasma Phe/LNAA ratio. In rats, addition of only 10% protein to the meal completely suppresses this carbohydrate-induced effect on the Phe/LNAA ratio (Yokogoshi and Wurtman 1986). The amounts of protein needed to suppress the carbohydrate response in people have not been calibrated, but are probably similar or slightly lower.

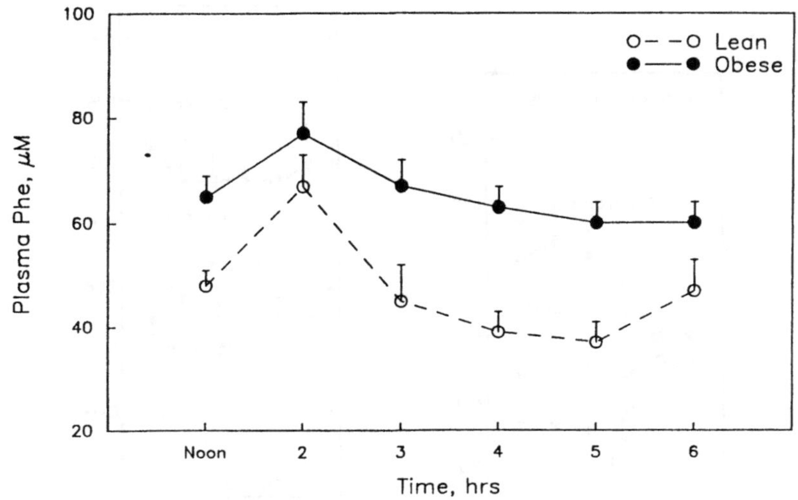

FIGURE 1.1. Effects of obesity on the plasma phenylalanine response to carbohydrate intake. Nine obese (over 130% of ideal body weight) and 7 lean controls consumed a standard meal at noon and a 30-g carbohydrate test meal at 2 P.M. Bars represent SEM.

Consumption of high-protein meals causes changes in plasma Phe levels that persist beyond the digestion and absorption of those meals. Consumption of 150 g of protein per day for five days produces higher fasting Phe levels, in normal subjects, than consumption of 0 or 75 g of protein per day (Fernstrom et al. 1979a). Long-term studies also reported significantly higher fasting plasma amino acid levels when 1.5 g/kg of protein per day was consumed than when 0.7 or 1 g/kg per day was consumed (Milsom et al. 1979).

Dietary protein content determines the shape of the diurnal variations in plasma Phe concentration. As shown in Figure 1.2, consumption of protein-rich meals increased mean plasma Phe levels throughout the day and caused daytime (3 to 7 P.M.) levels to be significantly higher than nighttime (3 to 7 A.M.) levels; there was a concurrent reduction during the daytime in the plasma Phe/LNAA ratio (Fernstrom et al. 1979a).

The introduction of the artificial sweetener aspartame prompted a number of studies to explore its effects on plasma amino acid concentrations (Stegink et al. 1977, 1979b, 1980, 1981, Yokogoshi et al. 1984). Of special interest were the potential effects of its phenylalanine content on the brain, since very high plasma phenylalanine concentrations were known to be associated with severe brain damage in phenylketonuria. The wide-scale use of aspartame in beverages and commercially prepared foods allowed, for the first time in history, the consumption of unrestricted amounts of

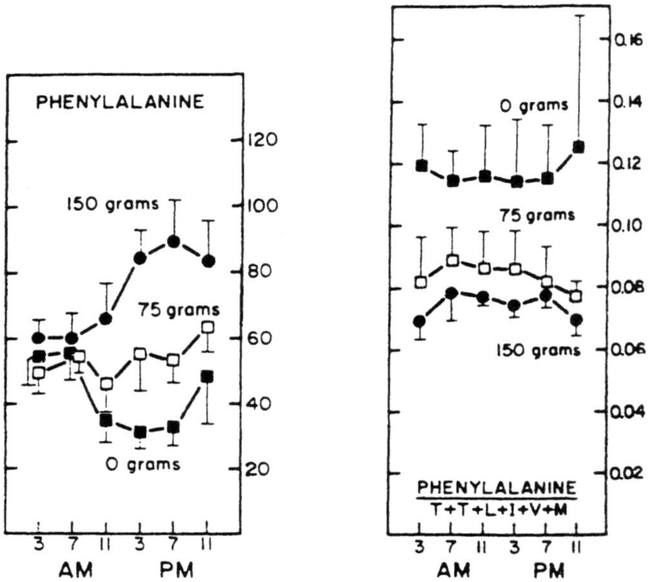

FIGURE 1.2. Diurnal variations in plasma Phe levels and in the plasma Phe/LNAA ratio among subjects consuming three meals per day containing 0, 25, or 50 g of protein per meal. Each test diet was consumed for five consecutive days, and blood samples were collected every four hours on the third, fourth, and fifth test days. The denominator in the plasma Phe/LNAA ratio included tryptophan, tyrosine, leucine, valine, isoleucine, and methionine. Error bars represent SD. Reprinted with permission from Am. J. Clin. Nutr. **32**:1912–1922 © American Society for Clinical Nutrition.

phenylalanine in the absence of the other LNAA. Studies by Stegink et al. (1977, 1979a) used an aspartame dose of 34 mg/kg, calculated by estimating the sugar consumption of the general population and assuming a replacement of half the calories by the artificial sweetener. This level of intake causes a rise in plasma Phe to a peak of 120 μM, about one hour after its ingestion. As pointed out by Stegink et al. (1977), this plasma Phe level is similar to that found after consumption of a high-protein meal. However, a very important difference between the effects of consuming aspartame and consuming protein becomes apparent when the plasma Phe/LNAA ratio is examined: while a high-protein meal causes a *fall* in the Phe/LNAA ratio (Fernstrom et al. 1979a), aspartame ingestion at the 34 mg/kg level causes an *increase* in this ratio, to a peak of 0.255 (Table 1.2). Proportionately higher increases in the Phe/LNAA ratio were reported at higher levels of sweetener intake or in persons who carry the gene for phenylketonuria (Stegink et al. 1981, Caballero et al. 1986).

As mentioned above, in assessing the possible effects of phenylalanine

TABLE 1.2. Changes in plasma Phe levels and in the plasma Phe/LNAA ratio in normal subjects after ingestion of different foods.

Food[a]	Mean Phe (μM)	Mean Phe/LNAA (μM)[b]	Reference
Fasting	56	0.100	Caballero et al. 1986
Protein			
0%	30	0.110	Maher et al. 1984
75%	61	0.087	
150%	87	0.085	
Carbohydrate			
30 g	36	0.110	Caballero et al. 1987
OGTT	33	1.120	
Aspartame			
15 mg/kg	60	0.170	Yokogoshi et al. 1984
34 mg/kg	120	0.255	Stegink et al. 1977
100 mg/kg	200	0.480	Stegink et al. 1981

[a]The "protein" diet also contained carbohydrate and fat; the "carbohydrate" and aspartame treatments contained only these compounds.
[b]The Phe/LNAA ratio refers to the plasma phenylalanine concentration divided by the sum of the plasma concentrations of valine, leucine, isoleucine, tyrosine, and tryptophan.

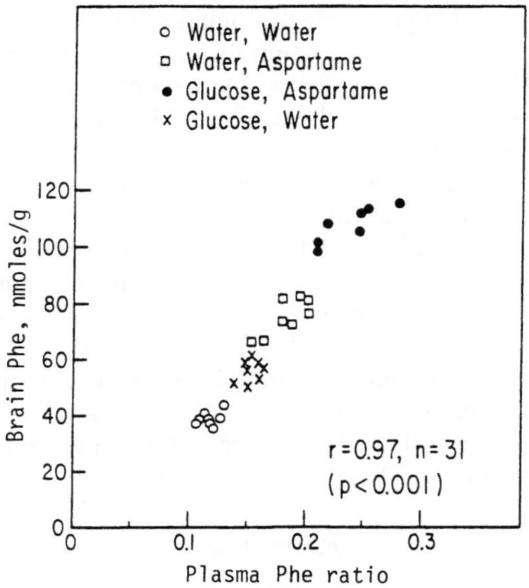

FIGURE 1.3. Correlation between the plasma Phe ratio and brain Phe levels in the rat. The denominator in the plasma ratio included tyrosine, tryptophan, valine, leucine, and isoleucine. Reprinted with permission from Am. J. Clin. Nutr. **40**:1–7 © American Society for Clinical Nutrition.

ingestion in the brain, the relative plasma concentration of all the large neutral amino acids should be considered, since the net brain uptake of each one of these amino acids will be determined by the competition with the others for transport through the L system at the blood-brain barrier (Pardridge 1977).

As shown in Figure 1.3, increasing the plasma Phe/LNAA ratio from 0.1 to 0.2 doubles the concentration of Phe in rat's brain (Yokogoshi et al. 1984). By comparison, ingestion of 34 mg of aspartame per kg by healthy men caused an estimated rise in the plasma Phe/LNAA ratio from 0.09 to 0.25 (Stegink et al. 1977). Although there are no direct in vivo human data on brain Phe levels after aspartame ingestion, studies by Choi and Pardridge (1986) using isolated human brain capillaries indicate that the phenylalanine transport system at the blood-brain barrier of rat and man are very similar. Moreover, simultaneous measurements of plasma Phe/LNAA ratios and cerebrospinal fluid Phe concentration in humans with phenylketonuria show also a good correlation (Berry et al. 1982). These data suggest that increases in the plasma Phe/LNAA ratio up to or above 0.20 are likely to be associated with significant rises in brain Phe concentrations in humans.

Summary and Conclusions

Plasma amino acids change widely throughout the day, in response to physiological variables such as diet composition, spacing of meals, and hormone output. They are also affected by a number of disease processes, particularly those involving the liver. Plasma Phe constitutes a small proportion of most proteins for human consumption, and its level increases moderately after protein intake, compared with the levels of the branched-chain amino acids (valine, leucine, and isoleucine), which are also present in all dietary proteins but in much larger proportion than Phe. This balance between the relative intake of Phe and the other neutral amino acids is the main factor limiting Phe's access into the brain. When consumed *without* these competing neutral amino acids (such as when aspartame is ingested), even relatively moderate amounts of Phe are likely to produce significant increases in brain uptake of this amino acid, raising its central nervous system concentration well above the levels attained when the same amino acid is consumed in naturally occurring proteins.

References

Berry, H.K., Bofinger, M.K., Hunt, M.M., Phillips, P.J., and Guilfoile, M.B. (1982). Reduction of cerebrospinal fluid phenylalanine after oral administration of valine, isoleucine and leucine. Pediatr. Res. **16**:751–755.

Bjerkenstedt, L., Edman, G., Hagenfeldt, L., Sedvall, G., and Wiesel, F.A. (1985). Plasma amino acids in relation to cerebrospinal fluid monoamine metabolites in schizophrenic patients and healthy controls. Br J. Psychiatr. **147**:276–282.

Caballero, B., Finer, N., and Wurtman, R.J. (1987). Plasma amino acid levels in obesity: effects of insulin resistance. *In* Kaufman, S. (ed.), Amino acids in health and disease: new perspectives, New York: Alan R. Liss, Inc., pp. 369–382.

Caballero, B., Mahon, B.E., Rohr, F.J., Levy, H.L. and Wurtman, R.J. (1986). Plasma amino acid levels after single-dose aspartame consumption in phenylketonuria, mild hyperphenylalaninemia, and heterozygous state for phenylketonuria. J. Pediatr. **109**:668–671.

Choi, T.B., and Pardridge, W.M. (1986). Phenylalanine transport at the human blood-brain barrier. Studies with isolated human capillaries. J. Biol. Chem. **261**:6536–6541.

Clarke, J.T.R., and Bier D.M. (1982). The conversion of phenylalanine to tyrosine in man. Metabolism **31**:999–1005.

Felig, P., and Wahren, J. (1971). Influence of endogenous insulin secretion on splanchnic glucose and amino acid metabolism in man. J. Clin. Invest. **50**:1702–1711.

Felig, P., Marliss, E., and Cahill, G.F. (1969). Plasma amino acid levels and insulin secretion in obesity. N. Engl. J. Med. **281**:811–816.

Fernstrom, J.D., and Faller, D.V. (1978). Neutral amino acids in the brain: changes in response to food ingestion. J. Neurochem. **30**:1531–1538.

Fernstrom, J.D., Wurtman, R.J., Hammarstrom-Wiklund, B., Rand, W.M., Munro, H.N., and Davidson, C.S. (1979a). Diurnal variations in plasma concentrations of tryptophan, tyrosine, and other neutral amino acids: effects of dietary protein intake. Am. J. Clin. Nutr. **32**:1912–1922.

Fernstrom, J.D., Wurtman, R.J., Hammarstrom-Wiklund, B., Rand, W.M., Munro, H.N., and Davidson, C.S. (1979b). Diurnal variations in plasma neutral amino acid concentrations among patients with cirrhosis: effect of dietary protein intake. Am. J. Clin. Nutr. **32**:1923–1933.

Gregory, D.M., Sovetts, D., Clow, C.L., and Scriver, C.R. (1986). Plasma free amino acid values in normal children and adolescents. Metabolism **35**:967–969.

Kaufman, S. (1986). Regulation of the activity of hepatic phenylalanine hydroxylase. Adv. Enzyme Reg. **25**:37–64.

Koh, E.T., and Cha, C.J. (1983). Comparison of plasma amino acids by race, sex and age. Nutr. Rep. Int. **28**:8–22.

Maher, T.J., Glaeser, B.S., and Wurtman, R.J. (1984). Diurnal variations in plasma concentrations of basic and neutral amino acids and in red cell concentrations of aspartate and glutamate: effects of dietary protein intake. Am. J. Clin. Nutr. **39**:722–729.

McIntosh, N., Rodeck, C.H., and Heath, R. (1984). Plasma amino acids of the mid-trimester human fetus. Biol. Neonate **45**:218–224.

Milsom, J.P., Morgan, M.Y., and Sherlock, S. (1979). Factors affecting plasma amino acid concentrations in control subjects. Metabolism **28**:313–319.

Moldawer, L.L., Kamamura, L., Bistrian, B.R., and Blackburn, G.L. (1983). The contribution of phenylalanine to tyrosine metabolism in vivo. Studies in the post-absorptive and phenylalanine-loaded rat. Biochem. J. **210**:811–817.

Moller, S.E. (1985). Effect of various protein doses on plasma neutral amino acid levels. J. Neural Trans. **61**:183–191.

Pardridge, W.M. (1977). Regulation of amino acid availability to the brain. *In* Wurtman, R.J., and Wurtman, J.J. (eds.), Nutrition and the Brain. New York: Raven Press, pp. 141–204.

Schauder, P., Scheder, K., Matthaei, D., Henning, H.V., and Langenbeck, U. (1983). Influence of insulin on blood levels of branched chain keto and amino acids in man. Metabolism **32**:323–327.

Scriver, C.R., Gregory, D.M., Sovetts, D., and Tissenbaum, G. (1985). Normal plasma free amino aid values in adults: the influence of some common physiological variables. Metabolism **34**:868–873.

Stegink, L.D., Filer, L.J., and Baker, G.L. (1977). Effect of aspartame and aspartate loading upon plasma and erythrocyte free amino acid levels in normal adult volunteeers. J. Nutr. **107**:1837–1845.

Stegink, L.D., Filer, L.J., and Baker, G.L. (1979a). Plasma, erythrocyte and human milk levels of free amino acids in lactating women administered aspartame or lactose. J. Nutr. **109**:2173–2181.

Stegink, L.D., Filer, L.J., Baker, G.L., and McDonnell, J.E. (1979b). Effect of aspartame loading upon plasma and erythrocyte amino acid levels in phenylketonuric heterozygotes and normal adult subjects. J. Nutr. **109**:708–717.

Stegink, L.D., Filer, L.J., and McDonnell, J.E. (1980). Effect of an abuse dose of aspartame upon plasma and erythrocyte levels of amino acids in phenylketonuric heterozygous and normal adults. J. Nutr. **110**:2261–2224.

Stegink, L.D., Koch, R., Blaskovics, M.E., Jr., L.J. Filer, Baker, G.L., and McDonnell, J.E. (1981). Plasma phenylalanine levels in phenylketonuric heterozygous and normal adults administered aspartame at 34 mg/kg body weight. Toxicology **20**:81–90.

Tewksbury, D.A., and Lohrenz, F.N. (1970). Circadian rhythm of human urinary amino acid excretion in fed and fasted states. Metabolism **19**:363–371.

Volz, V.R., Book, L.S., and Churella, H.R. (1983). Growth and plasma amino acid concentrations in term infants fed either whey-predominant formula or human milk. J Pediatr **102**:27–31.

Yokogoshi, H., Roberts, C., Caballero, B., and Wurtman, R.J. (1984). Effects of aspartame and glucose administration on brain and plasma levels of large neutral amino acids and brain 5-hydroxyindoles. Am. J. Clin. Nutr. **40**:1–7.

Yokogoshi, H., and Wurtman, R.J. (1986). Meal composition and plasma amino acid ratios: effect of various proteins or carbohydrates, and of various protein concentrations. Metabolism **35**:837–842.

CHAPTER 2

Adrenergic Influence on Plasma and Brain Concentrations of Phenylalanine and Other Large Neutral Amino Acids in Rats

Tomas Eriksson and Arvid Carlsson *

The β-adrenergic agonist terbutaline is demonstrated to cause an increase in the total concentration of large neutral amino acids (LNAA), as well as in the concentration of all the individual LNAA in rat brain. This increase in brain LNAA concentration is concomitant with a decrease in the plasma concentrations of the same amino acids. The increase in brain concentrations of LNAA could obviously not be explained by the competition among the LNAA in plasma for the carrier-mediated transport into the brain. These findings indicate that the transport of LNAA from plasma into the brain is regulated by a β-adrenergic mechanism.

Introduction

The transport of amino acids from plasma into the brain is carrier mediated. The large neutral amino acids (LNAA) (tyrosine, tryptophan, phenylalanine, valine, leucine, and isoleucine) are considered to be transported via the same saturable carrier system and to compete with each other for the carrier (Oldendorf 1973). The relation among the LNAA in plasma, rather than the concentrations of individual amino acids, seems to determine how much of each of the different amino acids will be transported into the brain (Fernstrom and Wurtman 1972, Wurtman and Fernstrom 1976).

We have previously reported that ethanol causes a decrease in most amino acids in plasma in man (Eriksson et al. 1983) and rat (Eriksson et al. 1980). This effect of ethanol could, at least partly, be inhibited by the β-adrenergic antagonist propranolol (Eriksson et al. 1981), indicating that the concentrations of amino acids in plasma could be influenced by adrenergic mechanisms. This hypothesis was supported by our finding that the β-adrenergic agonist isoprenaline also causes a marked decrease in most amino acids in rat plasma (Eriksson et al. 1984). Furthermore, iso-

*Department of Pharmacology, University of Göteborg, S–400 33 Göteborg, Sweden.

prenaline was found to cause a change in the relation among the LNAA in plasma in favor of the monoamine precursors tyrosine and tryptophan. This finding could explain the previous finding by Hutson and coworkers (Hutson et al 1980), later confirmed by us (Eriksson et al. 1984), that isoprenaline gives rise to increased concentrations of tyrosine and tryptophan in rat brain. The increase in the brain concentrations of these amino acids was, however, more pronounced than could be expected just on the basis of the change in the relation among the LNAA in plasma. Therefore, we deemed it necessary to investigate possible effects of adrenergic agonists on the transportation of LNAA from plasma into the brain. In the present study, the effect of the β_2-adrenergic agonist terbutaline on the concentrations of LNAA in plasma and brain was investigated.

Methods

Male rats of the Sprague-Dawley strain were housed before use for more than one week in a room maintained on a 14/10-h light/dark cycle where they had free access to food and water. In the experiment, the rats in one group were injected with terbutaline 6 mg/kg (0.6 mg/ml) intraperitoneally (i.p.). A control group received an equivalent volume of saline. Sixty minutes after the injections, the rats were killed by decapitation. About 5 ml of blood was collected in a tube containing 0.5 ml of a 1% EDTA solution and deproteinized. Immediately after death, the brains were taken out, frozen on dry ice, homogenized, and deproteinized. The concentrations of phenylalanine, valine, leucine, and isoleucine were determined by high-pressure liquid chromatography (HPLC) in a system using pre-column derivatization of the amino acids with *o*-phtaldialdehyde followed by fluorescence detection. The concentrations of tyrosine and tryptophan were determined in an HPLC system with electrochemical detection. Statistical significances were assessed by Student's *t* test.

Results

The β_2-agonist terbutaline caused a 32% decrease in the total amount of LNAA in plasma. The concentrations of all the individual LNAA were also decreased, though not significantly in the cases of tryptophan and phenylalanine (Figure 2.1). Thus, the individual LNAA in plasma were not decreased to the same extent; the relative concentrations of tyrosine, tryptophan, and phenylalanine were increased and the relative concentrations of valine, isoleucine, and leucine were decreased (Figure 2.2).

The effects of terbutaline on brain concentrations of the LNAA are demonstrated in Figure 2.3. Terbutaline gave rise to an increase in all LNAA in the brain; actually the total amount of LNAA in the brain increased by as much as 33%.

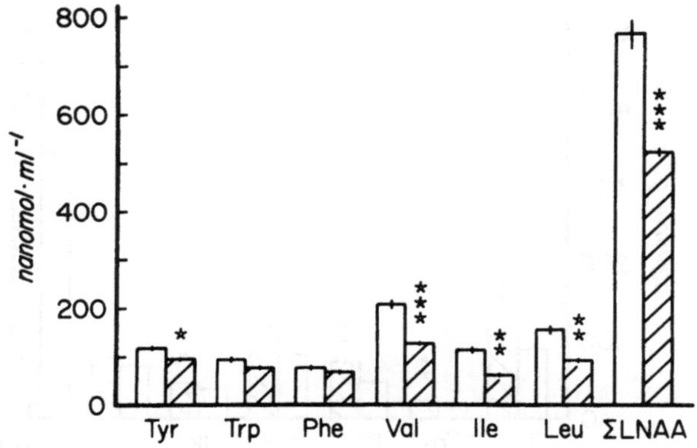

FIGURE 2.1. Effects of terbutaline (6 mg/kg, i.p.) on rat plasma concentrations of LNAA. Given are mean values and their SEM. Hatched bars: terbutaline treated; $n = 4$. Open bars: saline-treated controls; $n = 6$. ***$p < 0.001$; **$p < 0.01$; *$p < 0.05$.

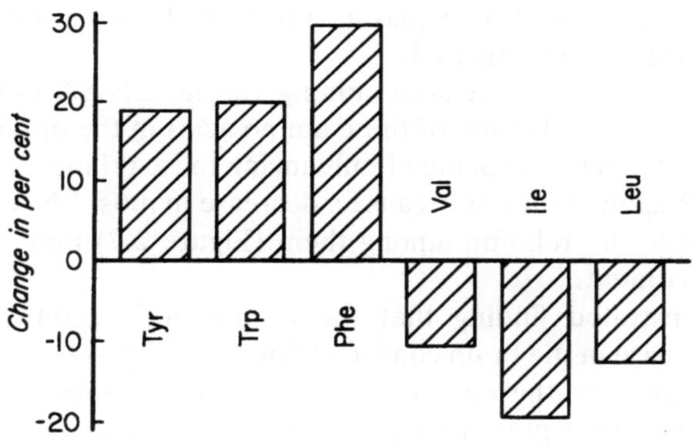

FIGURE 2.2. Effects of terbutaline (6 mg/kg, i.p.) on the relative concentrations of LNAA in rat plasma. The values given in the figure are calculated from the mean plasma concentration of each LNAA in the terbutaline-injected and in the saline-injected rats.

Discussion

Since β-adrenergic agonists are known to increase the blood perfusion of the liver (Greenway and Stark 1971), where amino acids are mainly metabolized, a possible explanation of the terbutaline-induced decrease in plasma concentrations of LNAA could be an increased liver metabolism. This suggestion is supported by the fact that α-adrenergic agonists, which reduce liver perfusion, give rise to increased concentrations of LNAA

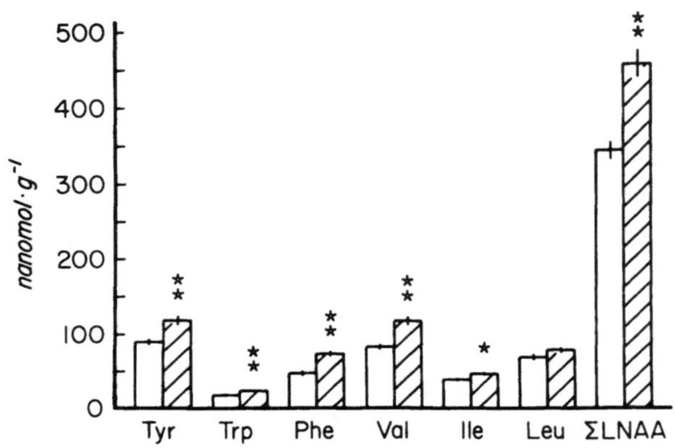

FIGURE 2.3. Effects of terbutaline (6 mg/kg, i.p.) on rat brain concentrations of LNAA. Given are mean values and their SEM. Hatched bars: terbutaline treated; $n = 4$. Open bars: saline-treated controls; $n = 6$. $^{**}p < 0.01$; $^{*}p < 0.05$.

in plasma (Eriksson and Carlsson 1982). The mechanism by which β-adrenergic agonists exert their plasma amino-acid-decreasing effect must, however, be further investigated.

From the view that the relation among various LNAA in plasma is of importance to the availability of these amino acids in the brain, it is necessary to elucidate factors capable of influencing these relations. Our present finding that β-agonists not only cause a decrease in most LNAA in plasma, but also change the relation among them (Figure 2.2) may be of importance in this context.

The demonstrated finding that the β-adrenergic agonist terbutaline causes an increase in the brain concentrations of all LNAA could obviously not be explained by the competition among these amino acids for the carrier-mediated transport from plasma into the brain. The fact that this increase occurs despite a concomitant decrease of the same amino acids in plasma indicates that β-adrenergic agonists are capable of influencing the distribution of LNAA between plasma and brain. We propose that this adrenergic effect is mediated via an influence on the amino acid carrier.

Acknowledgments. This work was supported by the Swedish Medical Research Council, Stiftelsen för Gamla Tjänarinnor and Åhlén-stiftelsen. The expert technical assistance of Mrs. Birgitta Holmgren is gratefully acknowledged.

References

Eriksson, T., and Carlsson, A. (1982). Adrenergic influence on rat plasma concentrations of tyrosine and tryptophan. Life Sci. **30**:1465–1472.

Eriksson, T., Carlsson, A., Hagman, M., and Jagenburg, R. (1984). β-Adrenergic influence on brain concentrations of monoamine precursors. *In* Usdin, E., Åsberg, M., Bertilsson, L., and Sjöqvist, F. (eds.), Frontiers in biochemical and pharmacological research in depression. Raven Press, New York: pp. 263–270.

Eriksson, T., Carlsson, A., Liljequist, S., Hagman, M., and Jagenburg, R. (1980). Decrease in plasma amino acids in rat after acute administration of ethanol. J. Pharm. Pharmacol. **32**:512–513.

Eriksson, T., Magnusson, T., Carlsson, A., Hagman, M., Jagenburg, R., and Edén, S. (1981). Effects of hypophysectomy, adrenalectomy and (−)-propranolol on ethanol-induced decrease in plasma amino acids. Naunyn-Schmiedeberg's Arch. Pharmacol. **317**:214–218.

Eriksson, T., Magnusson, T., Carlsson, A., Hagman, M., and Jagenburg, R. (1983). Decrease in plasma amino acids in man after an acute dose of ethanol. J. Stud. Alcohol. **44**:215–221.

Fernstrom, J.D., and Wurtman, R.J. (1972). Brain serotonin content: physiological regulation by plasma neutral amino acids. Science **178**:414–416.

Greenway, C.V., and Stark, R.D. (1971). Hepatic vascular bed. Physiol. Rev. **51**:23.

Hutson, P.H., Knott, P.J., and Curzon, G. (1980). Effect of isoprenaline infusion on the distribution of tryptophan, tyrosine and isoleucine between brain and other tissues. Biochem. Pharmacol. **29**:509–516.

Oldendorf, W.H. (1973). Stereospecificity of blood-brain barrier permeability to amino acids. Am. J. Physiol. **224**:967–969.

Wurtman, R.J., and Fernstrom, J.D. (1976). Control of brain neuro-transmitter synthesis by precursor availability and nutritional state. Biochem. Pharmacol. **25**:1691–1696.

CHAPTER 3

Effect of Aspartame on Plasma Phenylalanine Concentration in Humans

L.J. Filer, Jr., and Lewis D. Stegink *

We have determined the effects of aspartame (APM) ingestion by humans on plasma free phenylalanine concentration in a variety of feeding situations. These clinical studies have examined the plasma phenylalanine response when APM was given as a bolus dose in unsweetened beverage, orange juice, a sucrose solution, capsules, or a slurry or as part of a meal. The complexity of the meal feedings ranged from soup and a beverage to a hamburger and a milk shake.

The effect of aspartame on plasma phenylalanine concentration was also determined under conditions simulating repeated ingestion of an APM-containing beverage. The frequency of ingestion varied from a cycle of every two hours over a six-hour period to every hour over an eight-hour period. Aspartame intake in these studies was the equivalent of three 12-oz cans of diet carbonated soft drink per cycle.

These clinical studies have been carried out in normal young adults, lactating women, normal one-year-olds, known adult phenylketonuric (PKU) heterozygotes, adolescent PKU homozygotes, and young adults selected on the basis of an idiosyncratic response to monosodium L-glutamate (MSG).

All clinical trials were conducted using a randomized, crossover design to an appropriate placebo control. With the exception of studies involving one-year-olds, lactating women, PKU heterozygotes, and PKU homozygotes, equal numbers of male and female subjects were studied.

Heparinized blood samples were collected at frequent intervals following administration of APM and the plasma centrifuged and deproteinized within 30 minutes for analysis of plasma free amino acid concentrations.

In order to develop pharmacokinetic data to describe the rate of increase and disappearance of phenylalanine from plasma following APM ingestion, dose-response studies were conducted. The steady-state model predicted on the basis of these analyses was tested by clinical studies of response to repeated doses of APM.

* Departments of Pediatrics and Biochemistry, The University of Iowa, Iowa City, IA 52242, USA.

Introduction

This review is limited to considerations of the change in plasma phenylalanine concentration occurring as a result of aspartame (APM) ingestion and the factors affecting this response. In the course of these studies, APM intake was varied from 4 mg/kg body weight to 200 mg/kg body weight. These doses approximate the intake of APM following ingestion of 12 oz of a diet carbonated soft drink by a 50-kg person to an abuse level of intake equivalent to a 10-kg one-year-old ingesting 100 tablets of APM in tabletop sweetener form (20 mg per tablet).

Questions have been raised about the safety of APM as a food additive on the basis of its phenylalanine content. It has been speculated that APM ingestion at normal use levels would markedly elevate plasma phenylalanine concentration. Grossly elevated plasma phenylalanine concentrations, such as those found in children with classical phenylketonuria, are associated with mental retardation (Blaskovics 1974, Blaskovics et al. 1974, Koch et al. 1974).

Summary of the Iowa Studies

Projected Aspartame Intake

It is clear that phenylalanine toxicity requires gross elevation of plasma phenylalanine concentration. Thus, it is important to develop an understanding of APM intake under conditions of its use as a food additive. Projected levels of APM intake have been calculated by the Food and Drug Administration (FDA), the Market Research Corporation of America (MRCA), and our group. Table 3.1 summarizes these data (Anonymous 1974, 1976a, 1976b, Stegink et al. 1977). If APM totally replaces estimated mean daily sucrose intake on a sweetness basis, its intake will range between 3 and 11 mg/kg body weight per day. This is equivalent to 1.7 to 6.2 mg of phenylalanine per kg body weight. The highest daily APM ingestion, according to these calculations, would range from 22 to 34 mg/kg body weight.

More recently, Pardridge (1986a, 1986b) has interpreted an earlier study by Frey (1976) to indicate that 7- to 12-year-old children fed a normal diet

TABLE 3.1. Summary of projections for aspartame intake.

Source	Aspartame totally replaces mean sucrose sweetness (mg/kg body weight)	Maximum mg/kg body weight
FDA	Not calculated	22–28
MRCA	3–11	25–34
Iowa	7–9	23–25

TABLE 3.2. Fasting and high mean (± SD) plasma phenylalanine levels in subjects administered single bolus doses of APM.

APM dose	Plasma Phe concentration (μmol/dl) Fasting	Peak	Time of peak (min)
0	5.50 ± 0.54	5.61 ± 1.42	30
4	5.48 ± 0.85	5.67 ± 0.48	30
10	5.08 ± 0.82	6.73 ± 0.75	30
34	5.66 ± 1.21	11.1 ± 2.49	30
50	4.61 ± 1.72	16.2 ± 4.86	45
100	5.40 ± 1.05	20.2 ± 6.77	45
150	6.72 ± 1.93	35.0 ± 9.43	90
200	5.26 ± 0.67	48.7 ± 15.5	90

could ingest as much as 50 to 77 mg of APM per kg body weight. In our opinion, Pardridge has misinterpreted Frey's data. The children studied by Frey were deliberately given large doses of APM in addition to usual food since the purpose of the study was to investigate the effects of high doses of APM. The intake values noted in Frey's study cannot be extrapolated to calculate intake values of APM at the current market availability of APM-sweetened products. Indeed, the most recent measurements of APM intake in the United States (Abrams 1986), Canada (Kirkpatrick 1986), and England (Sherlock 1986) closely approximate the values calculated earlier (Anonymous 1974, 1976a, 1976b, Stegink et al. 1977).

Large Single Bolus Doses

Our initial studies focused on the effects of large single bolus doses of APM on plasma phenylalanine concentrations in normal adults ingesting APM at doses of 34, 50, 100, 150, and 200 mg/kg body weight (Stegink 1984). Plasma phenylalanine concentrations in subjects administered single bolus doses of APM at 0, 34, 50, 100, 150, and 200 mg/kg body weight are shown in Table 3.2. These data clearly demonstrate that the high mean plasma phenylalanine concentration is proportional to dose. The data also indicate a variation with dose in the time at which the high mean plasma phenylalanine concentration occurs.

Pharmacokinetic Studies

The dose-response curves plotted from these studies are shown in Figure 3.1. The curve for normal subjects predicts that peak plasma phenylalanine levels should increase approximately 1 μmol/dl above fasting level at an APM intake of 4 mg/kg body weight, and about 2 to 3 μmol/dl above fasting levels at an intake of 10 mg/kg body weight.

Pharmacokinetic analysis of these dose-response data enabled us to

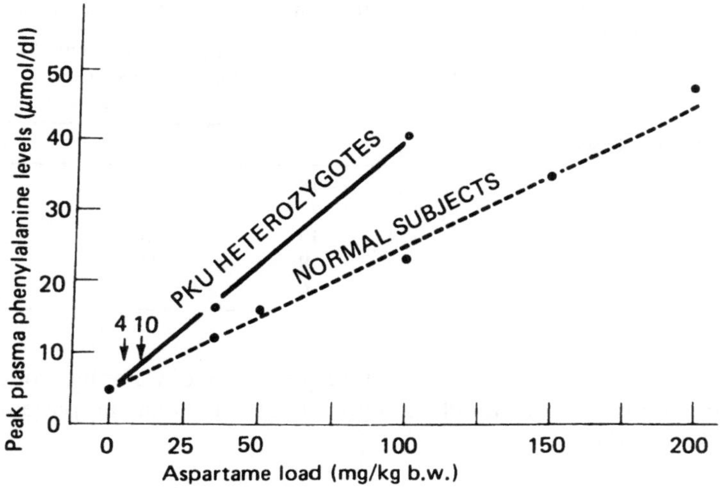

FIGURE 3.1. Correlation of high mean peak plasma phenylalanine concentrations with APM dose in normal adults and phenylketonuric heterozygotes. Reproduced with permission from Stegink (1984).

predict the average steady-state concentration of plasma phenylalanine following repeated doses of APM (Fischer 1978). The calculated half-life ($t_{1/2}$) for disappearance of phenylalanine from plasma under the conditions of study was 1.7 h (Table 3.3) with a first rate constant (K_e) of 0.41. The first-order input for phenylalanine in plasma (K_1) is dose dependent, ranging from 1.26 to 3.7 µmol/dl per hour. The variation in this constant might be due to the fact that the rate of hydrolysis of APM in the enterocyte is rate limiting.

Calculated average steady-state concentrations of plasma phenylalanine after oral doses of APM ranging from 34 to 200 mg/kg body weight, repeated at intervals of one to eight hours, are given in Table 3.4. Administration of APM at the 99th percentile of projected daily intake at two-hour intervals will produce an average steady-state concentration of 12 µmol/dl in normal adults. These calculations illustrate the abnormal situations re-

TABLE 3.3. Phenylalanine pharmacokinetics following aspartame dosing. Plasma phenylalanine half-life, K_e, and K_1.

Aspartame dose (mg/kg body weight)	Half-life ($t_{1/2}$) (h)	K_e (µmol/dl per hour)	K_1 (µmol/dl per hour)
34	1.65	0.420	3.47
100	1.7	0.408	2.72
200	1.7	0.408	1.26

TABLE 3.4. Plasma phenylalanine concentration in average steady state, C_{ss}.

Interval between doses (h)	Plasma Phe levels (μmol/dl) for aspartame dose (mg/kg body weight) of:		
	34	100	200
1	18	73	162
2	12	40	84
3	10	28	58
4	9	23	45
8	6	14	25

quired to achieve average steady-state plasma phenylalanine concentrations in excess of 60 μmol/dl, an upper limit for managing patients with phenylketonuria.

Small Bolus Doses

After exploring the effects of large single bolus doses of APM, we explored the effects of doses more likely to be ingested by normal adults. Probably the most common way that individuals ingest APM is in a diet beverage. A 12-oz serving of diet carbonated soft drink provides about 200 mg of APM; thus, a 50-kg person would receive 4 mg of APM per kg body weight. Similarly, a 20-kg four-year-old consuming this same size serving would receive 10 mg of APM per kg body weight. In normal adults, 10 mg of APM per kg body weight increased plasma phenylalanine concentration (Table 3.2) from a mean baseline concentration of 5.08 ± 0.82 μmol/dl to a high mean value of 6.73 ± 0.75 μmol/dl at 30 minutes after dosing (Stegink et al. 1987a). This plasma phenylalanine concentration is well within the normal fasting or postprandial range. Plasma phenylalanine concentration was not significantly affected (Table 3.2) in normal subjects ingesting 4 mg of APM per kg body weight (Wolf-Novak 1987).

Effects of Sampling Time

The data summarized in Table 3.2 clearly demonstrate the effects of sampling time on the observed high mean plasma phenylalanine concentration. High mean plasma phenylalanine concentrations were observed at 30 minutes after APM dosing at 4, 10, and 34 mg/kg body weight. High mean values were observed at 45 minutes after APM ingestion at 50 and 100 mg/kg body weight, and at 90 minutes after APM ingestion at 150 and 200 mg/kg body weight. Thus, clinical studies measuring plasma phenylalanine concentrations at a single time point after APM dosing should critically consider the time of blood sampling.

TABLE 3.5. High mean plasma phenylalanine levels in normal adults ingesting APM in solution or capsules.

APM ingested	High mean plasma Phe (μmol/dl)	Plasma Phe AUC [(μmol/dl) × min]
Solution	19.1 ± 6.54	1534 ± 482
Capsules	11.7 ± 3.95	847 ± 336

Solution Versus Capsule Administration

Many studies designed to investigate the putative effects of APM on a variety of functions must be carried out double-blind. This design requires that APM and placebo be administered in capsules because the sweetness of aspartame cannot be masked when administered in solution. However, test compounds administered in capsules may produce a different pharmacokinetic response than test compounds administered in solution. To determine this effect, we measured plasma phenylalanine concentrations in ten normal subjects given 3 g of APM in solution and in capsules (Stegink et al. 1987b). The results of these studies are shown in Table 3.5. Peak plasma phenylalanine concentrations were significantly higher and occurred significantly earlier when APM was given in solution. Similarly, the 4-h area under the plasma phenylalanine curve (AUC) was significantly greater after APM ingestion in solution than after ingestion in capsules.

These data do not rule out the use of capsules for administration of APM in clinical studies. They do, however, point out that certain variables must be controlled. For example, the plasma pharmacokinetic response must be determined at the dose to be studied prior to double-blind testing of the compound for putative side effects.

Solution Versus Slurry Administration

Aspartame is relatively insoluble; thus, some clinical studies may be compromised by the administration of APM as a slurry or in capsules with minimal quantities of fluid. To test the effects of slurry administration, six normal adults were administered an abuse dose of APM (100 mg/kg body weight) in both solution and slurry form (Stegink et al. 1979a). Aspartame was completely dissolved in 500 ml of orange juice when administered in solution and was mixed with 1.2 ml of orange juice per kg body weight when administered in slurry form. The results of this study are shown in Figure 3.2. Mean plasma phenylalanine concentrations were not significantly affected by the form of APM administration. The high mean plasma phenylalanine concentration was 20.3 ± 2.05 μmol/dl after administration of APM in solution and 26.0 ± 18.9 μmol/dl in slurry form.

Plasma phenylalanine levels, however, showed greater individual varia-

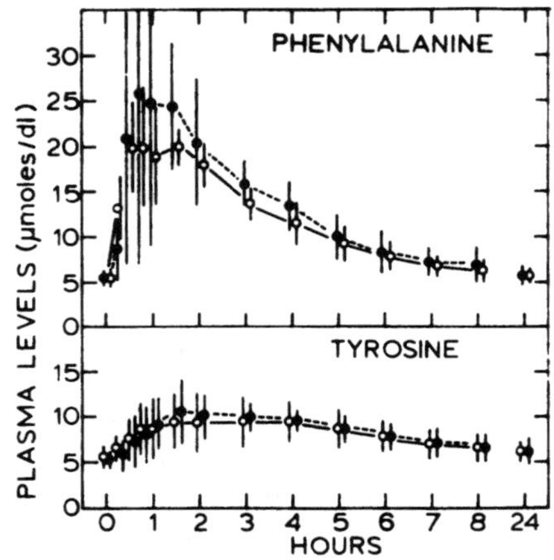

FIGURE 3.2. Mean (± SD) plasma phenylalanine and tyrosine levels in normal adults administered 100 mg APM per kg body weight either in solution (○) or slurry (●) form. Reproduced with permission from Stegink et al. (1979a).

tion after APM administration in slurry form than after administration in solution. The six subjects separated into three distinct groups of two subjects each (Figure 3.3). Two subjects showed a rapid rise in plasma phenylalanine concentration, reaching a high mean value of 49 μmol/dl. Two subjects showed a plasma phenylalanine response similar to the high mean values noted in subjects ingesting 100 mg of APM per kg body weight in solution. Two of the subjects showed a lower and slower rise in plasma phenylalanine concentration. The latter two subjects behaved as though the APM had been administered in capsules. Plasma aspartate and methanol concentrations showed a similar pattern. The precise reasons for differences in plasma phenylalanine concentrations in slurry versus solution ingestion are not known but probably involve differences in gastric emptying and/or luminal hydrolysis/absorption rates.

This study indicates that widely varying plasma phenylalanine concentrations will be obtained when large amounts of APM are administered in slurry form. We should expect similar variations in plasma phenylalanine concentration after administration of very large quantities of APM in capsules.

Multiple-Dose Studies

Since studies of single-dose administration of APM in those amounts likely to occur in foods showed only a small effect on plasma phenylalanine concentration, we modified our clinical trials to study individuals receiving multiple servings of APM-sweetened beverage. Two study designs were developed. The first involved ingestion of APM at 10 mg/kg body weight

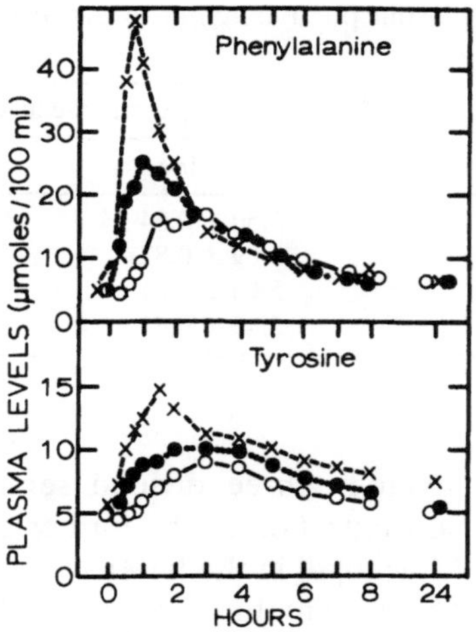

FIGURE 3.3. Plasma phenylalanine and tyrosine levels in six subjects administered APM in slurry form. Values appear to break down into three groups of two subjects each. Two individuals (×) showed plasma levels consistent with rapid gastric emptying; two others (○) showed values consistent with delayed gastric emptying; the other 2 subjects (●) showed levels similar to those noted after APM administration in solution. Reproduced with permission from Stegink et al. (1979a).

every two hours for three successive times. This model would simulate the response of a four-year-old child consuming three 12-oz beverage servings throughout the day. The second design involved consumption of APM at 600 mg per serving every hour for eight hours. This model simulates an adult drinking three 12-oz servings every hour or the equivalent of 24 bottles or cans of carbonated soft drink in eight hours. Although the studies to be described were carried out in adults, we assume they hold for children as well, since one-year-old infants absorb and metabolize APM as well as adults (Filer et al. 1983).

Mean ± SD fasting plasma phenylalanine concentrations and the high mean plasma phenylalanine values observed are listed in Table 3.6. Plasma phenylalanine concentrations did not differ significantly from baseline values after ingestion of three servings of unsweetened beverage but increased significantly over baseline values (5.09 ± 0.82 μmol/dl) after each serving of aspartame-sweetened beverage, with a high mean value of 8.10 ± 1.42 μmol/dl. These data clearly demonstrate that high mean plasma phenylalanine concentrations were markedly lower when an APM load was given in divided doses throughout the day as compared to values noted when an equivalent dose was given as a single bolus. Table 3.6 also compares high mean plasma phenylalanine concentrations in normal subjects ingesting 34 mg of APM per kg as a single bolus with the plasma phenylalanine concentrations observed in normal adults ingesting 30 mg of APM

TABLE 3.6. High mean plasma phenylalanine levels in normal subjects ingesting APM-sweetened beverage.

APM dose	Plasma Phe levels (μmol/dl)	
	Baseline	High mean
0	5.50 ± 0.54	5.61 ± 1.42
10 mg/kg × 3 doses	5.09 ± 0.82	8.10 ± 1.42
34 mg/kg	5.66 ± 1.21	11.1 ± 2.49
600 mg × 8 doses	5.45 ± 0.71	10.7 ± 1.35

per kg body weight given in three divided servings. The high mean plasma phenylalanine concentration in the former group was 11.1 ± 2.49 μmol/dl and 8.10 ± 1.42 μmol/dl in the latter.

We investigated the effects of multiple ingestion of APM in normal adult subjects consuming 600 mg of APM at hourly intervals over an 8-h period. Plasma phenylalanine concentration increased from a mean baseline value of 5.45 ± 0.71 μmol/dl to a high mean value of 10.7 ± 1.35 μmol/dl (Table 3.6). The total APM dose ingested was 4.8 g, or 70 mg/kg body weight. This APM dose ingested as a single bolus would be projected (using the data in Figure 3.1) to result in a high mean plasma phenylalanine concentration of approximately 19 μmol/dl in normal subjects.

Aspartame Ingestion with Meals

We have also examined the effects of APM ingestion with meals on plasma phenylalanine concentration (Stegink et al. 1982, 1983, 1987c). These studies investigated the interaction of aspartame with a hamburger/milk shake meal and a soup/beverage meal. The hamburger/milk shake meal provided protein at 1 g/kg body weight (42 mg/kg body weight of protein-bound phenylalanine), while the soup/beverage meal provided no protein-bound phenylalanine.

In our initial study, we examined the effect of adding 23 mg of APM per kg body weight (12 mg peptide-bound phenylalanine) to a hamburger/milk shake meal (Stegink et al. 1983). Plasma phenylalanine concentration increased from a fasting level of 5.09 ± 0.44 μmol/dl to a high mean value of 9.75 ± 2.55 μmol/dl after ingestion of the meal alone (Figure 3.4). The addition of 23 mg of APM per kg body weight to the meal resulted in a minimal change in plasma phenylalanine concentration, producing a high mean value of 10.2 ± 1.62 μmol/dl. This amount of APM is equivalent to a 50-kg person ingesting 5.75 12-oz cans (2000 ml) of diet beverage with the meal. A similar study was carried out with 34 mg of APM per kg body weight added to the hamburger/milk shake meal (Stegink et al. 1982). Plasma phenylalanine concentrations increased from a fasting level of 4.72 ±

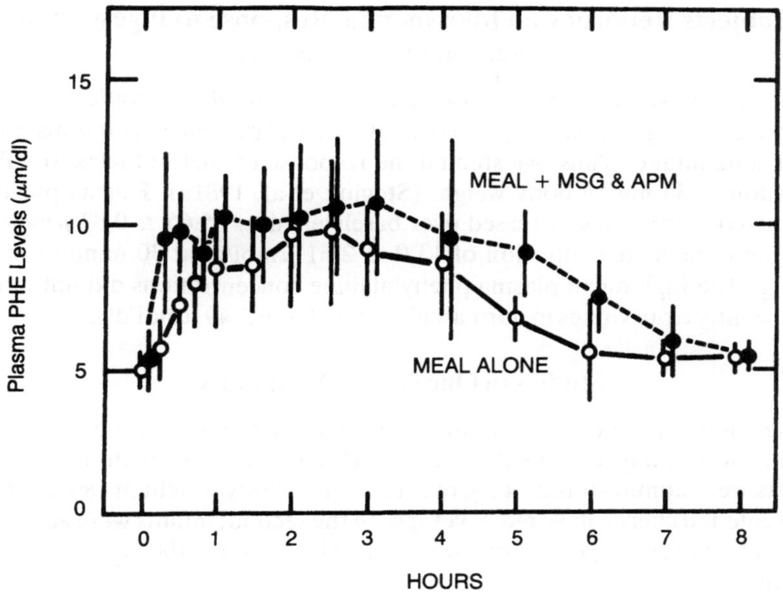

FIGURE 3.4. Mean (± SD) plasma phenylalanine concentrations in normal adults ingesting hamburger/milk shake meals with no additions (○) or with APM added at 23 mg/kg body weight (●). Reproduced with permission from Stegink (1984).

0.52 μmol/dl to a mean high value of 7.14 ± 0.82 μmol/dl after ingestion of the meal alone. With APM added to the meal (34 mg/kg body weight), the high mean plasma phenylalanine concentration was 9.34 ± 2.33 μmol/dl. This amount of aspartame is the projected 99th percentile of daily intake and is equivalent to a 50-kg person ingesting 8.5 12-oz cans of diet beverage (3000 ml) with the meal.

The results of the hamburger/milk shake meal study indicate that dietary protein modulates the rise in plasma phenylalanine concentration expected from the administered dose of APM. To test this hypothesis, we investigated plasma phenylalanine concentration in normal adults ingesting a soup/beverage meal with and without the addition of 34 mg of APM per kg body weight (Stegink et al. 1987c). The beef consommé provided 0.005 g of protein per kg body weight. Plasma phenylalanine concentration was not significantly changed from baseline values (5.81 ± 1.21 μmol/dl) after ingestion of the soup/beverage meal without APM. When 34 mg of APM per kg body weight was added, plasma phenylalanine concentration increased from a baseline value of 5.98 ± 1.08 μmol/dl to a high mean value of 14.5 ± 4.53 μmol/dl, a response similar to that noted when 34 mg of APM per kg body weight was ingested in orange juice (Stegink et al. 1977).

Subjects Reporting an Idiosyncratic Response to Ingestion of Monosodium L-Glutamate

It has been postulated that individuals reporting an idiosyncratic symptom response after glutamate ingestion might also experience symptoms after aspartame intake. Thus, we studied the response of such subjects to APM ingestion at 34 mg/kg body weight (Stegink et al. 1981a). Plasma phenylalanine concentrations increased over baseline values (5.66 ± 0.82 μmol/dl) to a high mean concentration of 13.0 ± 2.61 μmol/dl at 30 minutes after dosing. The high mean plasma phenylalanine concentrations did not differ significantly from values in normal subjects (11.1 ± 2.49 μmol/dl).

Studies in One-Year-Old Infants

These studies were carried out in three stages and all data were evaluated prior to proceeding to the next higher dose (Filer et al. 1983). In the first study, infants were administered 34 mg of APM per kg body weight in 180 ml of an unsweetened cherry-flavored beverage. In the second, infants were administered 50 mg of APM per kg body weight, and in the third, 100 mg per kg body weight.

For comparable doses of APM, the high mean value for plasma phenylalanine levels in infants was comparable to that in young adults (Table 3.7). These data indicate that infants metabolize the phenylalanine portion of APM as well as adults and that studies in adults can be extrapolated to infants and children.

Studies in Phenylketonuric Heterozygotes

Individuals heterozygous for phenylketonuria (PKU) represent a population that will metabolize the phenylalanine portion of APM less well than normal. The heterozygous state of phenylketonuria is estimated to occur in 1 in every 50 to 70 persons in the United States.

We studied the effects of APM loading in PKU heterozygotes and normal subjects administered 34 and 100 mg of APM per kg body weight in orange juice (Stegink et al. 1979b, 1980). Table 3.8 shows plasma phenylalanine concentrations in these subjects. Plasma phenylalanine concen-

TABLE 3.7. High mean plasma phenylalanine concentrations in one-year old infants and adults ingesting APM.

	High mean plasma Phe (μmol/dl)	
APM dose (mg/kg)	Adults	One-year olds
0	5.61 ± 1.42	4.92 ± 0.77
34	11.1 ± 2.49	9.37 ± 1.44
50	16.2 ± 4.86	11.6 ± 4.44
100	20.2 ± 6.77	23.3 ± 11.5

TABLE 3.8. High mean (± SD) plasma phenylalanine levels in normal and PKU heterozygotes ingesting APM.

APM dose	High mean plasma Phe ± SD (μmol/dl)	
	Normal subjects	PKU heterozygotes
0 mg/kg	5.61 ± 1.42	8.41 ± 2.05
10 mg/kg	6.73 ± 0.75	12.1 ± 2.08
34 mg/kg	11.1 ± 2.49	16.0 ± 2.25
100 mg/kg	20.2 ± 6.77	41.7 ± 2.33
10 mg/kg × 3 doses	8.10 ± 1.42	13.9 ± 2.15
600 mg × 8 doses	7.89 ± 1.36	16.5 ± 3.40

tration increased significantly after APM loading in both normal subjects and PKU heterozygotes, with levels in the PKU heterozygotes significantly higher. The high mean plasma phenylalanine concentration after ingestion of 34 mg of APM per kg body weight was 11.1 ± 2.49 μmol/dl and 16.0 ± 2.25 μmol/dl in normal subjects and PKU heterozygotes, respectively. When the number of normal subjects and heterozygous subjects was increased, similar results were noted (Stegink et al. 1981b). From the shape of the curves describing uptake and clearance of phenylalanine from plasma, it is evident that PKU heterozygotes clear phenylalanine less rapidly. In normal subjects and PKU heterozygotes ingesting APM at 100 mg/kg body weight, the high mean plasma phenylalanine concentration was 20.2 ± 6.77 μmol/dl and 41.7 ± 2.33 μmol/dl, respectively.

Based upon the observation that the plasma concentration-time curve of heterozygotes administered APM at 34 mg/kg body weight was similar to that of normal subjects administered APM at 50 mg/kg body weight, we concluded that heterozygotes metabolized the phenylalanine portion of aspartame approximately one-half as well as normal subjects. The possibility exists, however, that an aspartame dose of 34 mg/kg body weight represents the maximal quantity readily handled by the heterozygote. If this were true, the ingestion of large quantities of APM would exceed metabolic capacity and lead to a dramatic increase in plasma phenylalanine concentration. However, as shown in Figure 3.5, the high mean plasma phenylalanine concentration and area under the plasma phenylalanine concentration-time curve indicate that PKU heterozygotes ingesting APM at 100 mg/kg body weight metabolize the phenylalanine portion of APM at least as rapidly as normal adults metabolize and clear APM doses of 200 mg/kg body weight. Thus, it does not appear as though excessive intake of APM would overwhelm the uptake and clearance mechanisms for phenylalanine in phenylketonuric heterozygotes.

Our data in phenylketonuric heterozygotes are consistent with the data of Ford and Berman (1977). They measured the plasma phenylalanine response of 115 parent carriers of phenylketonuria, 40 offspring carriers, and 24 normal offspring administered 100 mg of phenylalanine per kg body

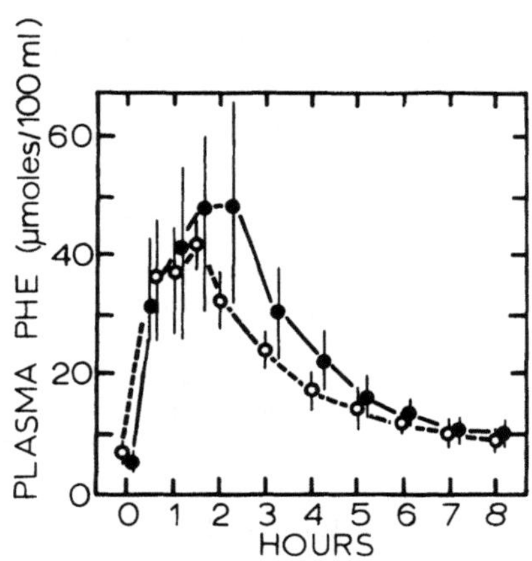

FIGURE 3.5. Mean (± SD) plasma phenylalanine levels in normal adults (●) administered APM at 200 mg/kg body weight and PKU heterozygotes (○) administered APM at 100 mg/kg body weight. Reproduced with permission from Stegink et al. (1980).

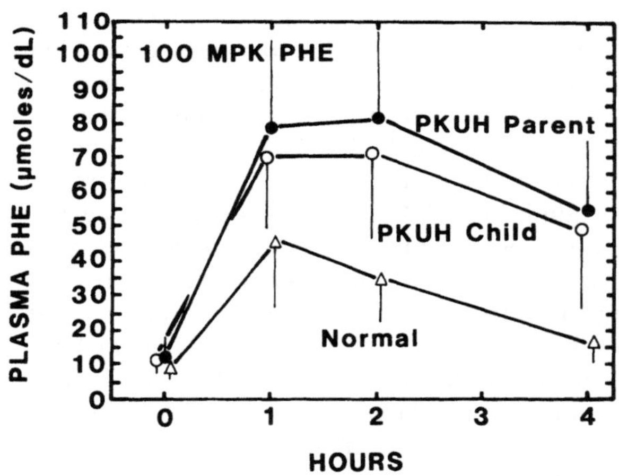

FIGURE 3.6. Mean (± SD) plasma phenylalanine concentrations in phenylketonuric heterozygous (PKUH) parents (●), PKUH children (○), and normal children (△) administered 100 mg L-phenylalanine per kg body weight. Graphic presentation of data from Ford and Berman (1977).

weight. Since 50% of the aspartame molecule is phenylalanine, their study was the equivalent of dosing with 200 mg of APM per kg body weight. Figure 3.6 presents the data of Ford and Berman (1977) in graphic form. In normal subjects administered phenylalanine at 100 mg/kg body weight, the high mean plasma phenylalanine concentration was 45.3 ± 18.5 μmol/dl. This value is in good agreement with the high mean plasma phenylalanine level that we observed in normal subjects administered APM at 200 mg/kg body weight (48.7 ± 15.5 μmol/dl; Table 3.2). Although we have not measured plasma phenylalanine concentrations in phenylketonuric heterozygotes administered 200 mg of APM per kg body weight, the data of Ford and Berman (Figure 3.6) permit us to predict such values. Ford and Berman (1977) reported high mean plasma phenylalanine concentrations of 82.0 ± 24.6 and 71.4 ± 24.6 μmol/dl, respectively, in parent and offspring heterozygotes ingesting 100 mg of phenylalanine per kg body weight. We would expect similar plasma phenylalanine concentrations in phenylketonuric heterozygotes ingesting 200 mg of APM per kg body weight.

After exploring the effects of large single bolus doses of APM in PKU heterozygotes, we investigated the effects of APM doses more likely ingested in a normal diet. In PKU heterozygotes, plasma phenylalanine concentrations (Table 3.8) increased from a baseline of 7.04 ± 1.71 μmol/dl to a high mean value of 12.1 ± 2.08 μmol/dl at 30 minutes after APM dosing at 10 mg/kg body weight. In contrast, normal subjects given the same quantity of APM increased their plasma phenylalanine concentration from 5.08 ± 0.82 μmol/dl to 6.73 ± 0.75 μmol/dl.

Studies in PKU Heterozygotes Ingesting Multiple Small Doses of Aspartame

We have evaluated the effects of successive ingestion of three beverage servings, each providing APM at 10 mg/kg body weight, on plasma phenylalanine concentrations in PKU heterozygotes (Baker et al. 1986). Beverage servings were ingested at 2-h intervals.

Table 3.8 lists the high mean plasma phenylalanine concentration observed in PKU heterozygotes during the six hours of study. Plasma phenylalanine concentration in these subjects increased from a baseline value of 7.04 ± 1.71 μmol/dl to a mean high value of 13.9 ± 2.15 μmol/dl. This value is lower than the mean high plasma phenylalanine concentration noted in PKU heterozygotes administered 34 mg of APM per kg body weight as a single serving (16.0 ± 2.25 μmol/dl).

We have also investigated the effects of multiple ingestion of APM in beverage by PKU heterozygotes given 600 mg of APM at hourly intervals over an 8-h period (Stegink et al. 1987d). The results of this study are also shown in Table 3.8. Plasma phenylalanine concentrations increased from a mean baseline level of 7.9 ± 1.4 μmol/dl to a high mean value of 16.5 ± 3.4 μmol/dl. The total APM dose was 4.8 g or 68 mg/kg body weight. This dose of APM ingested as a single bolus would be projected (Figure 3.1) to produce a high mean plasma phenylalanine concentration of 30 μmol/dl in PKU heterozygotes.

Phenylketonuric Homozygotes

APM-sweetened products are labeled "Phenylketonurics: Contains Phenylalanine," and homozygotes are discouraged from ingesting APM-sweetened products. However, adolescent homozygotes no longer on a phenylalanine-restricted diet often ask about the potential effect of inadvertent consumption of a 12-oz serving of APM-sweetened beverage on plasma phenylalanine concentration. To examine this point, plasma phenylalanine concentrations were measured in fasting adolescent phenylketonuric homozygotes after ingestion of 12 oz of APM-sweetened beverage providing 200 mg APM (Wolf-Novak 1987). Ingestion of the aspartame-sweetened beverage had no significant effect on baseline plasma phenylalanine concentrations (150 ± 23.0 μmol/dl). Thus, the ingestion of this quantity of APM had no significant effect on the already elevated plasma phenylalanine concentration produced by normal protein turnover.

Plasma Phenylalanine Response to Aspartame Loading Reported by Other Investigators

Plasma phenylalanine concentrations have been measured in several chronic feeding studies of APM. Frey (1976) reported no effect of APM on fasting plasma phenylalanine and tyrosine concentrations in 126 children and adolescents ingesting between 0.6 and 2.4 g of APM per day over a 13-week period. Knopp et al. (1976) measured plasma phenylalanine concentrations in 55 subjects ingesting 2.7 g of APM per day in capsules during a weight reduction program. In general, fasting plasma phenylalanine concentrations were not affected by dietary aspartame at this level. The maximum mean ± SD plasma phenylalanine concentration observed in the APM group was 7.08 ± 0.66 μmol/dl. Stern et al. (1976) measured fasting plasma phenylalanine concentrations in 43 adult diabetics ingesting 1.8 g of APM or placebo daily in capsules in a 90-day study. Fasting plasma phenylalanine concentrations after APM ingestion did not differ significantly from values in subjects ingesting placebo.

Koch et al. (1976a) measured fasting serum phenylalanine concentrations in 45 obligate phenylketonuric heterozygous adults ingesting 0.6 to 8.1 g of APM per day over a 27-week period. These authors reported no significant changes in serum phenylalanine concentration. Koch et al. (1976b) also measured plasma phenylalanine concentrations at selected time points in two PKU-homozygous adolescent boys ingesting 34 mg of APM per kg body weight or an equimolar quantity of phenylalanine. In subject 1, plasma phenylalanine increased from a baseline value of 112 μmol/dl to a value of 137 μmol/dl one hour after APM dosing. When subject 1 ingested an equimolar quantity of L-phenylalanine, plasma phenylalanine concentration increased from a baseline value of 105 μmol/dl to a value of 112 μmol/dl one hour after dosing. In subject 2, plasma phenylalanine concentration increased from a baseline value of 18 μmol/dl to a value of 27 μmol/dl one hour after dosing with 34 mg of APM per kg body weight. Ingestion of an equimolar dose of L-phenylalanine raised plasma phenylalanine concentration from a baseline value of 31 μmol/dl to 38 μmol/dl one hour after dosing. Güttler and Lou (1985) gave three 9-year-old children with classical phenylketonuria 34 mg of APM per kg body weight as a single bolus. They reported an increase in plasma phenylalanine concentration from a baseline value of 57 μmol/dl to a high mean value of 84 μmol/dl.

Yokogoshi et al. (1984) briefly described a preliminary study in normal adults given 15 mg of APM per kg body weight with 20 g of carbohydrate. Peak plasma phenylalanine and tyrosine concentrations were reported to be above the normal range, but no values or sampling times were given.

Garattini and colleagues (1986) recently reported mean plasma phenylalanine concentrations in normal adults administered APM at 0.83, 8.3, and 50 mg/kg body weight. They reported no change in plasma phenylalanine concentrations at APM doses of 0.83 and 8.3 mg/kg body weight. However, plasma phenylalanine concentrations increased from a mean baseline value of 5 μmol/dl to a high mean value of 14 μmol/dl after ingestion of an APM dose of 50 mg/kg body weight. These data agree well with the results of our earlier studies on APM dosing at 50 mg/kg body weight (Stegink 1984).

Caballero et al. (1986) recently reported plasma phenylalanine concentrations prior to and one hour after administering a single dose of APM at 10 mg/kg body weight to normal subjects, PKU homozygotes, mild hyperphenylalaninemics, and PKU heterozygotes. In normal subjects, plasma phenylalanine concentrations increased from a mean baseline value of 4.4 ± 1.3 μmol/dl to a value of 5.8 ± 0.9 μmol/dl at one hour. In PKU heterozygotes, plasma phenylalanine concentrations increased from a mean baseline value of 6.9 ± 1.4 μmol/dl to a value of 8.2 ± 1.7 μmol/dl at one hour. In mild hyperphenylalaninemics, plasma phenylalanine concentrations were not changed from baseline (41.2 ± 20.7 μmol/dl) one hour after APM dosing (41.3 ± 18.4 μmol/dl). Similarly, plasma phenylalanine concentrations in PKU homozygotes did not change from baseline (137 ± 24.0 μmol/dl) one hour after APM dosing (132 ± 21.0 μmol/dl) at 10 mg/kg body weight. However, the sampling time selected by these investigators (1 h) probably missed the peak plasma phenylalanine response to this dose of APM. Our data in normal subjects and PKU heterozygotes administered 10 mg of APM per kg body weight show peak plasma phenylalanine concentrations occur at 30 minutes after dosing, rather than one hour. In our studies, mean plasma phenylalanine concentrations at one hour postdosing were 84% of values seen at 30 minutes in normal subjects and were 88% of values seen at 30 minutes in PKU heterozygotes. Thus, the values observed by Caballero et al. (1986) are consistent with our data when the effect of sampling time is taken into consideration.

Matalon and colleagues reported plasma phenylalanine concentrations in normal adults and PKU heterozygotes administered APM as a slurry in a 12-week chronic feeding study (Matalon 1986, Matalon et al. 1987a, b). Unfortunately, the dose of APM administered was not clear. The first report (Matalon 1986) stated: "Each individual took 100 mg/kg of aspartame twice daily and random blood samples were taken every two weeks." The second report (Matalon et al. 1987a) did not list the dose. The third report (Matalon et al. 1987b) stated: "During the 12 weeks of aspartame intake (100 mg/kg/day) blood phenylalanine levels were determined every two weeks." A single blood sample was taken in the morning (presumably before dosing) and a second was taken at a random time in the afternoon. The investigators report (Matalon 1986) widely variable afternoon plasma phenylalanine concentrations in both normal subjects (range, 13.2 to 66 μmol/dl; mean $\pm$ SD, 28.1 $\pm$ 15.7 μmol/dl) and PKU heterozygotes (range, 28.8 to 103 μmol/dl; mean $\pm$ SD, 46.3 $\pm$ 23.9 μmol/dl). They suggested that their data indicated a wide variability in the ability of both normal subjects and PKU heterozygotes to metabolize aspartame and that this point was missed by previous investigators.

The high variability in plasma phenylalanine concentrations reported by Matalon and colleagues undoubtedly reflects slurry administration of aspartame rather than wide differences in metabolism. We have previously reported the high variability of the plasma phenylalanine concentration response when aspartame was administered as a slurry (Stegink et al. 1979a); these data have been reviewed earlier in this paper (Figures 3.2 and 3.3). These data indicate that the SD is approximately 73% of the mean when APM is administered as a slurry. Phenylketonuric heterozygotes administered 100 mg of APM per kg body weight in solution had a high mean plasma phenylalanine concentration of 41.7 μmol/dl. Although we do not have plasma phenylalanine data in phenylketonuric heterozygotes administered this dose of APM as a slurry, we can predict such values. Studies in normal subjects indicate that the high mean plasma phenylalanine response to administration of APM as a slurry or in solution was similar (Figure 3.2). These data also indicate that the SD after administration of APM as a slurry was approximately 73% of the mean. Thus, the high mean plasma phenylalanine level after administration of 100 mg of APM per kg of body weight as a slurry would approximate 42 $\pm$ 30 μmol/dl. A value 2 SD above the mean (96% of expected values fall within this range) would be 102.3 μmol/dl, a value in good agreement with the highest individual value (102.6 μmol/dl) reported by Matalon (1986).

Matalon et al. (1987a) also describe studies in normal adults and PKU heterozygotes given single loading doses of 50 and 100 mg of APM per kg body weight. There was no indication whether the aspartame was given in solution or as a slurry. They reported that peak plasma phenylalanine concentrations in normal subjects ranged from 11 to 19.3 μmol/dl at 50 mg of

APM per kg body weight and from 13.8 to 33 μmol/dl at 100 mg of APM per kg body weight. We reported a mean ± SD of 16.2 ± 4.86 μmol/dl at 50 mg of APM per kg body weight and 20.6 ± 6.70 μmol/dl at 100 mg of APM per kg body weight (Stegink 1984). Thus, a value 2 SD above the mean would be 26 μmol/dl at 50 mg/kg body weight and 34.3 μmol/dl at 100 mg/kg body weight. These values closely approximate the highest individual values reported by Matalon et al. (1987a). Similar calculations can be made for the results reported in PKU heterozygotes.

Potential for Adverse Effects at the Plasma Phenylalanine Concentrations Observed After APM Ingestion

The science of toxicology is based on the fact that all chemicals are toxic at some dose level. Thus, it is not surprising that high doses of phenylalanine produce adverse effects. The critical question is whether plasma phenylalanine concentrations are elevated to potentially harmful values at normal use levels of APM ingestion.

Table 3.9 summarizes the plasma phenylalanine concentrations in normal subjects, PKU heterozygotes, and patients with various forms of

TABLE 3.9. Plasma phenylalanine concentrations in various conditions.

Condition	Plasma Phe (μmol/dl)
Normal subjects	
Fasting	5.50 ± 2.50
Postprandial	9.75 ± 2.55
PKU heterozygotes	
Fasting	6.90 ± 1.40[a]
Hyperphenylalaninemics	
Classical PKU	137 ± 24.0[a]
Questionable variants	60 to 110
Mild hyperphenylalaninemics	41.2 ± 20.7[a]
Normal subjects	
70 mg of APM per kg as 8 divided doses	10.9 ± 2.49
50 mg of APM per kg as a single bolus	16.2 ± 4.86
34 mg of APM per kg as a single bolus	11.1 ± 2.46
30 mg of APM per kg as 3 divided doses	8.10 ± 1.42
10 mg of APM per kg as a single bolus	6.73 ± 0.75
4 mg of APM per kg as a single bolus	5.67 ± 0.48
PKU heterozygotes	
68 mg of APM per kg as 8 divided doses	16.5 ± 3.40
34 mg of APM per kg as a single bolus	16.0 ± 2.25
30 mg of APM per kg as 3 divided doses	13.9 ± 2.15
10 mg of APM per kg as a single bolus	12.1 ± 2.08

[a]Data from Caballero et al. (1986).

hyperphenylalaninemia. Normal fasting plasma phenylalanine concentrations are approximately 5.5 ± 2.5 μmol/dl. Normal postprandial plasma phenylalanine concentrations are about 9.75 ± 2.55 μmol/dl. Children with classical phenylketonuria have plasma phenylalanine concentrations ranging from 120 to 600 μmol/dl continually. These children are mentally retarded if not treated. Other children fall into the questionable variant grouping and have plasma phenylalanine concentrations ranging from 70 to 110 μmol/dl. These children may or may not become retarded if not treated with a phenylalanine-restricted diet, and they are usually treated. Some children have benign variant forms of hyperphenylalaninemia with plasma phenylalanine concentrations ranging from 24 to 60 μmol/dl. These children are not retarded and are usually not treated. The data shown in Table 3.9 indicate little reason to expect deleterious effects from the plasma phenylalanine concentrations produced by APM ingestion at normal use levels of the product.

Ratio of Plasma Phenylalanine Concentrations to the Sum of the Plasma Concentrations of the Large Neutral Amino Acids

Several investigators have suggested that simple evaluation of the high mean plasma phenylalanine concentrations may underestimate the potential for harmful effects at a given plasma phenylalanine concentration and suggest using the ratio of the plasma phenylalanine concentration to the sum of the plasma concentrations of the other large neutral amino acids (Phe/LNAA) sharing its transport site into brain (valine, methionine, isoleucine, leucine, tyrosine, and tryptophan). In the rat model, the transport of phenylalanine into brain was better reflected by the Phe/LNAA values than by plasma phenylalanine concentrations alone.

High mean plasma Phe/LNAA values in a typical cross section of our studies are shown in Table 3.10. For example, ingestion of a hamburger/milk shake meal does not raise the Phe/LNAA value above baseline values (0.100 ± 0.011) since the protein not only provides phenylalanine, but

TABLE 3.10. Plasma Phe/LNAA values in normal subjects ingesting APM.

Condition	High mean Phe/LNAA ± SD
Fasting	0.100 ± 0.02
Hamburger/milk shake meal	0.100 ± 0.015
34 mg of APM per kg in orange juice	0.230 ± 0.039
34 mg of APM per kg with hamburger/milk shake	0.130 ± 0.032
30 mg of APM per kg in 3 divided doses	0.161 ± 0.021
10 mg of APM per kg as a single dose	0.136 ± 0.020
4 mg of APM per kg as a single dose	0.111 ± 0.010

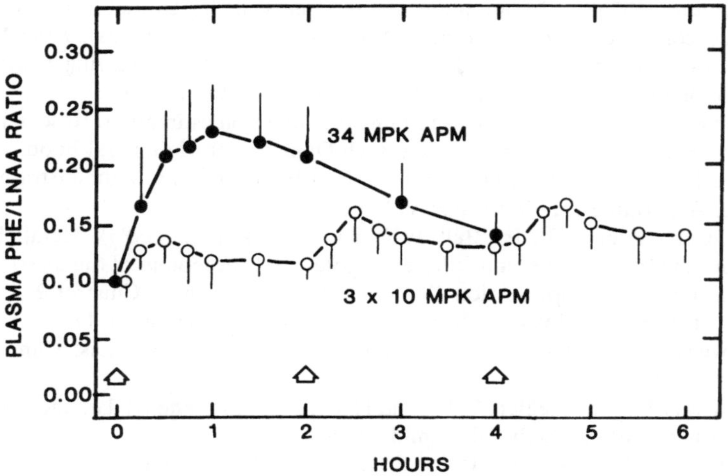

FIGURE 3.7. Mean (± SD) Phe/LNAA values calculated in eight normal adult subjects ingesting three 12-oz beverage servings, each providing aspartame at 10 mg/kg body weight (○). These values are compared with values in twelve normal adult subjects ingesting a single aspartame bolus of 34 mg/kg body weight (●). Data taken from Stegink et al. (1977, 1987a).

also valine, methionine, isoleucine, leucine, tyrosine, and tryptophan. The plasma levels of these amino acids also increase postprandially in proportion to their content in the meal, resulting in an unchanged Phe/LNAA value. In contrast, the ingestion of a single APM dose of 34 mg/kg body weight in orange juice increased the Phe/LNAA ratio from a baseline value of 0.102 ± 0.014 to a high mean value of 0.230 ± 0.039. When the same dose of APM was added to a hamburger/milk shake meal, the Phe/LNAA ratio increased from a baseline value of 0.100 ± 0.017 to a high mean value of 0.130 ± 0.032. As noted with free plasma phenylalanine concentrations, Phe/LNAA values are lower when a given dose is given in divided portions (Figure 3.7). Similarly, values are lower at lower doses of APM.

References

Abrams, I.J. (1986). Using the menu census survey to estimate dietary intake: post-market surveillance of aspartame. *In* Proceedings of the international workshop on aspartame, Session II, Chapter 2. International Life Sciences Institute-Nutrition Foundation, Washington, D.C.

Anonymous (1974). Title 21—Food and drugs. Chapter 1—Food and Drug Administration, Department of Health, Education, and Welfare. Subchapter B—Food and food products. Part 121—Food additives. Subpart D—Food additives permitted in food for human consumption. Aspartame. Fed. Reg. 39, 27317–27320.

Anonymous (1976a). Consumption of sweeteners in the United States and projected consumption of aspartame. A Report to General Foods by the Market Research Corporation. Food and Drug Administration, Hearing Clerk File, Administrative Record, Aspartame 75F–0355, File Volume 103.

Anonymous (1976b). Potential aspartame consumption estimation. Research Summary. Prepared by the Market Research Department of General Foods, March 1976. Food and Drug Administration, Hearing Clerk File, Administrative Record, Aspartame 75F–0355, File Volume 103.

Baker, G.L., Filer, L.J., Jr., Bell, E.F., and Stegink, L.D. (1986). Metabolism of aspartame by phenylketonuric heterozygotes. *In* Proceedings of the international aspartame workshop, Session IV, hyperphenylalaninemia, Chapter 2. International Life Science Institute-Nutrition Foundation, Washington, D.C.

Blaskovics, M.E. (1974). Phenylketonuria and phenylalaninaemias. Clin. Endocrinol. Metab. **3**:87–105.

Blaskovics, M.E., Schaeffler, G.E., and Hack, S. (1974). Phenylalaninaemias. Differential diagnosis. Arch. Dis. Child. **49**:835–843.

Caballero, B., Mahon, B.E., Rohr, F.J., Levy, H.L., and Wurtman, R.J. (1986). Plasma amino acid levels after single-dose aspartame consumption in phenylketonuria, mild hyperphenylalaninemia, and heterozgyous state for phenylketonuria. J. Pediatr. **109**:668–671.

Filer, L.J., Jr., Baker, G.L., and Stegink, L.D. (1983). Effect of aspartame loading upon plasma and erythrocyte free amino acid concentrations in one-year-old infants. J. Nutr. **113**:1591–1599.

Fischer, L.J. (1978). Personal communication.

Ford, R.C., and Berman, J.L. (1977). Phenylalanine metabolism and intellectual functioning among carriers of phenylketonuria and hyperphenylalaninemia. Lancet **i**:767–771.

Frey, G.H. (1976). Use of aspartame by apparently healthy children and adolescents. J. Toxicol. Environ. Health **2**:401–405.

Garattini, S., Caccia, S., and Salmona, M. (1986). Aspartame, brain amino acids and neurochemical mediators. *In* Proceedings of the international aspartame workshop, Session III, Central nervous system, Chapter 4. International Life Science Institute-Nutrition Foundation, Washington, D.C.

Güttler, F., and Lou, H. (1985). Aspartame may imperil dietary control of phenylketonuria. Lancet **i**:525–526.

Kirkpatrick, D.C. (1986). Canadian approaches to collecting consumption data for chemical substances. *In* Proceedings of the international workshop on aspartame, Session II, Chapter 3. International Life SciencesInstitute-Nutrition Foundation, Washington, D.C.

Knopp, R.H., Brandt, K., and Arky, R.A. (1976). Effects of aspartame in young persons during weight reduction. J. Toxicol. Environ. Health **2**:417–428.

Koch, R., Blaskovics, M., Wenz, E., Fishler, K., and Schaeffler, G. (1974). Phenylalaninemia and phenylketonuria. *In* Nyhan, W.L. (ed.), Heritable disorders of amino acid metabolism. Wiley: New York, pp. 109–140.

Koch, R., Shaw, K.N.F., Williamson, M., and Haber, M. (1976a). Use of aspartame in phenylketonuric heterozygous adults. J. Toxicol. Environ. Health **2**:453–457.

Koch, R., Schaeffler, G., and Shaw, K.N.F. (1976b). Results of loading doses of aspartame by two phenylketonuric (PKU) children compared with two normal children. J. Toxicol. Environ. Health **2**:459–469.

Matalon, R. (1986). Aspartame intake in carriers for phenylketonuria. *In* Proceedings of the international aspartame workshop, Session IV, Hyperphenylalaninemia, Chapter 3. International Life Science Institute-Nutrition Foundation, Washington. D.C.

Matalon, R., Michals, K., Sullivan, D., and Levy, P. (1987a). Aspartame intake and its effect on phenylalanine (Phe) and Phe metabolites. Pediatr. Res. **31**:344A (Abstract 1026).

Matalon, R., Michals, K., Sullivan, D., and Levy, P. (1987b). Aspartame consumption in normal individuals and carriers for phenylketonuria (PKU). *In* Wurtman, R.J., and Ritter-Walker, E. (eds.), Dietary phenylalanine and brain function, proceedings of the first international meeting on dietary phenylalanine and brain function. Cambridge, Massachusetts: Center for Brain Sciences and Metabolism Charitable Trust, pp. 81–93.

Pardridge, W.M. (1986a). Potential effects of the dipeptide sweetener aspartame on the brain. *In* Wurtman, R.J., and Wurtman, J.J. (eds.), Nutrition and the brain, Volume 7. New York: Raven Press, pp. 199–241.

Pardridge, W.M. (1986b). The safety of aspartame. J. Am. Med. Assoc. **256**:2678.

Sherlock, J.C. (1986). The estimation of intake. *In* Proceedings of the international workshop on aspartame, Session II, Chapter 4. International Life Sciences Institute-Nutrition Foundation, Washington, D.C.

Stegink, L.D. (1984). Aspartame metabolism in humans: acute dosing studies. *In* Stegink, L.D., and Filer, L.J., Jr. (eds.), Aspartame: physiology and biochemistry, New York: Marcel Dekker, pp. 509–553.

Stegink, L.D., Filer, L.J., Jr., and Baker, G.L. (1977). Effect of aspartame and aspartate loading upon plasma and erythrocyte free amino acid levels in normal adult volunteers. J. Nutr. **107**:1837–1845.

Stegink, L.D., Filer, L.J., Jr., Baker, G.L., and Brummel, M.C. (1979a). Plasma and erythrocyte amino acid levels of adult humans given 100 mg/kg body weight aspartame. Toxicology **14**:131–140.

Stegink, L.D., Filer, L. J., Jr., Baker, G.L., and McDonnell, J.E. (1979b). Effect of aspartame loading upon plasma and erythrocyte amino acid levels in phenylketonuric heterozygotes and normal adult subjects. J. Nutr. **109**:708–717.

Stegink, L.D., Filer, L.J., Jr., Baker, G.L., and McDonnell, J.E. (1980). Effect of an abuse dose of aspartame upon plasma and erythrocyte levels of amino acids in phenylketonuric heterozygous and normal adults. J. Nutr. **110**:2216–2224.

Stegink, L.D., Filer, L.J., Jr., and Baker, G.L. (1981a). Effect of aspartame and sucrose loading in glutamate-susceptible subjects. Am. J. Clin. Nutr. **34**:1899–1905.

Stegink, L.D., Koch, R., Blaskovics, M.E., Filer, L.J., Jr., Baker, G.L., and McDonnell, J.E. (1981b). Plasma phenylalanine levels in phenylketonuric heterozygous and normal adults administered aspartame at 34 mg/kg body weight. Toxicology **20**:81–90.

Stegink, L.D., Filer, L.J., Jr., and Baker, G.L. (1982). Effect of aspartame plus monosodium L-glutamate ingestion on plasma and erythrocyte amino acid levels in normal adult subjects fed a high protein meal. Am. J. Clin. Nutr. **36**:1145–1152.

Stegink, L.D., Filer, L.J., Jr., and Baker, G.L. (1983). Plasma amino acid concentrations in normal adults fed meals with added monosodium L-glutamate and aspartame. J. Nutr. **113**:1851–1860.

Stegink, L.D., Filer, L.J., Jr., and Baker, G.L. (1987a). Repeated ingestion of aspartame-sweetened beverage by normal adult subjects: effect on plasma amino acid concentration. Metabolism (in press).

Stegink, L.D., Filer, L.J., Jr., Bell, E.F., and Ziegler E.E. (1987b). Plasma amino acid concentrations in normal adults administered aspartame in capsules or solution: lack of bioequivalence. Metabolism **36**:507–512.

Stegink, L.D., Filer, L.J., Jr., and Baker, G.L. (1987c). Plasma amino acid concentrations in normal adults ingesting aspartame and monosodium L-glutamate as part of a soup-beverage meal. Metabolism **36**:1073–1079.

Stegink, L.D., Filer, L.J., Jr., Bell, E.F., and Ziegler, E.E. (1987d). Effects of repeated ingestion of aspartame-sweetened beverage on levels of plasma amino acids. Manuscript in preparation.

Stern, S.B., Bleicher, S.J., Flores, A., Gombos, G., Recitas, D., and Shu, J. (1976). Administration of aspartame in non-insulin-dependent diabetics. J. Toxicol. Environ. Health **2**:429–439.

Wolf-Novak, L. (1987). The effects of aspartame ingestion with and without carbohydrate on plasma phenylalanine and amino acid ratios in normal subjects and phenylketonurics. M.S. thesis, Department of Home Economics, The University of Iowa, Iowa City, Iowa.

Yokogoshi, H., Roberts, C.H., Caballero, B., and Wurtman, R.J. (1984). Effects of aspartame and glucose administration on brain and plasma levels of large neutral amino acids and brain 5-hydroxyindoles. Am. J. Clin. Nutr. **40**:1–7.

CHAPTER 4

Aspartame Consumption in Normal Individuals and Carriers for Phenylketonuria

Reuben Matalon, Kimberlee Michals, Debra Sullivan, Louise Wideroff, and Paul Levy *

Aspartame, a widely used sweetener, was studied in 53 adults, 28 of whom were carriers for phenylketonuria (PKU). The study was divided into a loading test followed by chronic intake for a period of 12 weeks. Two doses of aspartame were used in the loading test, 50 mg/kg and 100 mg/kg. Blood levels of phenylalanine and tyrosine were measured hourly, and the aromatic acid metabolites of phenylalanine were determined in urine. During the chronic intake, blood phenylalanine and tyrosine and urinary phenylalanine metabolites were determined every two weeks. The level of blood phenylalanine rose 5- to 10-fold when carriers for PKU ingested the 100-mg/kg load, as opposed to 2.5- to 6-fold in the normal controls. With a dose of 50 mg/kg, blood phenylalanine was increased from 2 to 7 times in the carriers, and from 2 to 3.5 times in the control group ($p < 0.02$). Blood tyrosine levels did not rise sharply, unlike blood phenylalanine. The normal blood phenylalanine-to-tyrosine ratio is 1.0. The ratio reached as high as 20.9 in carriers and 5.8 in normals. During 12 weeks of aspartame intake, blood phenylalanine levels were greater than 10 mg/dl in 12% of carriers for PKU and in 5% of the normal controls. Among the controls, the range of blood phenylalanine was 0.3 to 11.1 mg/dl. These data are of concern since blood phenylalanine levels are elevated above physiological levels even among normal individuals who take aspartame.

Introduction

Aspartame (L-aspartyl-L-phenylalanine methyl ester) is 180 to 200 times sweeter than sucrose (Nazur et al. 1960, Lelji et al. 1976, MacDonald et al. 1980, Kawai et al. 1980). The results of over 100 safety tests were submitted to the Food and Drug Administration (FDA) by G.D. Searle prior to the approval of aspartame in 1981 for consumption in the U.S. (Searle Food Resources 1979). Metabolic studies of aspartame in humans and laboratory animals have shown that following ingestion, the methyl ester is rapidly catalyzed by intestinal esterases to yield free methanol which is

* Departments of Pediatrics, Nutrition and Medical Dietetics, and Epidemiology and Biometry, University of Illinois at Chicago, Chicago, IL 60612, USA.

eventually oxidized (Opperman et al. 1973, Ranney et al. 1976). The free dipeptide is split at the mucosal surface by dipeptidases to aspartic acid and phenylalanine which are then absorbed.

Single loadings of normal adult humans (females and males) with aspartame at: (1) 34 mg/kg body weight, a dose suggested by Searle not to be exceeded by 99% of the population, and (2) "abuse" doses of up to 200 mg/kg represented no hazard with respect to the effects of the aspartyl residue and the released ester methanol or their metabolic products (Stegink et al. 1977, 1981). However, phenylalanine utilization is comparatively slower, and the ingestion of 34 mg of aspartame per kg body weight increased serum phenylalanine levels in normal human adults to 11.1 μmol/dl, and by twice this amount in carriers for phenylketonuria (PKU) (Stegink et al. 1979a, b). Elevated plasma phenylalanine concentrations ten times greater than normal were found after loading with doses of 100 to 200 mg/kg (Stegink et al. 1979a, 1981). These levels were considered "safe" based on the lack of acute deleterious effects during the short-term period of the test.

Population at Risk for Aspartame Consumption

Since its introduction in 1981, aspartame has gained a great deal of popularity among consumers. Consumer data from G.D. Searle show that in 1984 approximately 3,500 tons of aspartame were consumed. With this phenomenal rise in aspartame acceptability, it would not be an exaggeration to suggest that aspartame consumption may triple in the next decade. This means that the consumption data of "normal" use and "abuse" levels suggested early by G.D. Searle may not be relevant and that consumption of 50 to 100 mg of aspartame per kg body weight per day may not be that uncommon. Frey (1976) has indicated that children 7 to 12 years old may consume over 75 mg of aspartame per kg per day. Individuals consuming 50 to 100 mg of aspartame per kg per day may be at risk for increased levels of blood phenylalanine.

Heterozygotes for PKU with one normal allele for phenylalanine hydroxylase are usually not detected since they are clinically normal, maintaining about 15 to 50% (Tourian and Sidbury 1983) of the normal phenylalanine hydroxylase activity. Combinations of defective recessive genes show a spectrum of symptoms from moderate to mild severity, classified as hyperphenylalaninemia (Woolf 1971, Cotton 1977, Kaufman 1976), and have approximately 5 to 15% of the normal enzyme activity (Tourian and Sidbury 1983). A seven-week survey of 45 heterozygotes, given increasing amounts of aspartame, revealed no significant medical or biochemical changes (Koch et al. 1976a). The same was true for two adolescent PKU patients loaded once with 34 mg of aspartame per kg body weight and monitored for three days (Koch et al. 1976b). Studies on PKU heterozygotes given single aspartame loads of 34 mg/kg (Stegink et al. 1979c) or "abuse" doses of 100 mg/kg have shown no significant effects on erythrocyte aspartic acid and only slight increases of aspartic acid in plasma, but up to twice the normal plasma phenylalanine levels have been detected in these experiments. Since an intake of 50 to 100 mg of aspartame per kg body weight per day may not be that uncommon, heterozygotes for PKU are at risk for elevated blood phenylalanine levels.

Studies of Aspartame Safety

While the experiments regarding the safety of aspartame were encouraging in humans, they were usually restricted in scope and offered little guidance with respect to possible long-term effects. Investigators have only addressed themselves to plasma phenylalanine levels without taking into account that phenylalanine toxicity is not acute. Since the nature of the disease PKU is not acute, blood phenylalanine levels that are above physiological levels may not be the sole parameter of safety. Indeed, neuropsychological disturbances have been reported when blood phenylalanine rises above 10 mg/dl in PKU patients (Anderson and Siegel 1976, Brunner et al. 1983, Griffin et al. 1980, Krause et al. 1985, Low et al. 1985). No such studies have been performed on normals or carriers consuming aspartame. The data supporting the statement that aspartame ingestion is safe have been based primarily on (1) single loading studies, (2) blood phenylalanine levels, and (3) acceptable blood phenylalanine levels for PKU patients, which may be as high as 10 mg/dl. A blood phenylalanine level of 10 mg/dl should not be considered a safe level for normal school children or pregnant women.

Aspartame and Pregnancy

To the best of our knowledge, there are no data regarding the influence of aspartame, if any, on pregnant women where normal metabolism is dramatically altered and phenylalanine may rise to three times the preconception levels (Kang and Paine 1963). It is known that elevated blood phenylalanine levels in PKU women result in a high incidence of birth defects (Lenke and Levy 1980). Intermittent rise of blood phenylalanine following aspartame intake may be detrimental to the developing fetus. Indeed, fetuses of hyperphenylalaninemic mothers may have lower IQs and a higher incidence of developmental abnormalities (Bessman et al. 1978). The National Collaborative Study for Maternal PKU has recommended that during pregnancy blood phenylalanine levels should not exceed 6 mg/dl (Acosta et al. 1985).

Methods

Fifty-three healthy adults were enrolled in the aspartame study 28 of whom were obligate carriers for PKU. The study was divided into three subgroups: (1) aspartame load with 50 mg/kg; (2) aspartame load with 100 mg/kg; and (3) aspartame intake over 12-week period (100 mg/kg per day in two divided doses). During the loading experiments, blood phenylalanine and tyrosine levels were determined at baseline and at 1 hour and 3 hours. The organic acid metabolites of phenylalanine and tyrosine in urine were assayed at baseline and 3, 8, and 24 hours following the aspartame load. During 12 weeks of chronic aspartame consumption, blood samples were obtained every two weeks for determination of phenylalanine and tyrosine. The phenylalanine and tyrosine analyses were done fluorometrically (McCaman and Robins 1962, Udenfriend and Cooper 1952). Urine samples (A.M. and P.M.) were collected every two weeks for extraction of the organic acid metabolites and phenylethylamine. The organic acid metabolites and phenylethylamine were determined by gas chromatography-mass spectrometry (Goodman and Markey 1981, Blau and Claxton 1979).

Results

Aspartame Loading

Blood phenylalanine levels were determined on 28 carriers for PKU and 25 normal individuals following aspartame loading. The following loading doses were given: 50 mg/kg (16 PKU carriers and 18 normal individuals) and 100 mg/kg (12 PKU carriers and 7 normal individuals). Blood phenylalanine was determined at baseline and at 1 hour and 3 hours following loading. Phenylalanine loading data are summarized in Table 4.1. The differences between carriers and normals are statistically significant at 1 hour and 3 hours. The rise in blood phenylalanine in both groups is also statistically significant. The mean increases in blood phenylalanine from baseline to 1 hour is seen in Table 4.2. The increase from baseline is also statistically significant.

Aspartame Intake for 12 Weeks

During the 12 weeks of aspartame intake (100 mg/kg per day), blood phenylalanine levels were randomly determined every two weeks. The re-

TABLE 4.1. Mean levels of phenylalanine following aspartame loading.

Group (aspartame dose)	Phenylalanine level[a] (mg/dl) at: Baseline	1 hour	3 hours
Normals (50 mg/kg)	1.63 ± 0.32 (18)	2.59 ± 0.22 (18)	1.82 ± 0.11 (18)
Carriers for PKU (50 mg/kg)	1.58 ± 0.11 (16)	3.66 ± 0.36 (18)	2.54 ± 0.22 (18)
Normals (100 mg/kg)	1.41 ± 0.19 (9)	3.60 ± 0.51 (9)	2.80 ± 0.40 (7)
Carriers for PKU (100 mg/kg)	1.74 ± 0.12 (12)	7.78 ± 0.92 (12)	4.76 ± 0.44 (12)

[a] Values are mean ± SE for number of subjects given in parentheses.

TABLE 4.2. Increase in mean blood phenylalanine levels from baseline to 1 hour after aspartame loading.

Group (aspartame dose)	n	Mean ± SE phenylalanine level (mg/dl)	p
Normals (50 mg/kg)	18	0.96 ± 0.31	
Carriers for PKU (50 mg/kg)	16	2.13 ± 0.39	<0.03
Normals (100 mg/kg)	9	2.29 ± 0.51	
Carriers for PKU (100 mg/kg)	12	6.03 ± 0.89	<0.01

TABLE 4.3. Ranges of blood phenylalanine levels at biweekly intervals during 12 weeks of aspartame intake (100 mg/kg).[a]

	Number of individuals with phenylalanine levels (mg/dl) in range:			
Random blood sample	0.0–5.9	6.0–7.9	8.0–9.9	10+
1st				
Normals	21	0	0	1
Carriers for PKU	13	3	0	2
2nd				
Normals	21	0	0	0
Carriers for PKU	16	0	1	0
3rd				
Normals	22	0	0	1
Carriers for PKU	11	2	2	2
4th				
Normals	16	6	0	0
Carriers for PKU	13	2	1	1
5th				
Normals	18	2	0	1
Carriers for PKU	13	1	1	1

[a]Mantel extension analysis: $X^2_{\text{m-Est}} = 11.02$, $p < 0.01$.

sults are shown in Table 4.3. The data are summarized according to blood phenylalanine ranges. The group of individuals with blood levels below 6 mg/dl are not subdivided. The group above 6 mg/dl is further subdivided into individuals in the range of 8 to 10 mg/dl and those above 10 mg/dl. Blood phenylalanine concentrations greater than 6 mg/dl were chosen because in pregnant females with PKU (and therefore in any pregnant female), the level of blood phenylalanine should be below 6 mg/dl (Acosta et al. 1985).

These data indicate that the prevalence of elevated blood phenylalanine levels is higher in carriers than in controls. Of concern is the fact that six carriers had phenylalanine levels of 10.8 to 18.8 mg/dl. Such levels have not been reported in any previous studies. In the noncarrier group, two individuals had levels above 10 mg/dl (11.0 to 11.1 mg/dl). These data indicate

TABLE 4.4. Summary statistics on the random blood phenylalanine levels during 12 weeks of aspartame intake.

	Blood sample				
	1st	2nd	3rd	4th	5th
Proportion > 6.0 mg/dl					
Normals	0.05	0.00	0.00	0.11	0.14
Carriers for PKU	0.28	0.06	0.35	0.24	0.19
Proportion > 8.0 mg/dl					
Normals	0.05	0.00	0.00	0.00	0.05
Carriers for PKU	0.11	0.06	0.24	0.12	0.13
Proportion > 10.0 mg/dl					
Normals	0.05	0.00	0.00	0.00	0.05
Carriers for PKU	0.11	0.00	0.12	0.06	0.06

TABLE 4.5. Mean levels of tyrosine following aspartame loading.

	Tyrosine level[a] (mg/dl) at:		
Group (aspartame dose)	Baseline	1 hour	3 hours
Normals (50 mg/kg)	1.76 ± 0.27 (13)	1.68 ± 0.16 (13)	1.93 ± 0.22 (13)
Carriers for PKU (50 mg/kg)	1.42 ± 0.16 (9)	1.41 ± 0.14 (11)	1.60 ± 0.16 (11)
Normals (100 mg/kg)	1.49 ± 0.20 (9)	2.10 ± 0.22 (9)	2.17 ± 0.41 (9)
Carriers for PKU (100 mg/kg)	1.54 ± 0.11 (11)	1.85 ± 0.18 (11)	2.14 ± 0.20 (11)

[a] Values are mean ± SE for number of subjects given in parentheses.

that it is important to study chronic intake of aspartame and not only single loading experiments. The proportion of individuals with blood phenylalanine above 6 mg/dl during the 12 weeks of aspartame intake is summarized in Table 4.4.

The data in Table 4.4 are significant and important for both carriers and normals who take aspartame at the 100 mg/kg per day level. The data indicate that 14% of normals exceed 6 mg/dl and 35% of carriers exceed that level. Of even greater concern are the levels exceeding 10 mg/dl which were found in 5% of normals and 12% of carriers. In terms of absolute numbers, the 5% of normals constitute a large proportion of the population.

Tyrosine blood levels changed only slightly following aspartame loading. The data on blood tyrosine levels are summarized in Table 4.5. The mean ratios of phenylalanine to tyrosine are indicated in Table 4.6. The ratios in the carriers are higher than in the normals. The individual ratios in certain carriers reached 21.0 on 100 mg of aspartame per kg and 8.0 on 50 mg per kg. In the normal individuals, the ratios of phenylalanine to tyrosine were also abnormal and reached 3.4 on 50 mg of aspartame per kg and 6.0 on 100 mg per kg.

In summary, the data on the effect of a loading dose of aspartame of 50

TABLE 4.6. Mean ratio of phenylalanine to tyrosine following aspartame load.

Group (aspartame dose)	Phenylalanine/tyrosine ratio[a] at: Baseline	1 hour	3 hours
Normals (50 mg/kg)	1.19 ± 0.41 (13)	1.77 ± 0.37 (13)	1.30 ± 0.27 (13)
Carriers for PKU (50 mg/kg)	1.25 ± 0.26 (9)	3.26 ± 0.66 (11)	2.14 ± 0.52 (11)
Normals (100 mg/kg)	1.38 ± 0.24 (9)	1.92 ± 0.18 (9)	1.71 ± 0.39 (9)
Carriers for PKU (100 mg/kg)	1.22 ± 0.06 (11)	5.75 ± 1.56 (11)	1.85 ± 0.56 (11)

[a]Values are mean ± SE for number of subjects given in parentheses.

mg/kg per day or 100 mg/kg indicate a statistically significant rise in blood phenylalanine levels in both the normals and carriers. There is also a significant difference in blood phenylalanine between the normals and carriers. Due to the fact that blood tyrosine levels did not rise significantly, the normal ratio of phenylalanine to tyrosine becomes abnormally elevated. During the 12 weeks of chronic aspartame intake of 100 mg/kg, there was a significant rise in blood phenylalanine in normal individuals and carriers for PKU, with 5% and 12% of the subjects, respectively, greater than 10 mg/dl.

Aromatic Acids of Phenylalanine and Tyrosine

The aromatic acid metabolites of phenylalanine—phenylpyruvate, phenyllactate, and phenylacetate—are excreted in large amounts in urine of patients with PKU. The interesting finding in this study was that neither carriers for PKU nor normal subjects excreted large amounts of these compounds. Thus, the major aromatic acid metabolites of phenylalanine were the *hydroxy* derivatives, which are most likely the result of non-enzymatic hydroxylation of the aromatic acid derivatives of phenylalanine. The data from the 50-mg/kg aspartame load are shown in Table 4.7 and the data

TABLE 4.7. Mean levels of aromatic acid metabolites of phenylalanine/tyrosine in urine following 50mg/kg aspartame load.

			Mean level mg/gm creatinine of: 3-hydroxy phenylacetate		4-hydroxy phenylacetate		4-hydroxy phenyllactate	
Subjects	n	Time	$\bar{X}$	S.E.	$\bar{X}$	S.E.	$\bar{X}$	S.E.
Normals	11	0 min	167	81	387	121	102	47
		3 h	281	66	393	67	97	57
		8 h	243	57	591	119	194	29
Carriers for PKU	15	0 min	271	52	533	67	123	45
		3 h	312	49	527	76	80	24
		8 h	599	326	1181	504	183	56

TABLE 4.8. Mean levels of aromatic acid metabolites of phenylalanine/tyrosine in urine following 100mg/kg aspartame load.

			Mean level mg/gm creatinine of:					
			3-hydroxy phenylacetate		4-hydroxy phenylacetate		4-hydroxy phenyllactate	
Subjects	*n*	Time	$\bar{X}$	S.E.	$\bar{X}$	S.E.	$\bar{X}$	S.E.
Normals	7	0 min	286	105	531	131	282	154
		3 h	584	166	1171	370	446	144
		8 h	481	187	847	241	399	152
Carriers for PKU	9	0 min	237	146	389	217	173	98
		3 h	951	372	1644	659	413	160
		8 h	353	66	821	222	219	45

TABLE 4.9. Mean levels of aromatic acid metabolities of phenylalanine/tyrosine in random urine samples during 12 weeks of aspartame.

		mg/gm creatinine					
		carriers for PKU			normals		
Sample		n	$\bar{X}$	S.E.	n	$\bar{X}$	S.E.
Baseline	AM	23	730	242	18	892	325
(0)	PM	22	1355	395	18	930	372
Week 2	AM	22	1728	534	22	1270	283
(1st)	PM	18	1331	203	20	1231	195
Week 4	AM	19	1324	260	21	1217	187
(2nd)	PM	13	974	123	18	1430	330
Week 6	AM	18	1143	199	20	1341	198
(3rd)	PM	15	1220	164	17	1190	165
Week 8	AM	18	1887	315	20	1438	284
(4th)	PM	15	1376	195	20	1436	256
Week 10	AM	18	1994	383	19	1469	334
(5th)	PM	16	1381	234	18	1553	329

* Sum = 3-hydroxy phenylacetate + 4-hydroxy phenylacetate + 4-hydroxy phenyllactate.

from the 100-mg/kg load are shown in Table 4.8. The differences between carriers and normals seen in these tables are not statistically significant due to the large standard deviations. There are considerable variations in the values for each individual. Nevertheless, a sharp rise in these metabolites is not considered normal.

During the 12 weeks of chronic aspartame consumption, random urine samples were collected for analysis of the aromatic acid derivatives of phenylalanine and tyrosine. The levels of these metabolites are summarized in Table 4.9. The differences between the carriers for PKU and normals are not statistically significant. However, the surprising finding is that both carriers and normals excreted a wide range of these metabolites. Five of the 13 normals had levels that exceeded the baseline level by 8 times and of the 15 carriers, 11 had random levels that exceeded the baseline level by 8 times. Since these metabolites are very low in normal urines, intermittent surges of high levels should be of concern.

Phenylethylamine

Phenylethylamine (PEA) is a biogenic amine that is primarily derived from phenylalanine. It is a labile compound and is degraded by monoamine oxidase (MAO). As in patients with PKU there are increased concentrations of PEA, this compound was examined in urines of the individuals taking aspartame.

The random urine samples collected during the 12 weeks were analyzed for PEA. Urine from 10 carriers for PKU and 11 normal individuals were analyzed. There was no statistically significant difference between carriers and normals, but all 21 individuals had intermittently high levels of PEA during the 12-week experimental period, exceeding by a factor of 5 the baseline levels of 2.2 to 5.8 mg creatinine. Since an intermittent rise in PEA may be of concern regarding behavior, concentration, and learning, such studies need to be extended.

Discussion and Conclusions

The data presented in this study are of importance to carriers for PKU and normal individuals who consume aspartame. Since the incidence of PKU carriers in the general population is 1/50, and since most carriers are unknown, this study is of significance to that population. In our studies, we have found that blood phenylalanine significantly rises above physiological levels in both normals and carriers for PKU. This rise in concentration is significant even with a dose of 50 mg/kg. The idea that blood phenylalanine levels of 10 mg/dl or higher are "safe" has no physiological basis. These data are generated primarily from literature dealing with the treatment of PKU and should not be used to extend "safety" to above physiological levels. Another important issue is the question of allowed daily intake (ADI) for aspartame consumption. Currently, the FDA has suggested an ADI of 50 mg/kg. In our studies, this level of intake leads to significant rises in blood phenylalanine, in both PKU carriers and normal subjects. Therefore, the ADI should be scaled down so that the intake of aspartame does not lead to above-physiological levels of blood phenylalanine and distort the ratio of phenylalanine to tyrosine or other essential amino acids.

It is also important to realize that the data suggesting that aspartame consumption of 34 mg/kg per day would constitute the 99th percentile may not be realistic. For example, if a 7- or 8-year-old child were to consume 1 milk shake, 1 slice of cheesecake, and 1 can of a beverage that were all sweetened with aspartame, his consumption of aspartame would be approximately 50 mg/kg. Therefore, relying on unrealistic consumption data may lead to a false sense of security that no one will consume high levels of aspartame.

The biochemical data obtained thus far following aspartame ingestion show statistically significant differences when phenylalanine levels in carriers for PKU are compared to those in normal individuals. The surprising finding is that normal controls are at risk for having significantly high blood phenylalanine levels. Since the tyrosine levels did not significantly rise as did the phenylalanine levels, the ratio of phenylalanine to tyrosine became abnormally high. The increased acceptance of aspartame by the general population suggests that intakes of 50 mg/kg per day or higher may not be a rare occurrence. In our study, 5% of normal individuals had blood levels of phenylalanine exceeding 10 mg/dl while consuming 100 mg of aspartame per kg per day. On the same dose, 12% of the carriers for PKU exceeded 10 mg/dl. If the level of 6 mg/dl is adopted as safe for pregnancy, then 14% of the normal population and 35% of the carriers for PKU may be at risk to exceed such a level with an aspartame intake of 100 mg/kg per day.

Studies regarding the metabolites of phenylalanine and tyrosine need to be extended. These metabolites were elevated above baseline with considerable individual variations. Since the role of increased levels of these metabolites is unclear, they should be further studied.

Based on these data, two immediate conclusions can be made:

1. The ADI should be calculated so that blood phenylalanine does not rise above physiological levels, taking into account the heterozygote population for PKU in this country.
2. Aspartame content and the ADI should be specified exactly on each product containing aspartame.

Acknowledgments. This study was supported in part by grant HD-191-50-01A1 from the National Institutes of Health, USA.

References

Acosta, P.B., Michals, K., Castiglioni, L., Rohr, F., and Wenz, E. (1985). Protocol for nutrition support of maternal PKU. Supported by DHHS and NICHHD. Department of Health and Human Services, National Institute of Child Health and Human Development.

Anderson, V.E., and Siegel, F.S. (1976). Behavioral and biochemical correlates of diet change in phenylketonuria. Pediatr. Res. **10**: 10–17.

Bessman, S.P., Williamson, M.L., and Koch, R. (1978). Diet, genetics, and mental retardation interaction between phenylketonuric heterozygous mother and fetus to produce nonspecific diminution of IQ: evidence in support of the justification hypothesis. Proc. Natl. Acad. Sci. USA **75**: 1562–1566.

Blau, K., and Claxton, I.M. (1979). Urinary phenylethylamine excretion: gas chromatographic assay with electron-capture detection of the pentafluorobenzoyl derivative. J. Chromotography **163**: 135–142.

Brunner, R.L., Jordan, M.K., and Berry, H.K., (1983). Early-treated phenylketonuria: neuropsychologic consequences. J. Pediatr. **102**: 831–835.

Cotton, R.G.H., (1977). The primary molecular defects in phenylketonuria and its effects. Int. J. Biochem. **8**: 333–341.

Frey, G.H. (1976). Use of aspartame by apparently healthy children and adolescents. J. Toxicol. Environ. Health **2**: 401–415.

Goodman, S., and Markey, S.P. (1981). Diagnosis of organic acidemias by gas chromatography. *In* Goodman, S., and Markey, S. (eds.), Lab and research methods in biology and medicine. pp. 1–43. New York, Alan R. Liss.

Griffin, F.D., Clarke, J.T.R., and d'Entremont, D.M. (1980). Effect of dietary phenylalanine restriction on visual attention span in mentally retarded subjects with phenylketonuria. Can. J. Neurol. Sci. **7**: 127–131.

Kang, E., and Paine, R.S. (1963). Elevation of plasma phenylalanine during pregnancies of women heterozygous for phenylketonuria. J. Pediatr. **63**: 283–289.

Kaufman, S., (1976). The phenylalanine hydroxylating system in phenylketonuria and its variants. Biochem. Med. **15**: 42–52.

Kawai, M., Chorev, M., Marin-Rose, J., and Goodman, M. (1980). Peptide sweeteners 4-hydroxy and methoxy substitution of the aromatic ring in L-aspartyl-L-phenylalanine methyl ester. Structure-taste relationships. J. Med. Chem. **23**: 420–424.

Koch, R., Shaw, K.N.F., Williamson, M., and Haber, M. (1976a). Use of aspartame in phenylketonuric heterozygous adults. J. Toxicol. Environ. Health **2**: 453–457.

Koch, R., Schaeffler, G., and Shaw, K.N.F. (1976b). Results of loading doses of aspartame by two phenylketonuric (PKU) children compared with two normal children. J. Toxicol. Environ. Health **2**: 459–469.

Krause, W., Halminski, M., McDonald, L., Dembure, P., Salvo, R., Freides, D., and Elsas, L. (1985). Biochemical and neuropsychological effects of elevated plasma phenylalanine in patients with treated phenylketonuria. J. Clin. Invest. **75**: 40–48.

Lelji, F., Tancredi, T., Temussik, P.A., and Tonioloc, C. (1976). Interaction of alpha-aspartyl-L-phenylalanine methyl ester with the receptor site of a sweet taste bud. J. Am. Chem. Soc. **98**: 6669–6675.

Lenke, R.R., and Levy, J.L. (1980). Maternal phenylketonuria and hyperphenylalaninemia. N. Engl. J. Med. **303**: 1202–1208.

Lou, H.C., Güttler, F., Lykkelund, C., Bruhn, P., and Niederwieser, A. (1985). Decreased vigilance and neurotransmitter synthesis after discontinuation of dietary treatment for phenylketonuria in adolescents. Eur. J. Pediatr. **144**: 17–20.

MacDonald, S.A., Willson, C.G., Chorev, M., Varenacchia, F.S., and Goodman, M. (1980). Peptide sweeteners 3: Effect of modifying the peptide bond on the sweet taste of L-aspartyl-L-phenylalanine methyl ester and its analogues. J. Med. Chem. **23**: 413–420.

McCaman, M.W., and Robins, E. (1962). Fluorometric method for the determination of phenylalanine in serum. J. Lab. Clin. Med. **59**: 885–890.

Nazur, R.H., Schlatter, J.M., and Goldkamp, A.H., (1960). Structure-taste relationships of some dipeptides. J. Am. Chem. Soc. **91**: 2684–2691.

Opperman, J.A., Muldoon, E., and Ranney, R.E. (1973). Metabolism of aspartame in monkeys. J. Nutr. **103**: 1454–1459.

Ranney, R.E., Opperman, J.A., Muldoon, E., and McMahon, F.G. (1976). Comparative metabolism of aspartame in experimental animals and humans. J. Toxicol. Environ. Health **2**: 441–451.

Searle Food Resources (1979). Thoroughly tested food additive: results of aspartame safety tests. Searle Food Resources, Inc., Information Letter No. 564, Sept. 1979.

Stegink, L.D., Filer, L.J., Jr., and Baker, G.L. (1977). Effects of aspartame and aspartate loading upon plasma and erythrocyte free amino acid levels in normal adult volunteers. Toxicology **14**: 131–140.

Stegink, L.D., Filer, L.J., Jr., and Baker, G.L. (1981). Plasma and erythrocyte concentrations of free amino acids in adult humans administered abuse doses of aspartame. J. Toxicol. Environ. Health **7**: 291–305.

Stegink, L.D., Filer, L.J., Jr., Baker, G.L., and Brummel, M.C. (1979a). Plasma and erythrocyte amino acid levels of adult humans given 100 mg/kg body weight aspartame. Toxicology **14**: 131–140.

Stegink, L.D., Filer, L.J., Jr., Baker, G.L. (1979b). Plasma phenylalanine levels in phenylketonuric heterozygous and normal adults administered aspartame 34 mg/kg body weight. Toxicology **20**: 81–90.

Stegink, L.D., Filer, L.J., Jr., Baker, G.L., and McDonnell, J.E. (1979c). Effects of aspartame loading upon plasma and erythrocyte amino acid levels in phenylketonuric heterozygotes and normal adult subjects. J. Nutr. **109**: 708–717.

Tourian, A.Y., and Sidbury, J.B. (1983). *In* Stanbury, J.B., Wyngaarden, J.B., and Frederickson, D.S. (eds.), The metabolic basis of inherited disease. New York: McGraw-Hill, Ch. 12.

Udenfriend, S., and Cooper, J.T. (1952). The enzymatic conversion of phenylalanine to tyrosine. J. Biol. Chem. **194**: 503–511.

Woolf, L.I. (1971). *In* Bickel, H., Hudson, F.P., and Woolf, L.I. (eds.), Phenylketonuria and some other inborn errors of metabolism. Stuttgart: Georg Thieme Verlag p. 103.

Part II Blood-Brain Barrier Transport of Phenylalanine

CHAPTER 5

Phenylalanine Transport at the Human Blood-Brain Barrier

William M. Pardridge *

Phenylalanine transport through the brain capillary wall, i.e., the blood-brain barrier (BBB), is characterized by a very high affinity (low K_m) transport system in laboratory rats. To test whether phenylalanine transport at the human BBB is also mediated by a very high affinity (low K_m) transport system, brain capillaries were isolated from fresh autopsy human brain, as well as from fresh rat brain or from fresh or 42-hour postmortem rabbit brain. The results show that the K_m values for phenylalanine transport are virtually identical at either the rat or human BBB. Therefore, the marked derangements in brain amino acid metabolism caused by hyperphenylalaninemia in rats may also occur in humans at comparable increases in blood phenylalanine.

Phenylketonuria (PKU) is an inborn error of phenylalanine metabolism in humans that is characterized by mental retardation, seizures, and spasticity (Scriver and Clow 1980). Investigations reported by a number of laboratories over the last several years have indicated that an important cause of the derangements of brain amino acid metabolism in PKU is a selective saturation of neutral amino acid transport through the human brain capillary wall, i.e., blood-brain barrier (BBB) (Oldendorf et al. 1971, Andersen and Avins 1976, Pardridge and Oldendorf 1977; Pardridge and Choi, 1986). In experimental hyperphenylalaninemia, the impairment of brain protein synthesis can be normalized by the coadministration of other large neutral amino acids (Binek-Singer and Johnson 1982) that compete with phenylalanine at the neutral amino acid carrier sites within the brain capillary (Oldendorf et al. 1971, Pardridge and Oldendorf 1977). The BBB neutral amino acid transport system has a very high affinity (very low K_m) for the large neutral amino acids. Phenylalanine has the highest affinity for this transport system among the neutral amino acids in blood (Pardridge and Oldendorf 1977, Miller et al. 1985). Studies in the laboratory rat have shown that the very low K_m, e.g., 25 to 100 μM, of neutral amino acid

*Department of Medicine, UCLA School of Medicine, Los Angeles, CA 90024, USA.

transport at the BBB is unique compared to neutral amino acid transport at cell membranes in nonbrain organs (Pardridge and Oldendorf 1977). In peripheral tissues, the neutral amino acid K_m parameters are in the 1 to 10 mM range, i.e., values that are 1 to 2 log orders higher than the existing plasma amino acid concentration. The very low K_m values at the rat BBB explain the selective vulnerability of the rat brain to hyperaminoacidemias (Pardridge and Oldendorf 1977). The human brain is also known to be selectively vulnerable to hyperaminoacidemias based on the clinical findings in PKU (Scriver and Clow 1980). Moreover, studies using either brain scanning (Oldendorf et al. 1971) or positron emission tomography (Comar et al. 1981) have shown that the brain uptake of neutral amino acids such as methionine is greatly inhibited by increased blood concentrations of phenylalanine in humans. To test the hypothesis that the human BBB, like the rat BBB, has a neutral amino acid transport system with a very low K_m (very high affinity) for neutral amino acids, the present studies were performed. It has recently been shown that brain capillaries isolated from fresh autopsy brain may be used as a model system for biochemical studies of the human BBB (Pardridge et al. 1985). This preparation has been used to characterize receptors on the human BBB for circulating peptides such as insulin, insulin-like growth factors, or transferrin (Pardridge et al. 1985, Duffy et al. 1988, Pardridge et al. 1987).

Methods

Microvessel Isolation

Capillaries were isolated with a mechanical homogenization technique from fresh autopsy human brain (Choi and Pardridge 1986). Brains were obtained from six autopsies over a one-year period. The patient ages ranged from 43 to 76 years (3 men and 3 women), and final diagnoses included esophageal carcinoma, barbiturate overdose, pulmonary carcinoma, and breast carcinoma. None of the patients had neurological disease. The brains were obtained from the pathologist 20 to 45 hours after death. Brains were also obtained from laboratory Sprague-Dawley male rats and from New Zealand white rabbits at either 0 or 42 hours postmortem. The capillaries were shown to be intact and free of cellular contaminants by visualization with light microscopy. In addition, human brain capillaries are enriched in high concentrations of factor VIII antigen, gamma-glutamyl transpeptidase (a BBB-specific enzyme), insulin receptor, transferrin receptor, or insulin-like growth factor receptor (Pardridge et al. 1985, Duffy et al. 1988, Pardridge et al. 1987). In addition, previous studies have shown that the membrane proteins separated by sodium dodecyl sulfate-polyacrylamide gel electrophoresis (SDS-PAGE) from human brain capillaries are qualitatively comparable to those from freshly isolated bovine brain capillaries (Pardridge and Choi 1986). Moreover, when capillaries isolated from rabbit brain after either 0 or 42 hours of autolysis were separated by SDS-PAGE, there were no significant differences between the protein patterns comprising the microvessels isolated from either the 0- or 42-hour postmortem rabbit brain.

Uptake Studies

Uptake studies were performed by resuspending capillaries in physiologic buffer containing [^{3}H] phenylalanine and [^{14}C] sucrose, an extracellular space marker, for incubation times ranging from 5 seconds to 5 minutes. Uptake studies were terminated by the addition of 4 ml of cold physiologic buffer followed by rapid filtration on a Millipore filtration apparatus attached to a vacuum pump (Choi and Pardridge 1986). Capillaries were solubilized in NaOH for protein determination, and the uptake data were expressed as volume of distribution (V_D, μl per mg of protein), which was calculated from the cell/medium DPM (disintegrations per minute) ratio corrected for the sucrose uptake. V_D was then converted to clearance (μl/min per mg of protein) by dividing by incubation time (2 minutes). Clearance (V/S) was then plotted against substrate concentration(s), and these saturation plots were fit to a five-parameter transport function. The preliminary fitting of the saturation data to a transport system composed of a single saturable route and a nonsaturable mechanism gave poor data fits. Therefore, the model was expanded to five parameters: two saturable mechanisms, one high affinity (K_m^{H}), low capacity ($V_{\mathrm{max}}^{\mathrm{H}}$), and another low affinity (K_m^{L}), high capacity ($V_{\mathrm{max}}^{\mathrm{L}}$), and a nonsaturable system (K_D). The function used in nonlinear regression analysis was

$$\frac{V}{S} = \frac{V_{\mathrm{max}}^{\mathrm{H}}}{K_m^{\mathrm{H}} + (S)} + \frac{V_{\mathrm{max}}^{\mathrm{L}}}{K_m^{\mathrm{L}} + (S)} + K_D$$

where V/S represents the experimentally determined clearance and S represents the medium phenylalanine concentration. Nonlinear regression analysis was done on an IBM 360-91 mainframe computer with the use of Subroutine P3R of BMPD (Biomedical Computer P Series) Programs developed at the Health Sciences Computing Facility, UCLA, Los Angeles, California. Because the standard error was roughly proportional to the means, the data were weighted using weights equal to 1/(clearance)^2.

Results

Capillaries were isolated intact and in high yield from postmortem human brain, and micrographs of these capillaries have been published previously (Pardridge et al. 1985, Choi and Pardridge 1986). The uptake of [^{3}H] phenylalanine by human brain capillaries increased with time and reached equilibrium in approximately five minutes. Initial studies showed that similar transport saturation curves were obtained from either 5-second or 2-minute incubation periods; therefore, the 2-minute incubation period was used for all saturation studies. The uptake of [^{3}H] phenylalanine by isolated human brain capillaries was not inhibited by an acidic amino acid, e.g., glutamic acid, or by a basic amino acid, e.g., arginine. Amino acids such as proline or glycine did not inhibit human brain capillary uptake of [^{3}H] phenylalanine at concentrations of 50 μM; 50 μM alanine caused a modest inhibition of [^{3}H] phenylalanine uptake, whereas 50 μM concentrations of large neutral amino acids (e.g., valine, leucine, tyrosine, isoleucine, phenylalanine, or tryptophan) markedly suppressed the uptake of [^{3}H] phenylalanine by isolated human brain capillaries (Choi and Pardridge 1986).

The uptake of [^{3}H] phenylalanine by the human brain microvessels was characterized by two saturable components. The K_m values of phenylalanine transport into human brain capillaries via the two saturable systems averaged 0.26 ± 0.08 and 22.3 ± 7.1 μM for five different human subjects. These studies provided the first evidence for a very high affinity ($K_m = 0.26$ μM) neutral amino acid transport system at the human BBB. Uptake via the high-K_m transport system was assumed to represent pathways of neutral amino acid flux through the human brain capillary that occur in vivo, since this K_m (22.3 μM) was not significantly different from the K_m of phenylalanine transport through the rat BBB in vivo, 26 ± 6 μM (Miller et al. 1985). Moreover, the K_m of phenylalanine uptake by human brain capillaries, 22 ± 7 μM, is not different from the K_m found for rabbit brain capillaries (at either 0 or 42 hours postmortem), 29 ± 6 μM, and approximates the phenylalanine transport K_m for rat brain capillaries, 11 ± 2 μM (Choi and Padridge 1986).

Discussion

These studies provide evidence that the K_m of neutral amino acid transport at the human BBB, $K_m = 22 \pm 7$ μM (Choi and Pardridge 1986), approximates the usual range of plasma phenylalanine concentrations, 50 to 100 μM. This close parallel between human BBB transport K_m and plasma amino acid concentration indicates the human BBB is heavily saturated by physiological concentrations of phenylalanine, similar to a situation known to occur at the rat BBB (Pardridge and Oldendorf 1977). These results provide the quantitative basis for the selective vulnerability of the human brain to changes in plasma amino acid concentrations that occur in the hyperaminoacidemias. The hyperaminoacidemias may be severe, as in the case of inborn errors of metabolism, e.g., PKU, or may be mild as in the case of such nutritional factors as high-dose aspartame intake (Pardridge 1986). For example, the ingestion of 50 mg of aspartame per kg per day would be expected to cause an increase in blood phenylalanine to 125 μM in normal subjects and an increase in blood phenylalanine to 250 μM in subjects with phenylalanine intolerance (Pardridge 1986). It is estimated that there may be as many as 20 million individuals in the United States with phenylalanine intolerance (Levy and Waisbren 1983). Since 7 to 12-year-old children are known to consume up to 77 mg of aspartame per day when aspartame-sweetened products are provided liberally in their diet (Frey 1976), any child consuming aspartame products freely may have substantial increases in blood phenylalanine that approximate 200 μM in the postprandial state, particularly if these individuals have phenylalanine intolerance.

A plasma concentration of 200 μM phenylalanine is ninefold greater than the K_m of phenylalanine transport at the human BBB (Choi and Pardridge 1986), and phenylalanine increases in blood of this magnitude would be expected to readily saturate neutral amino acid transport sites at the BBB and possibly cause derangements in amino acid availability in human brain. For example, tryptophan, which is an essential amino acid precursor to serotonin, a neurotransmitter (Fernstrom and Wurtman 1972), enters brain from blood via transport through the neutral amino acid carrier system (Pardridge and Oldendorf 1977). Inhibition of brain tryptophan uptake in the presence of hyperphenylalaninemia can be predicted by computation of the plasma tryptophan ratio, e.g.,

$$\text{Regular Trp ratio} = \frac{\text{Trp}}{\sum \text{LNAA}}$$

where Trp = total tryptophan concentration and $\sum$ LNAA = sum of large neutral amino acid concentrations in plasma. The computation of the regular tryptophan ratio makes a crucial assumption: that all neutral amino acids have the same affinity for the BBB neutral amino acid transport system (Fernstrom and Faller 1978). However, the neutral amino acids have markedly different K_m values for BBB transport (Pardridge and Oldendorf 1977, Miller et al. 1985). For example, the K_m values for phenylalanine, tryptophan, methionine, tyrosine, leucine, isoleucine, histidine, and valine transport through the BBB in the conscious rat are 32 ± 9, 52 ± 14, 83 ± 16, 86 ± 17, 87 ± 11, 145 ± 29, 164 ± 28, and 168 ± 72 μM. Thus, the K_m values can range over a spectrum of 500%. Because of this variability, it is useful to also compute K_m-normalized amino acid ratios. The K_m-normalized ratio for tryptophan is

$$K_m\text{-normalized Trp ratio} = \frac{\text{Trp}}{\sum \frac{K_m^{\text{Trp}}\,(\text{LNAA})}{K_m^{\text{LNAA}}}}$$

where K_m^{Trp} is the K_m of tryptophan and K_m^{LNAA} is the K_m of the competing large neutral amino acid. Using this formulation, the plasma amino acid concentration of phenylalanine is multiplied by the ratio of the K_m for tryptophan divided by the K_m for phenylalanine, which is 52 μM/32 μM

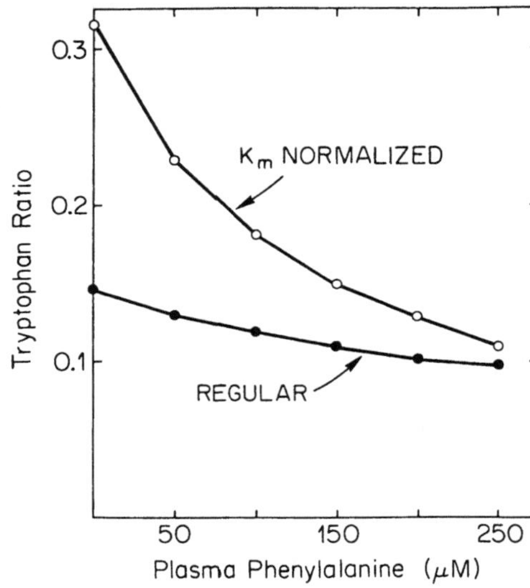

FIGURE 5.1. Comparison of regular tryptophan (TRP) amino acid ratio [per Fernstrom and Wurtman (1972)] and the K_m normalized ratio. The ratios, which predict brain uptake of TRP, are inversely related to the plasma phenylalanine concentration. The K_m normalized ratio was determined according to equation in the text, using the absolute K_m values for large neutral amino acid transport through the rat blood-brain barrier. Reproduced with permission (Pardridge, 1987).

= 1.6. Thus, as shown in Figure 5.1, the K_m-normalized tryptophan ratio falls precipitously when plasma phenylalanine is selectively increased to 250 μM, whereas the fall predicted by the regular tryptophan ratio is relatively modest. These considerations indicate that when there is a discrepancy between the K_m's of the competing neutral amino acids, the computation of the regular ratio for a particular large neutral amino acid markedly underestimates the inhibition of brain uptake of the amino acid in the face of selective hyperphenylalaninemia.

In summary, the transport of phenylalanine and other large neutral amino acids through the BBB in humans, as in rats, is mediated by a carrier that has a high affinity (low K_m) for the amino acids (Choi and Pardridge 1986). The functional significance of this finding is that the human brain is acutely (and uniquely, compared to other organs) sensitive to the selective increase in the blood concentration of single amino acids. This sensitivity may be quantitated by calculation of K_m-normalized amino acid ratios. Hyperphenylalaninemia in the 200-μM range inhibits brain tryptophan uptake (Figure 5.1). Therefore, the likelihood of generating plasma phenylalanine concentrations of this magnitude in humans following liberal aspartame intake raises serious questions on the wisdom of injecting 4 thousand tons of phenylalanine into the food supply every year (as of 1986, USDA) in the form of aspartame (Pardridge 1986).

Acknowledgments. The author is indebted to Thomas B. Choi for many valuable discussions, to Jing Yang and Jody Eisenberg for valuable technical assistance, and to Dawn Brown who skillfully prepared the manuscript. This research was supported by grant RO1-NS-19271 from the National Institutes of Health, by a grant from the American Diabetes Association, Southern California Affiliate, and by a grant from the International Life Sciences Institute.

References

Andersen, A.E., and Avins, L. (1976). Lowering brain phenylalanine levels by giving other large neutral amino acids. Arch. Neurol. **33**:686–686.

Binek-Singer, P., and Johnson, T.C. (1982). The effects of chronic hyperphenylalaninaemia on mouse brain protein synthesis can be prevented by other amino acids. Biochem. J. **206**:407–414.

Choi, T., and Pardridge, W.M. (1986). Phenylalanine transport at the human blood-brain barrier. Studies in isolated human brain capillaries. J. Biol. Chem. **261**:6536–6541.

Comar, D., Saudubray, J.M., Duthilleul, A., Delforge, J., Maziere, M., Berger, G., Charpentier, C., Todd-Pokropek, A., with Crouze, M., and Depondt, E. (1981). Brain uptake of ^{11}C-methionine in phenylketonuria. Eur. J. Pediatr. **136**:13–19.

Duffy, K.R., Pardridge, W.M., and Rosenfeld, R.G. (1988). Human blood-brain barrier insulin-like growth factor (IGF) receptor. Metabolism **37**:136–140.

Fernstrom, J.D., and Faller, D.V. (1978). Neutral amino acids in the brain: changes in response to food ingestion. J. Neurochem. **30**:1531–1538.

Fernstrom, J.D., and Wurtman, R.J. (1972). Brain serotonin content: physiological regulation by plasma neutral amino acids. Science **178**:414–416.

Frey, G.H. (1976). Use of aspartame by apparently healthy children and adolescents. J. Toxicol. Environ. Health **2**:401–415.

Levy, H.L., and Waisbren, S.E. (1983). Effects of untreated maternal phenylketonuria and hyperphenylalaninemia on the fetus. N. Engl. J. Med. **309**:1269–1274.

Miller, L., Braun, L.D., Pardridge, W.M., and Oldendorf, W.H. (1985). Kinetic constants for blood-brain barrier amino acid transport in conscious rats. J. Neurochem. **45**:1427–1432.

Oldendorf, W.H., Sisson, W.B., and Silverstein, A. (1971). Brain uptake of selenomethioine Se 75. II. Reduced brain uptake of selenomethionine Se 75 in phenylketonuria. Arch. Neurol. **24**:524–528.

Pardridge, W.M. (1986). Potential effects of the dipeptide sweetener aspartame on the brain. *In* Wurtman, R.J., and Wurtman, J.J. (eds.), Nutrition and the brain, Volume 7. New York: Raven Press, pp. 199–241.

Pardridge (1987): Phenylalanine transport at the human blood-brain barrier. *In* Kaufman, S. (ed.), Amino acids in health and disease: new perspectives. New York: Alan R. Liss, Inc., pp. 43–64.

Pardridge, W.M., and Choi, T. (1986). Amino acid transport at the human blood-brain barrier. Fed. Proc. **45**:2073–2078.

Pardridge, W.M., and Oldendorf, W.H. (1977): Transport of metabolic substrates through the blood-brain barrier. J. Neurochem. **28**:5–12.

Pardridge, W.M., Eisenbergy J., and Yang, J. (1985). Human blood-brain barrier insulin receptor. J. Neurochem. **44**: 1771–1778.
Pardridge, W.M., Eisenberg, J., and Yang, J. (1987). Human blood-brain barrier transferrin receptor. Metabolism **36**: 892–895.
Scriver, C.R., and Clow, C.L. (1980). Phenylketonuria: epitome of human biochemical genetics. N. Engl. J. Med. **303**: 1336–1342.

CHAPTER 6

Regional Transport of Phenylalanine and Other Neutral Amino Acids Across the Blood-Brain Barrier

Richard A. Hawkins,*† Anke M. Mans,* and Julien F. Biebuyck*

The supply of phenylalanine to individual brain structures of normal, anesthetized rats and of rats with hepatic encephalopathy was studied using quantitative autoradiography and state-of-the-art computer techniques. Influx varied considerably between structures, ranging from 5 nmol/min per g in the globus pallidus to 12 nmol/min per g in the inferior colliculus, in normal rats. Anesthesia reduced influx by altering the plasma amino acid spectrum. There was no effect on the activity of the neutral amino acid transport system in any structure. In hepatic encephalopathy, influx was markedly increased by both changes in plasma amino acids and increases in the transport activity. These changes may be of etiologic significance to cerebral dysfunction.

The availability of essential nutrients to the brain is determined by the properties of the transport systems of the blood-brain barrier as well as the plasma concentrations. In normal animals, the blood-brain barrier mediates a delicate balance between supply and need, but in metabolic disorders such as portal-systemic shunting, starvation, or diabetes, the permeability characteristics of several transport systems can be altered, and this may be of etiologic significance to the development of cerebral dysfunction. Transport across the blood-brain barrier is believed to be the rate-limiting step for the penetration of amino acids into brain cells since the maximal velocities of neuronal membrane transport systems are much greater than those of endothelial cells. Studies of amino acid transport in vivo showed that most amino acids could be assigned to one of three carriers: one for acidic amino acids, one for basic amino acids, and one for neutral amino acids. The neutral amino acid system carries nine large neutral amino acids at similar rates. Because the affinity of the carrier for each amino acid is similar to its plasma concentration, the rate of entry of any particular amino acid is influenced by the presence of the competing amino acids (Pardridge 1983). For example, alterations in the plasma concen-

---

Departments of Anesthesia* and Physiology,† The Pennsylvania State University, College of Medicine, The Hershey Medical Center, Hershey, PA 17033 USA.

trations of the neutral amino acids that compete with tryptophan can influence brain tryptophan content in normal rats.

We have been interested in determining substrate movement into individual brain structures as it occurs in vivo in normal and pathological situations and have devised appropriate methods to study this. A tracer quantity of ^{14}C-labeled substrate is infused into a rat in such a way as to establish and maintain a steady plasma level of [^{14}C]substrate. Initially, the accumulation of [^{14}C]substrate in brain is linear with time. Influx and the permeability-to-surface-area product (PA) are calculated from the brain radioactivity, the integral of blood radioactivity, and the plasma phenylalanine concentration. This method, combined with quantitative autoradiography of brain sections, enables the measurement of flux into any cerebral structure with an intact blood-brain barrier (Hawkins et al. 1982).

The autoradiographic images can be analyzed manually by taking measurements in individual structures or they can be reconstructed with the aid of a computer (Hibbard and Hawkins 1984). Computer reconstruction involves placing the sections in register with each other by a set of programs which reorients the images on the basis of their intrinsic mathematical properties. Once the data have been reassembled in a three-dimensional matrix, rates of transport in the brain can be viewed in a varie-

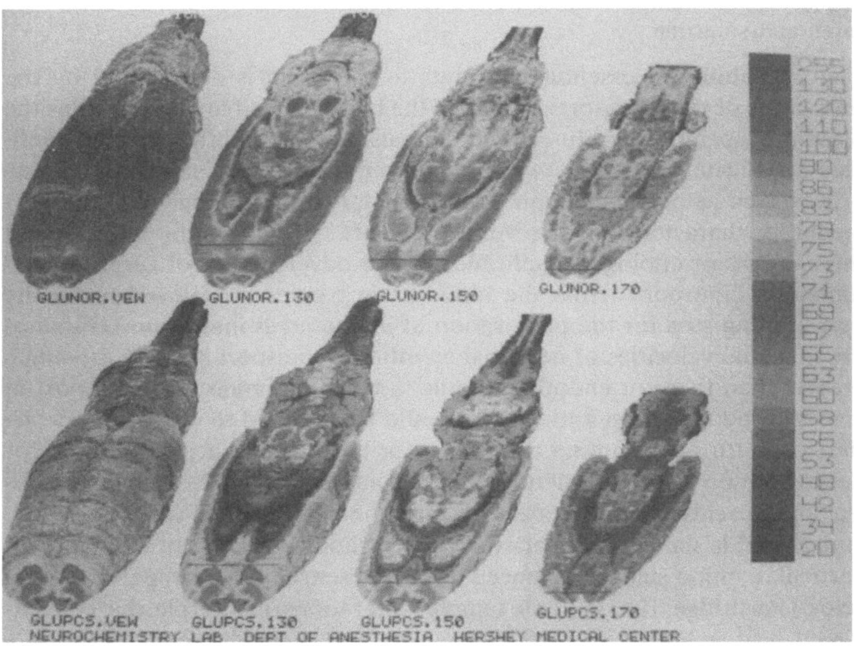

FIGURE 6.1. Autoradiographs of coronal sections were automatically aligned with respect to a fixed coordinate system and put together by computer. The stereograms shown represent glucose use in whole brain of a normal rat (top row) and a rat with a portacaval shunt (bottom row) and at various horizontal sections through the brain.

ty of ways, such as in the form of stereograms (whole brains or cut along axial or sagittal planes) or as transverse sections (see Figure 6.1). It is also possible to combine several whole brains together, thereby forming an average brain in which each point has a mean and standard deviation. This enables the comparison of experimental and control brains for global patterns of statistically significant changes.

Using the infusion-autoradiographic method, the transport of essential neutral amino acids has been studied in conscious, anesthetized, and portacaval-shunted rats (Hawkins et al. 1982, Mans et al. 1982). These studies, summarized below, have made it possible to quantify regional influx in detail as well as to identify the location and magnitude of permeability changes.

Initially, we studied the blood-brain barrier transport of phenylalanine into brain regions of conscious and anesthetized rats. There was considerable variation in transport between the structures in both groups. The highest rate of influx was found in the inferior colliculus and the lowest in the globus pallidus (Table 6.1). Phenylalanine influx was lower in

TABLE 6.1. Phenylalanine influx into brain regions in normal and anesthetized rats.

	Phenylalanine influx[a] (nmol/min per g)			
Brain region	Control	N_2O	Halothane	Pento-barbital
Gray tissue				
Frontal cortex	8.4 (80)	7.7 (76)	6.5 (71)	6.2 (73)
Caudate	6.6 (63)	6.6 (65)	5.5 (60)	5.4 (63)
Globus pallidus	5.0 (48)	4.7 (47)	4.1 (45)	4.2 (49)
Amygdala	5.9 (56)	5.6 (56)	5.2 (57)	5.0 (59)
Hippocampus	5.7 (54)	5.4 (53)	4.7 (51)	5.0 (58)
Hypothalamus	7.0 (67)	6.2 (62)	5.7 (62)	5.7 (67)
Thalamus				
Anterior nucleus	9.2 (88)	8.3 (82)	7.0 (77)	6.9 (81)
Ventrolateral nucleus	7.9 (76)	7.6 (75)	6.6 (72)	6.7 (79)
Lateral geniculate	7.4 (71)	7.2 (71)	5.8 (64)	6.2 (73)
Medial geniculate	8.5 (81)	8.1 (79)	6.8 (74)	6.8 (79)
Substantia nigra	7.1 (68)	6.5 (64)	5.7 (62)	5.6 (65)
Red nucleus	7.8 (74)	7.2 (71)	6.5 (71)	5.6 (65)
Reticular formation	6.6 (63)	5.9 (59)	5.3 (58)	5.3 (62)
Superior colliculus	8.5 (81)	8.3 (82)	6.9 (76)	6.8 (80)
Inferior colliculus	12.0 (114)	11.2 (110)	8.9 (97)	9.5 (111)
Pons	8.6 (82)	8.9 (87)	7.6 (83)	6.4 (75)
Cerebellum-granular	9.2 (88)	8.1 (80)	6.5 (71)	6.7 (78)
White tissue				
Internal capsule	4.9 (47)	4.9 (49)	4.9 (54)	4.4 (52)
Corpus callosum	5.1 (49)	5.1 (50)	5.3 (58)	4.3 (51)
Cerebellum	5.8 (55)	5.2 (52)	4.8 (53)	3.8 (44)

[a]Influx values are means for six rats. The values in parentheses are calculated values of maximal transport (T_{max}) based on Michaelis-Menten kinetics, measured concentrations of amino acids, and kinetic constants reported by Pardridge and Oldendorf (1977).

TABLE 6.2. Plasma and brain amino acids in chronic liver disease.

Amino acid	Level[a] Plasma	Brain
Tryptophan	70	300
Phenylalanine	160	400
Tyrosine	160	390
Methionine	80	140
Histidine	120	170
Leucine	60	110
Isoleucine	60	100
Valine	60	80
Threonine	50	90

[a] All values are expressed as percent of control value.

pentobarbital-anesthetized rats, but this did not seem to be a consequence of changes in the transport system itself. Pentobarbital altered the plasma neutral amino acid spectrum, including a lowering in the concentration of phenylalanine. The maximal transport capacity of the neutral amino acid system for phenylalanine can be estimated from the concentrations of all the competing amino acids and their kinetic constants. When this was done, it became apparent that there was no change in the total transport capacity; all changes in influx were due to shifts in the neutral amino acid spectrum.

We have also studied the transport of phenylalanine across the blood-brain barrier in the metabolic disease hepatic encephalopathy as a consequence of portal-systemic shunting. This disease is interesting because of the characteristic changes in the plasma neutral amino acid spectrum as well as the consequences to brain function. In hepatic encephalopathy caused by a portal-systemic shunting, there is a marked decrease in brain function. In rats, cerebral energy metabolism, an indirect measure of total nerve cell activity, is depressed by 20 or 25% throughout the brain, and the rats become very sensitive to any metabolic stress. There is a characteristic change in the plasma amino acid spectrum; aromatic amino acids, which are precursors of neurotransmitters, rise in concentration, whereas branched-chain amino acids fall (Table 6.2). There are also marked changes in the brain content of many of the amino acids, some expected and some not. For instance, the concentrations of aromatic amino acids are much higher than would be expected on the basis of changes in the plasma concentrations, whereas the branched-chain amino acids are approximately normal. This observation raised the possibility that changes in the blood-brain barrier were occurring, which was confirmed by autoradiographic analysis. It was found that the neutral amino acid transport was markedly stimulated in hepatic encephalopathy. These changes occurred throughout the brain, but they were most severe in several limbic structures and the reticular formation, whereas the hypothalamus was least affected. Of the neutral amino acids studied, the PA of tryptophan was increased by about 200%, that of phenylalanine and tyrosine by about 80%, and that of leucine by about 30% (Mans et al. 1982). This variation among the amino acids was unexpected, because all four amino acids are believed to be transported by the same transport carrier. As yet, the mechanism for the changes is unknown. In any event, in this disease there is both a change in plasma amino acid concentrations as well as increased regional transport capacity, which causes the brain to be flooded with aromatic amino acids. This may have important effects on the metabolism of neurotransmitters and neuromodulators for which these amino acids are the precursors (e.g., serotonin, norepinephrine, octopamine).

In summary, the rate of phenylalanine transport across the blood-brain barrier varies considerably among the brain structures; this variability seems to be correlated with capillarity. The main factors determining the entry of phenylalanine into normal brain regions are its concentration, the concentrations of competing neutral amino acids, and the activity of the blood-brain barrier transport system. In at least one metabolic disease (hepatic encephalopathy), the transport capacity increases rather remarkably, and there is an excess availability of amino acids, some of which are precursors of important neurotransmitters. Whether these changes are of etiologic significance to the altered state of cerebral function remains to be determined.

Acknowledgments. This study was supported by the National Institutes of Health (grants NS16389 and NS16737), the National Science Foundation (grant BNS8506479), and the American Diabetes Association.

References

Hawkins, R.A., Mans, A.M., and Biebuyck, J.F. (1982). Amino acid supply to individual cerebral structures in awake and anesthetized rats. Am. J. Physiol. **242**:E1–E11.

Hibbard, L.S., and Hawkins, R.A. (1984). The three-dimensional reconstruction of metabolic data from quantitative autoradiography of rat brain. Am. J. Physiol. **247**:E412–E419.

Mans, A.M., Biebuyck, J.F., Shelley, K. and Hawkins, R.A. (1982). Regional blood-brain barrier permeability to amino acids after portacaval anastomosis. J. Neurochem. **38**:705–717.

Pardridge, W.M. (1983). Brain metabolism: a perspective from the blood-brain barrier. Physiol. Rev. **63**:1481–1535.

Pardridge, W.M., and Oldendorf, W.H. (1977). Kinetic analysis of blood-brain barrier transport of neutral amino acids. Biochim. Biophys. Acta **401**:128–136.

CHAPTER 7

Dual Role of Transport Competition in Amino Acid Deprivation of the Central Nervous System by Hyperphenylalaninemia

Halvor N. Christensen and Carlos de Céspedes*

Evidence is presented that an amino acid such as phenylalanine accumulating in the organism to excessive concentrations may occasion not only a direct inhibition of the passage of endogenous amino acids across the blood-brain barrier, but also muscle and liver sequestration of certain amino acids. These are not precisely the same amino acids whose passage to the brain undergoes direct inhibition. This action explains, we believe, the lowering of the plasma levels of the affected-amino acids and may further handicap the nutritional flow of amino acids to the brain. We propose that the total effect of these two actions determines which amino acids will be depleted in the brain in hyperphenylalaninemia (both experimental and phenylketonuric) or in leucinemia or, possibly, other amino acid accumulations in the body.

When phenylketonuria (PKU) first came to be called by that name, the biomedical world found it natural to suppose that formation of an abnormal product of metabolism, as suggested by that name, caused the recognized damage to neurological development. The idea that the accumulation of a normal metabolite could severely disturb other metabolic events has in the meantime, however, gained greatly enhanced acceptance. We now appreciate that virtually every metabolic sequence falls under the influence of numerous pertinent modulators or effector molecules, whose excess rather than foreign character may cause the injurious aspect. In the meantime, however, the idea of poisoning by phenylpyruvate or a subsequent product has gained a somewhat persistent foothold.

The understanding that membrane processes are associated with such metabolic sequences, sharing the risk of perturbation by abnormal concentrations of effector metabolites, has grown even more slowly. The transport step first intruded into the tracing of metabolic sequences in 1913, specifically for the amino acids, with a discovery that should have been breathtaking. In the very words of their paper's title, Van Slyke and Meyer

*Department of Biological Chemistry, The University of Michigan Medical School, Ann Arbor, MI 48109, USA.

(1913–1914) discovered "the absorption of amino acids from the blood by the tissues." This finding astounded me half a century ago, and, even today, I maintain that its authors should be famous for it, if for no other contribution. The pattern of only partial specificity of the amino acid transport systems that has emerged during these past five decades made it seem obvious to us already in 1953 that these systems were a locus at which the accumulating phenylalanine in PKU must impair the nutrition of the developing nervous system (Christensen 1953).

Giving support to this view is the important subject of interorgan amino acid nutrition, which we've taken occasion to summarize elsewhere (Christensen 1982). Initially, the natural tendency had been to perceive the amino acids as flowing from their source in the alimentary canal to their points of use in the various tissues. We now know, as we've stressed elsewhere, that this is a vastly oversimplified picture of organismal nutrition. Not only do the various tissues compete for the amino acids "at a common feed trough," so to speak, but they also feed each other. To mention only a few from a myriad of examples, the muscles feed the liver with alanine, and the kidneys and the intestines with glutamine; the liver also tends to feed the muscle with creatine and the brain with tyrosine. I have suggested elsewhere (Christensen 1987b) that we should extend our understanding of homeostasis, sometimes overly restricted to the maintenance of constancy of concentrations, to include, perhaps equally, a "homeorhysis," whereby concentrations may be sacrificed to maintain the stability of interorgan flows. The rather low specificity of most of the amino acid transport systems makes that step in the total flow vulnerable to competition among amino acids in ultimately predictable ways.

Some of these transport steps occur at the plasma membrane, some at the membrane of an intracellular organelle, as we and others have shown for the lysosome (Pisoni et al. 1985, 1986). Therapeutic efforts in cystinosis, for example, are aimed at allowing proteolytically formed cystine to escape from the cystinotic lysosome via an alternative transport system unaffected in cystinosis. The lysosome appears to have as extensive a set of amino acid transport systems as the plasma membrane, although apparently under independent genetic determination. To discover possible evolutionary relations between such systems, we must now look for sequence homologies between the proteins characterizing transport systems of the plasma membrane and of the organelle membranes.

First, two important principles need to be considered when analyzing the complexities inherent in perturbations of amino acid transport:

1. Unless the transport process is perturbed, e.g., by accumulation of a competitive analogue or by a transport-regulatory change, the plasma and the tissue levels of an amino acid tend to change together (Christensen et al. 1948). Likewise, the gradient across the blood-brain barrier, unless perturbed, shows a tendency toward constancy.

2. Wherever more than one transport system participates in a parallel way in catalyzing the transport of a molecular species across a biomembrane, the steady state obtained is not likely to be the one characteristic of either system operating alone (Christensen 1975, 1987a, b). One of the systems is likely to contribute disproportionately to influx, the other to efflux.

I was trying to solve the puzzle of the relative roles of influx and efflux across biological membranes in establishing the steady states of amino acid distribution forty years ago, when we secured the results shown in Figure 7.1. The panels of this figure show how we were able to perturb the glycine gradient between the liver and the plasma, on the one hand, and the muscle and the plasma, on the other hand. We did this by feeding by stomach tube enough of one or another amino acid to secure competitive inhibition of transport. At the right of each panel, we see that each of a group of amino acids tended to lower the tissue-to-plasma glycine gradient in the guinea pig, somewhat in proportion to the degree to which we had succeeded in elevating the circulating level of the amino acids. At the upper right, we see the opposite effect of a different group of amino acids. After some delay, we realized that two different populations of amino acids were producing these opposite perturbations: amino acids with bulky side chains tended to steepen the glycine gradient (more examples being provided in panel B than panel A); amino acids with smaller, often polar side chains tended to decrease the glycine gradient. Some years later, we came to understand the difference between these two populations. The first of them included amino acids transported to greater degrees by a transport system we designated System L, e.g., leucine, phenylalanine, and tryptophan. The second population, which lowered the glycine gradients, was found to show characteristically vigorous transport by System A (Oxender and Christensen 1963), the relative affinities of methionine and histidine being not quite the same for the two tissues. How could such a division as to transport reactivity with Systems L and A lead to these opposite effects?

Figure 7.2 is a version of an explanatory diagram we have presented in various forms (Christensen 1962, 1975). It shows a given amino acid, G, pumped into a cell by one rather steeply uphill, scarcely reversible route, A, and escaping by another route, L, which tends to operate only weakly uphill. Because of the difference in the reversibility of the two systems, System A tends to establish for several amino acids gradients too steep for System L to sustain, with the consequence that such amino acids would show net entry by System A, and net exodus by System L. From this relation, we drew the conclusion that an amino acid supplied in excess can inhibit either the net tissue uptake or the net tissue release of an endogenous amino acid, depending on the directional role exerted by the participating transport system suffering the larger competitive influence.

Thus, we were led to the hypothesis that members of the group of amino

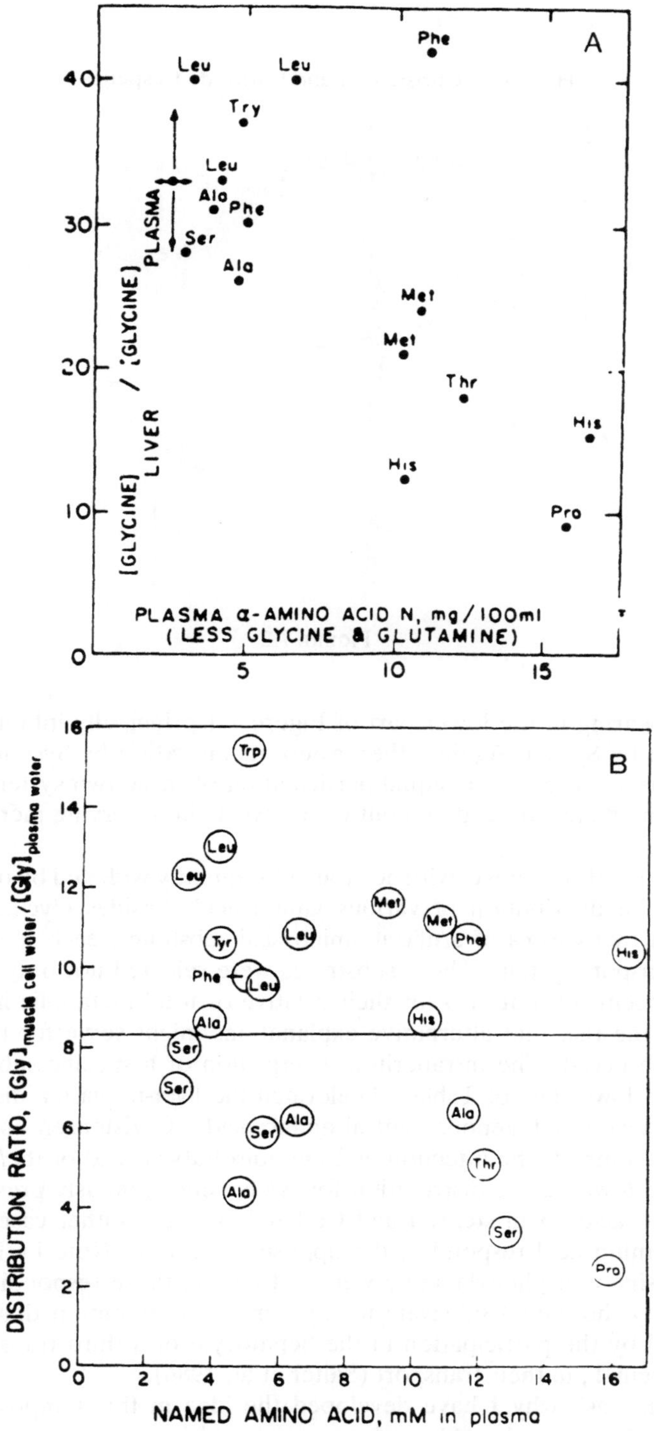

A
Phe
Leu
Leu
Try
Leu
Ala
Phe
Ser
Ala
Met
Met
Thr
His
His
Pro
[GLYCINE] LIVER / [GLYCINE] PLASMA
PLASMA α-AMINO ACID N, mg/100ml
(LESS GLYCINE & GLUTAMINE)
0
10
20
30
40
5
10
15
B
Trp
Leu
Leu
Met
Met
Phe
Tyr
Leu
His
Phe
Leu
Ala
His
Ser
Ser
Ala
Ser
Ala
Thr
Ala
Ser
Pro
DISTRIBUTION RATIO, [Gly]muscle cell water/[Gly]plasma water
NAMED AMINO ACID, mM in plasma
0
2
4
6
8
10
12
14
16

FIGURE 7.1

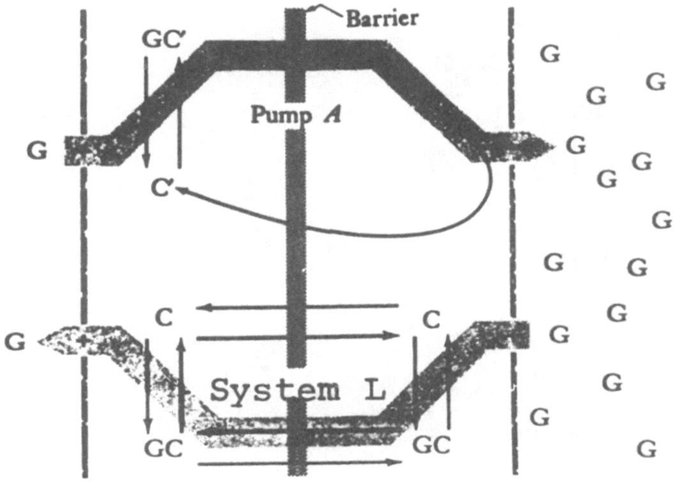

FIGURE 7.2

acids appearing in the lower part of Figure 7.1 principally inhibit glycine migration by System A; the other group, its migration by System L. The results suggest then the unequal participation of these two systems in the two glycine fluxes, inward and outward, glycine here serving merely as an example.

The scope of the above evidence was subsequently widened by our showing that the distribution of various amino acids besides glycine can be perturbed by excesses of artificial amino acid substrates, each specific to a given transport system. These substrates were selected not only for their system specificity, but also for their relative or total metabolic inertness, to lower the risk that alternative explanations might serve for the interactions observed. The intraperitoneal injection of a specific substrate of System L (lower line of Table 7.1) elevated the liver-to-plasma distribution ratio for seven endogenous neutral amino acids (Christensen and Cullen 1981). In contrast, the injection of 2-aminoisobutyric acid or its *N*-methyl derivative lowered the distribution for twelve endogenously present neutral amino acids (Christensen and Cullen 1968). In neither case did any neutral amino acid respond in the opposite direction. Note in Table 7.1 that tyrosine and phenylalanine were not among those responding to inhibition by the model substrate for System L, an exception that may be explained by the participation in the hepatocyte of a third transport system, System T, in their transport (Salter et al. 1986).

One may ask why I have developed the idea in this symposium that either the influx or the efflux of an amino acid into or from tissues may suffer the greater competitive inhibition by the accumulation in the organism of another amino acid in excess? We concluded in 1953 that in phenylketonuria, the excess phenylalanine must cause competition with the entry

TABLE 7.1. Hepatic amino acid levels in the rat are sharply decreased or increased, relative to plasma levels, by a transport-specific analogue, injected intraperitoneally 2 to 4 hours previously.

Perturbing analogue	Amino acids sharply decreased	Amino acids sharply increased	Reference
AIB, MeAIB	Thr, Ser, (Asn + Gln), Pro, Gly, Ala, Val, Met, Ileu, Leu, Tyr, Phe, His	No neutral amino acid	Christensen and Cullen 1968
BCH	None	Pro, Ala, Met, Leu, Val, Ileu, Orn	Christensen and Cullen 1981

of various amino acids across the blood-brain barrier. The subsequent history of this concept has been reviewed recently (Christensen 1987b). The flow across that barrier was identified by Wade and Katsman (1975) largely with transport System L, a system favoring especially amino acids with bulky side chains, sometimes spoken of as "large neutral amino acids." This system operates mainly inwardly across that barrier, except apparently for the case of brain-generated glutamine, which can be exchanged across the barrier for entering amino acids (James et al. 1979). If transport System A contributes at all to inward amino acid transport across this barrier, its participation must be small (Betz and Goldstein 1978). Hence, the directionality of the competitive action of phenylalanine on transport across this barrier appears largely unambiguous.

Another important change in amino acid distribution in phenylketonuria and in experimental hyperphenylalaninemia has, however, heretofore escaped satisfactory explanation. A lowering of the plasma levels of several amino acids, first observed by Linneweh and Ehrlich (1962) and then by Efron et al. (1969), has remained a paradox. Since satisfactory evidence that these amino acids are depleted by their movement into the urine or into the intestinal lumen has not been obtained, we have tended to suspect a second mode of perturbation of amino acid transport by excess of an amino acid, beyond competition at the blood-brain barrier. This is a stimulation of the net movement of amino acids into various tissues, an action which could be produced by stimulation of influx or inhibition of efflux. The results of Table 7.2 show that within two hours intraperitoneal phenylalanine does, indeed, cause the accumulation of several amino acids into the skeletal muscle and the livers of weanling rats. Similar results were observed for amino acid distribution between the plasma and the livers of rats fed 5% phenylalanine for 6 or 7 days. Therefore, given that this second perturbation of transport lowers the circulating level of amino acids in hyperphenylalaninemia, interfering in a second way with the amino acid nutrition of the brain, we next face the question of the origin of that perturbation. We have shown above how an excess of phenylalanine, through

TABLE 7.2. Amino acids whose relative tissue levels in the rat rose while the plasma level fell after phenylalanine treatment.

	Plasma level[a] (μmol/l)		Change(%) in distribution ratio	
Amino acid	Control	+ Phe	Liver/plasma	Muscle/plasma
Leu	254.4 ± 26.6	139.0 ± 37.0[b]		+ 54[d]
Ala	418.9 ± 119.7	218.5 ± 127.9[c]	+277[c]	+160[d]
Pro	118.0 ± 37.8	99.3 ± 41.8[c]		+292[b]
Gly	624.8 ± 120.7	390.6 ± 133.8[c]	+ 87[d]	+115[d]
His	82.8 ± 19.6	58.6 ± 11.1[c]		+ 53[d]
Orn	82.1 ± 22.8	52.8 ± 26.9[d]		+ 83[c]
Arg	166.5 ± 25.8	99.9 ± 20.9[b]		+ 65[c]

[a] Values are mean ± SD of 8 experimental male weanling rats vs. 8 controls.
[b] $p < 0.001$.
[c] $p < 0.01$.
[d] $p < 0.05$.

its greater inhibition of their efflux *from*, rather than of the influx *into*, various tissues, might act to cause tissue sequestration of various endogenously present amino acids. Both as a direct consequence of that sequestration and through a resultant acceleration of the catabolism of the amino acids sequestered, a lowering of their plasma levels would be expected. The shorter list of amino acids showing changed distribution for the liver rather than for muscle in the right-hand column of Table 7.2 may arise from a masking of some of the increases through a more widely accelerated catabolism in liver than in muscle.

A somewhat different interaction between Systems A and L known as *competitive stimulation* could also conceivably allow an overloading of the organism with a given amino acid to intensify rather than weaken the intracellular:extracellular gradients of certain other amino acids. In the initial observation of this phenomenon, the simultaneous presentation of external methionine and tryptophan was seen to intensify the accumulation of tryptophan into the Ehrlich ascites tumor cell within a minute or two, rather than cause the expected competitive inhibition of tryptophan uptake (Jacquez 1963). Oxender and Christensen (1963) showed that this phenomenon had its logical origin in (a) the rapid cellular accumulation of methionine, followed by (b) an exchange of the internalized methionine for external tryptophan via System L. Other pairs of amino acids met the kinetic criteria needed for this effect, in that the *stimulated* amino acid, and not the *stimulating* amino acid, showed a high influx by exchange, relative to its net influx (Christensen and Handlogten 1979; in this study, complications from transport by System ASC were minimized by limiting the study to relatively weak substrates of that system). As another example of an interacting pair of amino acids, the concurrent presence of "cycloleucine" intensifies the accumulation of the model amino acid BCH (± 2-aminoendo bicyclo [3.2.1.] heptane-2-carboxylic acid) into the Ehrlich cell. The phenomenon of competitive stimulation can be described also by reference to Figure 7.2 and indeed resembles the phenomenon described earlier, differing only in the detail that in the present instance the exodus of the exiting amino acid is balanced by a coupled uptake of the *stimulated* amino acid. Furthermore, because steps (a) and (b) are consecutive, the stimulating effect disappears as the observation time is shortened. Whether a persistent hyperphenylalaninemia or leucinemia can cause the tissue sequestration of a given circulating amino acid by competitive stimulation of its uptake depends on the kinetic criteria already described, and should be subject to direct test.

A third, more important possible explanation of the results of Figure 7.1 and Tables 7.1 and 7.2 has, in the meantime, however, proved not a likely cause of the tissue sequestration of various plasma amino acids than either of these two competitive phenomena. In 1928, Luck and his associates showed that insulin administration lowers the blood level of the amino acids collectively. This phenomenon became understandable when the endocrine stimulation of the accumulation of amino acids in tissues, especially by System A, came to be recognized (see Noall et al. 1957, Kipnis and Noall 1958). Its possible applicability to hyperleucinemia and hyperphenylalaninemia became clear with the finding of Fajans and his associates (1967) that these two fell conspicuously among the neutral amino acids showing in man an insulin secretogogue action of a particular glucose-independent, diazoxide-inhibitable type. The extension of this type of insulin secretogogue action to the nonmetabolizable leucine analogue, levorotatory 2-amino *endo*-bicycloheptane-2-carboxylic acid (Christensen and Cullen 1969), Fajans et al. 1971) showed that their secretogogue action does not depend on any metabolic modification of these amino acids.

Only one of the four isomers of the bicycloheptane amino acid, namely, the levorotatory form in which the amino group is in the *endo* orientation, shows an insulin secretogogue action, whereas all four isomers are rather similarly active (although not equally specific) in inhibiting System L transport (Christensen et al. 1969, Tager and Christensen 1971a, b). This circumstance gave us an opportunity to see whether insulin secretion plays a mediating role in the perturbation of transport in the rat. The non-secretogogue amino *exo* isomer of BCH caused significant increases in the muscle/plasma distribution ratios for alanine and glycine (marginally significant for proline); the liver/plasma distribution ratios tended to rise also, but with p slightly above 0.05 (results to be published). Supporting results had been obtained with extended feeding of 5% phenylalanine (Christensen et al. 1987a). Considering also that phenylalanine has failed to stimulate insulin release from perfused rat pancreas (Landgraf et al. 1974), we believe that inhibition of amino acid exodus via System L may be regarded as a main cause for the tissue sequestration of certain amino acids produced by hyperphenylalaninemia (although perhaps the insulin-secretogogue action of leucine may also play a role in hyperleucinemia) and that this mechanism adds to the consequences of competitive inhibition of amino acid transport in producing the types of amino acid starvation within the central nervous system characteristic of phenylketonuria and clinical conditions that include hyperleucinemia. We plan to examine this hypothesis for the case of leucinemia by testing the effects of additional amino acids (i.e., to look for dissociation between their effects on insulin release and on transport via System L), and also by monitoring plasma insulin levels. The degree to which hypersecretion of insulin (results to be published) presents a problem to both young and older PKU patients (see Antonozzi et al. 1987), even after termination of phenylalanine restriction, remains an interesting question.

Other authors (Huether et al. 1984) have recently concluded, despite their finding of intense tissue sequestration of various amino acids in very young rats, that competition of amino acids for transport plays no significant physiological role in the disturbed amino acid distribution of hyperphenylalaninemia. However, that study failed to take into account the competitive action of phenylalanine and various analogues on amino acid transport *from* as well as *into* tissues, as discussed here.

We plan to continue the study of chronic hyperphenylalaninemia using the α-methyl derivative of phenylalanine to inhibit hepatic phenylalanine hydroxylase, which is well known to have a much higher activity in rodents than in man.

Acknowledgments. These studies received support from grants HD01233 and DK32281 from the National Institutes of Health, U.S. Public Health Service. Collaboration by associates as cited should be noted.

References

Antonozzi, I., Carducci, C., Vestri, L., Manzari, V., and Dominici, R. (1987). Plasma amino acid values and pancreatic β-cell function in phenylketonuria. J. Inher. Metab. Dis. **10**:66–72.

Betz, A.L., and Goldstein, G.W. (1978). Polarity of the blood-brain barrier: neutral amino acid transport into isolated brain capillaries. Science **202**:225–227.

Christensen, H.N. (1953). Metabolism of amino acids and proteins. *Annu. Rev.* Biochem. **22**:233–260.

Christensen, H.N. (1962). Biological transport, 1st ed. New York: W.A. Benjamin.

Christensen, H.N. (1975). Biological transport, 2nd ed. Reading, Massachusetts: W.A. Benjamin.

Christensen, H.N. (1982). Interorgan amino-acid nutrition. Physiol. Rev. **62**:1193–1233.

Christensen, H.N. (1987a). Hypothesis: where do the depleted amino acids go in phenylalaninemia and why? Persp. Biol. Med. **30**:186–196.

Christensen, H.N. (1987b). Role of membrane transport in interorgan amino acid flows: where do the depleted amino acids go in phenylketonuria? J. Cell. Biochem. **35** Supplement 11E, p 172.

Christensen, H.N., and Cullen, A.M. (1968). Effects of non-metabolizable analogs on the distribution of amino acids in the rat. Biochim. Biophys. Acta **150:**237–252.

Christensen, H.N., and Cullen, A.M. (1969). Behavior in the rat of a transport-specific bicyclic amino acid. Hypoglycemic action. J. Biol. Chem. **244:**1521–1526.

Christensen, H.N., and Cullen, A.M. (1981). Intensified gradients for endogenous amino acid substrates for transport system L on injecting a specific competitor for that system. Life Sci. **29:**749–753.

Christensen, H.N., and Handlogten, M.E. (1979). Interaction between parallel transport systems examined with tryptophan and related amino acids. J. Neural Transm., Suppl. **15:**1–13.

Christensen, H.N., Streicher, J.A., and Elbinger, R.L. (1948) Effects of feeding individual amino amino acids upon the distribution of other amino acids between cells and extracellular fluid. J. Biol. Chem. **172:**151–524.

Christensen, H.N., Handlogten, M.E., Lam, I., Tager, H.S., and Zand, R. (1969). A bicyclic amino acid to improve discriminations among transport systems. J. Biol. Chem. **244:**1510–1520.

Efron, M.L., Song Kong, E., Visakorpi, J., and Feller F.X. (1969). Effects of elevated plasma phenylalanine levels on other amino acids in phenylketonuric and normal subjects. J. Pediatr. **74:**399–405.

Fajans, S.S., Floyd, J.S., Jr., Knopf, R.F., Conn, J.W. (1967): Effects of amino acids and proteins on insulin secretion in man. Recent Prog. Hormone Res. **23:**617–662.

Fajans, S.S., Quibrera, R., Pek, S., Floyd, J.C., Jr., Christensen, H.N., and Conn, J.W. (1971). Stimulation of insulin release in the dog by a nonmetabolizable amino acid. Comparison with leucine and arginine. J. Clin. Endocrinol. **33:**35–41.

Huether, G., Schott, K., Sprotte, U., Thoemke, F., and Neuhoff, V. (1984). Regulation of the amino acid availability in the developing brain. No physiological significance of amino acid competition in experimental hyperphenylalaninemia. Int. J. Dev. Neurosci. **2:**43–54.

Jacquez, J.A. (1963). Carrier-amino acid stoichiometry in amino acid transport in Ehrlich ascites cells. Biochim. Biophys. Acta **71:**15–33.

James, J.H., Jeppson, V., Ziparo, V., and Fischer, J.E. (1979). Hyperammonaemia, plasma aminoacid imbalance, and blood-brain aminoacid transport: a unified theory of portal-systemic encephalopathy. Lancett **ii:**772–775.

Kipnis, D.M., and Noall, M.W. (1958). Stimulation of amino acid transport by insulin in the isolated diaphragm. Biochim. Biophys. Acta **28:**226–227.

Landgraf, R., M.M.C. Landgraf-Leurs and Hörl, R. (1974). L-Leucine and Phenylalanine induced insulin release and the influence of *D*-glucose. Kinetic studies with perfused rat pancreas. Diabetologia **10:**415–420.

Linneweh, F., and Ehrlich, M. (1962). Zur Pathogenese des Schwachsinns bei Phenylketonuria. Klin. Wochenschr. **40:**225–226.

Luck, M.M., Morrison, G., and Wilbur, L.F. (1928). The effect of insulin on the amino acid content of blood. J. Biol. Chem. **77:**151–156.

McKean, C.M., Boggs, D.E., and Peterson, N.A. (1968). The influence of high phenylalanine and tyrosine on the concentration of essential amino acids in the brain. J. Neurochem. **15**:235–241.

Noall, M.W., Riggs, T.R., Walker, L.M., and Christensen, H.N. (1957). Endocrine control of amino acid transfer. Science **126**:1002–1005.

Oxender, D.L., and Christensen, H.N. (1963): Distinct mediating systems for the transport of neutral amino acids by the Ehrlich cell. J. Biol. Chem. **238**:3686–3699.

Pisoni, R.L., Thoene, J.G., and Christensen, H.N. (1985). Detection and characterization of carrier-mediated cationic amino acid transport in lysosomes of normal and cystinotic human fibroblasts. Role in therapeutic cysteine removal? J. Biol. Chem. **260**:4791–4798.

Pisoni, R.L., Flickinger, K.S., Thoene, J.G., and Christensen, H.N. (1986). Specific lysosomal transport of small neutral amino acids. Fed. Proc. **45**:1759.

Salter, M., Knowles, R.G., and Pogson, C.J. (1986). Transport of aromatic amino acids into isolated liver cells. Biochem. J. **233**:1499–1506.

Tager, H.S., and Christensen, H.N. (1971a). Hypoglycemic action of 2-aminonorbornane-2-carboxylic acid in the rat. Biochem. Biophys. Res. Commun. **44**:185–191.

Tager, H.S., and Christensen, H.N. (1971b). Transport of the four isomers of 2-aminonorbornane-2-carboxylic acid in selected mammalian systems and in Escherichia coli. J. Biol. Chem. **246**:7572–7580.

Van Slyke, D.D., and Meyer, G.M. (1913–14). The fate of protein digestion products in the body. III. The absorption of amino-acids from the blood by the tissues. J. Biol. Chem. **16**:197–212.

Wade, L.A., and Katzman, R. (1975). Synthetic amino acids and the nature of L-DOPA transport at the blood-brain barrier. J. Neurochem. **25**:837–842.

Part III Effects of Aspartame on Brain Monoamines and Seizure Thresholds in Experimental Animals

CHAPTER 8

An In Vivo Study of Dopamine Release in Striatum: The Effects of Phenylalanine

Matthew J. During, Ian N. Acworth, and Richard J. Wurtman*

We used intracerebral dialysis to monitor extracellular levels of dopamine and its major metabolites dihydroxyphenylacetic acid (DOPAC) and homovanillic acid (HVA) in the striatum of chloralose/urethane-anesthetized rats. Levels of these compounds were determined after intraperitoneal administration of phenylalanine (200, 500, and 1000-mg/kg). A dose of 200 mg phenylalanine per kg (which increases brain tyrosine by more than phenylalanine) *increased* basal dopamine release by 59%, peaking at 75 min. There was *no change* in basal dopamine release after the 500-mg dose, whereas the 1000-mg/kg dose (which increases brain phenylalanine more than tyrosine) significantly *reduced* (26%) dopamine release. No corresponding changes were observed in the concentrations of DOPAC and HVA with any of the treatments, indicating that changes in brain phenylalanine and tyrosine levels may selectively affect production of the dopamine molecules that are preferentially released into synapses.

Introduction

There is good in vitro evidence and some indirect in vivo evidence that brain phenylalanine levels can influence the synthesis of dopamine and the other monoamine neurotransmitters (McKean 1972, Milner et al. 1986). Such data have suggested that phenylalanine can substitute for tyrosine at low concentrations, but inhibits dopamine synthesis at higher doses. When rodents receive low phenylalanine doses, most of the amino acid is converted to tyrosine by the hepatic enzyme phenylalanine 4-monooxygenase; hence, blood and brain tyrosine levels increase by more than those of phenylalanine. At higher doses, however, this enzyme becomes saturated, and the excess phenylalanine might be expected to suppress catecholamine synthesis by (1) competitively inhibiting tyrosine 3-monooxygenase (Ikeda et al. 1967) and (2) effectively competing with tyrosine for attachment to

* Department of Brain and Cognitive Sciences, Massachusetts Institute of Technology, Cambridge, MA 02139, USA.

their common large neutral amino acid (LNAA) transporter, at the blood-brain barrier (Oldendorf 1973, Pardridge 1977, Fernstrom and Faller 1978).

In the present study, we have examined the relationship between phenylalanine dose, in vivo, and brain catecholamine release.

Methods

Dopamine in brain perfusates was collected and measured using a microdialysis system coupled to a highly sensitive high-pressure liquid chromatograph (HPLC) with coulometric detector (ESA, Inc.). [The use of brain dialysis was pioneered by Delgado et al. (1972) and elegantly adapted for studying rat brain neurons by Ungerstedt et al. (1982).]

Dialysis probes were obtained from Carnegie Medicin (4 mm membrane length, 0.5 mm o.d., 5000 MW cutoff). These were perfused with a Krebs-Ringer buffer (1 m*M* Ca, pH 7.35) at a flow rate of 1.5 μl/min using a CMA 100 pump. Fifteen-minute samples were collected into 5 μl of 0.5 *M* perchloric acid. The probe was calibrated by measuring the recovery of standards of known concentration (100 n*M*). Male Sprague-Dawley rats (300 to 350g) were anesthetized using chloralose/urethane (50/500 mg)/kg intraperitoneally (i.p.). Rats were placed in a Kopf stereotaxic frame; the skull was exposed, a hole drilled, and the probe implanted in the right striatum at coordinates: bregma +0.5; right 2.5; ventral −7.0 mm (Paxinos and Watson 1986). The rats were maintained at a temperature of 37°C and kept stable under anesthesia by administering additional doses of anesthetic as required.

Dialysates were injected directly onto the HPLC, and the concentrations of dopamine, dihydroxy phenylacetic acid (DOPAC), and homovanillic acid (HVA) were measured. After an initial period of "injury release" (60 to 90 minutes post probe implantation), stable concentrations of dopamine were obtained. Following a minimum of three stable collections (45 min), treatments were administered. Rats received either 0.9% NaCl i.p. (control) or phenylalanine (as a suspension in saline) in doses of 200, 500, or 1000 mg/kg i.p. All volumes were given at 4 ml/kg. Dialysates were injected immediately on collection; there was no sample preparation nor storage. Chromatograms were completed within 11 minutes.

Results

The basal dopamine concentrations in dialysates were similar in all groups, the mean concentration being 4.62 ± 0.31 n*M*. As the in vitro probe calibration gave a recovery of 16% ± 1%, the basal dialysates therefore reflected an extracellular dopamine level of 29 ± 3 n*M*. The dopamine release following treatment was determined for every animal by recording

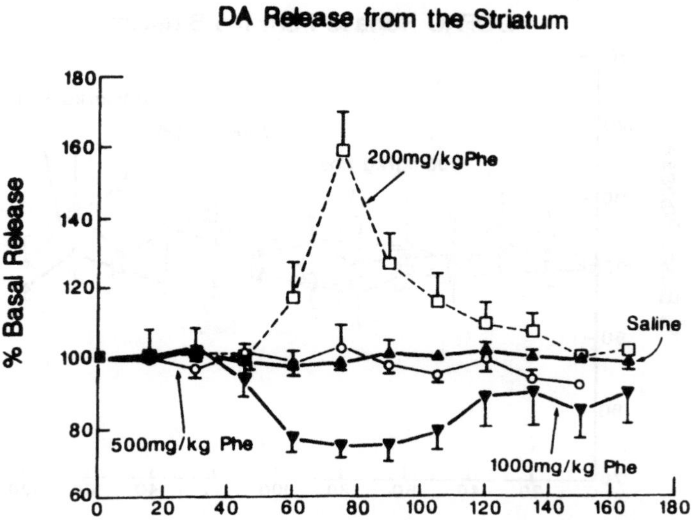

FIGURE 8.1. The effects of phenylalanine on the release of dopamine from the rat striatum, in vivo. Probes were placed acutely in the right striata, under chloralose/urethane anesthesia, and were perfused with artificial CSF (1.5 μl/min) for 3 hours; CSF samples, collected for 15-min periods, were assayed for dopamine by HPLC. After a basal (60 to 90 min) period, animals received phenylalanine (200, 500, or 1000 mg/kg) i.p. Vertical bars represent standard error of the mean.

the percent change from stable baseline during a minimum of three 15-minute collections.

Rats receiving normal saline ($n = 4$) showed no change in dopamine release during the 120 minutes following its i.p. administration (Figure 8.1). When rats received 200 mg of phenylalanine per kg ($n = 4$), the mean dopamine concentration in the dialysate was increased to a maximum of 159% ± 11% of basal levels at 75 minutes post administration. Rats receiving 500 mg of phenylalanine per kg ($n = 4$) showed no significant change in dopamine release (103% ± 3% of basal levels at 75 minutes). Rats receiving the highest phenylalanine dose (1000 mg/kg) ($n = 4$) exhibited, in contrast to those receiving the lower doses, a significant *inhibition* of dopamine release by 26% ± 3.6%.

Both DOPAC (Figure 8.2) and HVA (Figure 8.3) levels in the dialysate remained unchanged in all of the treatment groups. Basal DOPAC levels were 2092 ± 135 n*M*, and those of HVA were 1388 ± 79 n*M*, reflecting extracellular levels of 8046 ± 572 n*M* for DOPAC and 4477 ± 254 n*M* for HVA. (The in vitro recoveries for DOPAC and HVA were 26% and 31%, respectively.) The basal extracellular concentrations of these metabolites were approximately 200 times greater than those of dopamine, confirming findings of others (Sharp et al. 1986).

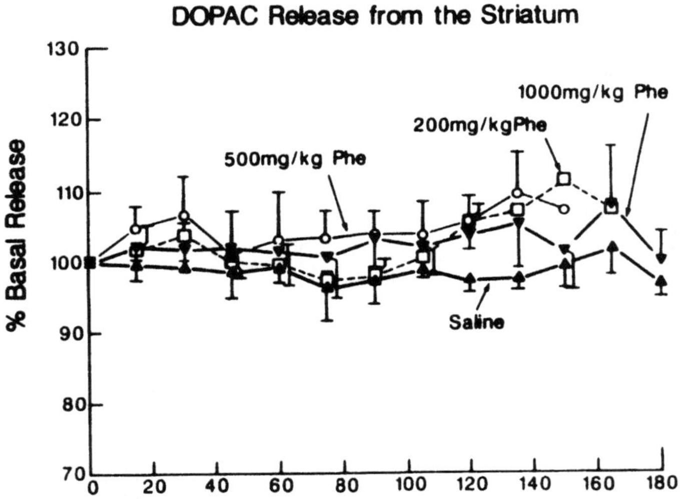

FIGURE 8.2. The effect of phenylalanine on the release of DOPAC from The Rat Striatum, in vivo. Animals were prepared as described in the legend of Figure 8.1. CSF samples, collected for 15-min periods, were assayed for DOPAC by HPLC. After a basal (60 to 90 min) period, animals received phenylalanine (200, 500, or 1000 mg/kg) i.p. Vertical bars represent standard error of the mean.

Discussion

These results provide, to our knowledge, the first direct in vivo evidence that high doses of phenylalanine *decrease* striatal dopamine release, whereas low doses, which, in rats, selectively increase brain tyrosine, increase dopamine release. This reduction in extracellular dopamine levels (as seen after the 1000-mg/kg dose) is likely to reflect at least two processes, i.e., the reduction in the plasma tyrosine/LNAA ratio and the consequent reduction in brain tyrosine levels (Fernstrom and Faller 1978); and the inhibition of tyrosine 3-monooxygenase (Milner et al. 1986, Katz et al. 1976). This effect of exogenous phenylalanine on brain catecholamines is complicated, in rodents, by their highly efficient hepatic conversion of phenylalanine to tyrosine; this causes the net effect of small phenylalanine doses to be a greater increase in plasma tyrosine than in plasma phenylalanine levels. At somewhat higher phenylalanine doses, the hydroxylation of phenylalanine becomes less efficient, so that the plasma tyrosine/phenylalanine ratio is unchanged from baseline, with proportionate increases in brain phenylalanine and tyrosine. Still higher doses increase plasma phenylalanine levels more than those of tyrosine. (It should be noted that, in humans, apparently *all* doses raise phenylalanine by more than tyrosine, since the human liver is relatively inefficient at hydroxylating phenylalanine.)

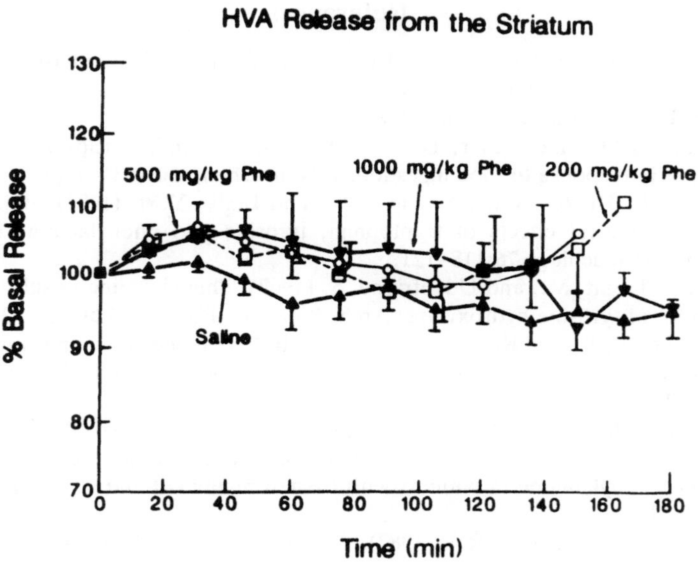

FIGURE 8.3. The effect of phenylalanine on the release of HVA from the rat striatum, in vivo. Animals were prepared as described in the legend of Figure 8.1. CSF samples, collected for 15-min periods, were assayed for HVA by HPLC. After a basal (60 to 90 min) period, animals received phenylalanine (200, 500, or 1000 mg/kg) i.p. Vertical bars represent standard error of the mean.

Although the changes in dopamine release (the increase by 59% and the decrease by 26%) were small compared with the changes seen after a drug like amphetamine (2 mg/kg increasing release by 1600%), they may still be sufficient to cause behavioral effects, particularly in animals or humans with impaired nigrostriatal neurons. For example, phenylalanine decreased immobility in a swim test (Gibson et al. 1982) (200 mg had a greater effect than 400 mg) and also increased open field behavior in mice (Thurmond et al. 1980). Doses of 1000 mg/kg or greater lowered seizure threshold to subcutaneous pentylenetetrazole, in mice, whereas lower doses had no effect (Pinto and Maher 1988, Maher and Wurtman 1987). In humans with phenylketonuria, urinary dopamine excretion is reduced, and this reduction can be reversed by lowering plasma phenylalanine levels (Nadler and Hsia 1961, Krause et al. 1985). The results of the present study are likely to be relevant to humans who, for genetic (phenylketonuria) or dietary reasons (high dietary intake of phenylalanine), may have impaired striatal dopamine release.

Acknowledgments. These studies were supported in part by grants from The Center For Brain Sciences and Metabolism Charitable Trust, the National Institutes of Health, and The United States Air Force.

References

Delgado, J.M.R., De Feudis, F.V., Roth, R.H., Ryugo, D.K., and Mitruka, B.M. (1972). Dialytrode for long term intracerebral perfusion in awake monkeys. Arch. Int. Pharmacodyn. Ther. **198:**9.

Fernstrom, J.D., and Faller, D.V. (1978). Neutral amino acids in the brain: changes in response to food ingestion. J. Neurochem. **30:**1531–1538.

Gibson, C.J., Deikel, S.M., Young, S.N., and Binik, Y.M. (1982). Behavioural and biochemical effects of tryptophan, tyrosine and phenylalanine in mice. Psychopharmacology **76:**118–121.

Ikeda, M., Levitt, M., and Udenfriend, S. (1967). Phenylalanine as substrate and inhibitor of tyrosine hydroxylase, Arch. Biochem. Biophys. **120:**420–427.

Katz, I., Lloyd, T., and Kaufman, S. (1976). Studies on phenylalanine and tyrosine hydroxylation by rat brain tyrosine hydroxylase. Biochim. Biophys. Acta **445:**567–578.

Krause, W., Halminski, M., McDonald, L., Dembure, P., Salvo, R., Friedes, D., and Elsas, L. (1985). Biochemical and neuropsychological effects of elevated plasma phenylalanine in patients with treated phenylketonuria. J. Clin. Invest. **75:**40–48.

Maher, T., and Wurtman, R.J. (1987). Possible neurological effects of aspartame, a widely used food additive. *In* Environmental health perspectives, Twentieth anniversary of the National Institute of Environmental Health Sciences **75:**53–58.

McKean, C.M. (1972). The effects of high phenylalanine concentrations on serotonin and catecholamine metabolism in the human brain. Brain Res. **47:**469–476.

Milner, J.D., Irie, K., and Wurtman, R.J. (1986). Effects of phenylalanine on the release of endogenous dopamine from rat striatal slices. J. Neurochem. **47:**1444–1448.

Nadler, H.L., and Hsia, D.Y. (1961). Epinephrine metabolism in phenylketonuria. Proc. Soc. Exp. Biol. **107:**721–723.

Oldendorf, W.H. (1973). Stereospecificity of the blood brain barrier to amino acids. Am. J. Physiol. **224:**967–969.

Pardridge, W. (1977). Regulation of amino acid availability to the brain. *In* Wurtman, R.J., and Wurtman, J.J. (eds.), Nutrition and the brain, Vol. 1. New York: Raven Press, pp. 142–204.

Paxinos, G. and Watson, C. (1982). The rat brain in stereotaxic coordinates. New York: Academic Press.

Pinto, J.M.B. and Maher, T.J. (1988). Aspartame administration potentiates fluorothyl- and pentylenetetrazole-induced seizure in mice. Neuropharmacol. **27:**51–5.

Sharp, T. Zetterstrom, T., and Ungerstedt, U. (1986). An in vivo study of dopamine release and metabolism in rat brain regions using intracerebral dialysis. J. Neurochem. **47:**113–122.

Thurmond J.B., Kramacy, N.R., Lasley, S.M., and Brown, J.W. (1980). Dietary amino acid precursors: effect on central monoamines, aggression and locomotor activity in the mouse. Pharmacol. Biochem. Behav. **12:**525–532.

Ungerstedt, U., Herrera-Marschitz, M., Jungnelius, U., Stahle, L., Tossman, U., and Zetterstrom, T. (1982). Dopamine synaptic mechanisms reflected in studies combining behavioural recordings and brain dialysis. *In* Kohsaka, M. et al. (eds.), Advances in the biosciences, Vol. 37: Advances in dopamine research. Oxford: Pergamon Press, pp. 219–231.

CHAPTER 9

Effects of Aspartame Ingestion on Large Neutral Amino Acids and Monoamine Neurotransmitters in the Central Nervous System

John D. Fernstrom*

The acute ingestion of aspartame in large doses by rats leads to rapid increments in serum tyrosine and phenylalanine levels. Such increments should enhance the uptakes of these amino acids into brain and also antagonize the entry into brain of other large neutral amino acids. This article reviews evidence that such amino acid effects do occur in brain following aspartame ingestion by rats and then evaluates their impact on the rates of synthesis of the monoamine neurotransmitters.

Aspartame (APM) is a phenylalanine-containing dipeptide (L-aspartyl-L-phenylalanine methyl ester). Following its ingestion by rats, serum levels of both phenylalanine (Phe) and tyrosine (Tyr) rapidly rise (Fernstrom et al. 1983, 1986, Yokogoshi and Wurtman 1986). (Serum Tyr levels rise because of rapid postprandial hydroxylation of Phe in the liver.) Because of these increments, the issue has been raised as to whether APM ingestion could enhance the transport into brain of Tyr and Phe and simultaneously inhibit the transport of tryptophan (Trp) and other large neutral amino acids (LNAA). Such effects might occur because Phe and Tyr are LNAA, and LNAA compete among themselves for transport into brain via a shared carrier, located at the blood-brain barrier (Pardridge and Oldendorf 1975). A rapid increase in serum levels of Phe and Tyr would thus be expected to enhance the uptake of these amino acids into brain, and at the same time diminish the uptake of the other LNAA. These other LNAA include Trp and the branched-chain amino acids (BCAA). The consequence of this competitive challenge would presumably be to reduce the brain levels of Trp and the BCAA, while raising those of Phe and Tyr.

Such changes in brain are of interest because a number of brain neurotransmitters are synthesized from the LNAA: serotonin (5-HT) from Trp, and the catecholamines [e.g., dopamine (DA)] from Tyr and Phe. Under appropriate conditions, their rates of synthesis in brain neurons are directly

*Departments of Psychiatry and Behavioral Neuroscience, University of Pittsburgh School of Medicine, Pittsburgh, PA 15213, USA.

influenced by the local availability of substrate (Fernstrom 1983, Wurtman et al. 1980). Such changes in the synthesis of transmitters could potentially modify brain function, since these substances are critical for information transfer among neurons. Hence, if following APM ingestion, changes in the brain levels of one or more of the LNAA were to occur, and this led to alterations in the formation of one or more neurotransmitters, then the ingestion of the sweetener might be predicted to influence brain functions.

This article reviews published data that evaluate the above notions, and focuses principally on three issues. The first is whether APM-induced increments in serum Phe and Tyr levels affect steady-state brain Trp levels and the increments in brain Trp level that accompany the ingestion of carbohydrates. The second is whether APM-induced increments in serum Phe and Tyr levels might be expected to influence the brain uptakes of LNAA drugs (DOPA, aldomet) that produce their therapeutic action following brain uptake. And the third is whether APM-induced increases in brain Phe and Tyr levels influence their rates of conversion to catecholamines.

Issue 1: Aspartame, Tryptophan, and Serotonin

In the earliest studies, APM was simply administered by gavage (p.o.) to fasting rats, and the effects on serum and brain LNAA levels measured. The sweetener produced dose-related (50 to 200 mg/kg) increments in the serum and brain levels of both Tyr and Phe that peaked within 60 minutes of administration (Fernstrom et al. 1983). Despite such changes, however, no remarkable effects were noted on the brain levels of either Trp or the BCAA. Furthermore, in keeping with the absence of an effect on brain Trp level, and the known relationship between brain Trp levels and 5-HT synthesis rate (Fernstrom 1983, Wurtman et al. 1980), no changes were noted in either the level of 5-HT (or its principal metabolite) or in the rate of 5-HT synthesis (Fernstrom et al. 1983).

Although APM administration had no effect on the levels of Trp or 5-HT in fasting rats, perhaps APM would affect the rise in brain Trp levels and 5-HT synthesis that is produced by the ingestion of carbohydrates (Wurtman 1983). Carbohydrate ingestion is thought to raise brain Trp uptake and levels, and thus 5-HT synthesis via insulin secretion. Insulin release causes the serum levels of Trp's competitors to fall (mainly the BCAA), thus giving Trp a competitive edge for access to the transport carrier into brain (Fernstrom 1983). Brain Trp uptake and levels accordingly increase. Since the carbohydrate-induced rise in brain Trp is dependent on the decline in blood of Trp's competitors, the notion was that APM ingestion along with carbohydrates would antagonize the carbohydrate-induced rise in brain Trp levels. The antagonism would occur because of the APM-induced increments in the serum levels of Trp's transport competitors, Phe and Tyr (Wurtman 1983). Subsequently, data were reported by Yokogoshi et al. (1984) regarding such an antagonistic effect of APM at an oral dose of 200 mg/kg.

Recently, we have reinvestigated this issue using an experimental design in which rats ingested carbohydrates by mouth in combination with APM (0 to 1500 mg/kg). The animals were fasted overnight, and the next morning given free access to a source of food containing carbohydrates and a given level of APM. They were killed 2 hours later. We noted dose-related increases in the serum levels of Phe and Tyr and identified a threshold dose of APM above which the carbohydrate-induced rise in brain Trp was suppressed. However, we found this dose to be in excess of 500 mg/kg (Fernstrom et al. 1986). This result fits with data obtained by Carlsson and Lindqvist (1978), who measured brain Trp levels soon after rats received an intraperitoneal injection of Phe. They found that a Phe dose of 100 mg/kg produced no effect on brain Trp level, while 300 mg/kg elicited a 33% reduction in brain Trp. A higher dose, 1000 mg/kg, reduced brain Trp levels by 50%. In other studies, Tyr was found to be similar in potency to Phe in reducing brain Trp levels. A dose of APM of 500 mg/kg represents a Phe dose of about 300 mg/kg; our results are thus consistent with those of Carlsson and Lindqvist (1978) and illustrate the need for a very large dose of Phe (APM) to modify Trp levels in brain. It is not obvious to us why our data differ from those of Yokogoshi et al. (1984). However, we have used as our paradigm that developed originally by Fernstrom and Wurtman (1971) to demonstrate carbohydrate effects on brain Trp and 5-HT levels, and we have consistently obtained the same dose-response results in about 15 experiments, utilizing a wide range of APM doses (Fernstrom et al. 1986). We are thus confident that with this meal paradigm, a dose of APM in excess of 500 mg/kg is required to block the carbohydrate-induced rise in brain Trp levels and 5-HT synthesis. In humans, if a metabolic rate adjustment of 5 is used (Munro 1969), the theoretical dose level needed to produce the same effect would be 100 mg/kg or more. This dose would have to be ingested in a single load, as was the case in rat experiments. Such a practice is not likely to occur. For example, if a 30-kg child consumed this amount, the total load would be 3000 mg or more (considerably more for a 70-kg adult). Given that APM is 200 times sweeter than sucrose (Mazur 1984), this dose represents a substantial level of sweetness (the equivalent of 600 g or 1.3 lb of sucrose). It seems improbable that this amount of APM would ever knowingly be consumed in a single sitting, much less in combination with sucrose. It therefore does not seem likely in man that enough APM would ever knowingly be ingested to block the rise in brain Trp levels and 5-HT synthesis that might follow carbohydrate ingestion by this species.

It should also be noted in this context that the difference between man and rats regarding the rate of Phe conversion to Tyr is not of particular importance, since the k_i (the concentration that produces 50% inhibition) for Phe's inhibition of Trp uptake into brain should be roughly the same as that for Tyr (Pardridge 1977). Hence, though the rat experiences a larger rise in serum Tyr than Phe after APM gavage, and the human a larger rise in Phe than Tyr (Fernstrom et al. 1986, Stegink et al. 1979), the impact of this difference on brain Trp uptake following the ingestion of APM should be minimal. Hence, correcting for overall metabolic rate differences, the rat should continue to be a good model for predicting APM's effects on brain Trp uptake and 5-HT synthesis in humans.

Issue 2: Aspartame and LNAA Drugs

The competitive carrier for transporting the LNAA across the blood-brain barrier not only transports natural LNAA, but also LNAA drugs (Markovitz and Fernstrom 1977, Pardridge 1977). Two LNAA drugs are of particular interest because of their therapeutic utility: L-DOPA and L-α-methyldopa (aldomet). DOPA is naturally occurring (as the hydroxylation product of tyrosine), but is used as a pharmacologic agent in large doses to treat Parkinson's disease. Aldomet is not a natural LNAA, but differs from L-DOPA only by the addition of a methyl group to the alpha-carbon. This is apparently not an important difference to the transport carrier, for aldomet has been shown to compete with natural LNAA for uptake into brain (Markovitz and Fernstrom 1977). Aldomet is used in the treatment of hypertension. Both drugs exert their therapeutic effects in the brain. Since both gain access to the brain via the LNAA transport carrier, it might be asked if APM ingestion should interfere with the therapeutic action of these drugs.

The answer is probably no. First, some data already exist suggesting that the coadministration of APM (200 mg/kg p.o.) with aldomet does not antagonize the antihypertensive action of the drug (Fernstrom 1984). Second, if a dose were found that antagonized aldomet's ability to lower blood pressure in rats, this dose would most likely be about that required to suppress the carbohydrate-induced rise in brain Trp and 5-HT (i.e., very high, >500 mg/kg). Such would be the case since the brain uptake of aldomet should probably be equally inhibited by Phe and Tyr, much like Trp. Third, given the above, the rat should thus be a valid model for extrapolating results to man. This view reinforces the notion that a very high dose of APM would probably be needed in man [>100 mg/kg, based on the fivefold metabolic difference between rats and man (Munro 1969)] to inhibit brain aldomet uptake (and thus possibly its antihypertensive actions). The issue for L-DOPA would probably be even less interesting, all of the above being true, since this amino acid is typically given in gram quantities, often in combination with a decarboxylase inhibitor to minimize DOPA metabolism outside the brain (Bianchine 1980). Very high doses of APM would therefore probably be required to inhibit brain DOPA uptake. This view is supported by the finding that an oral Phe dose of 100 mg/kg will antagonize the therapeutic efficacy of intravenously administered L-DOPA in Parkinsonian patients (Nutt et al. 1984). This is such a large dose of Phe that it alone will raise the total plasma LNAA level by as much as the ingestion of a meal containing 30 grams of protein (Nutt et al. 1984). This amount of protein is considerable, representing in some Western diets 40% to 50% of the total daily intake of protein (Milsom et al. 1979). From a sweetness perspective, if APM had been ingested instead of Phe, the equivalent dose would have been 167 mg/kg, and thus the subjects would have to have knowingly consumed the sweetness equivalent of 2.34 kg (5.15 lb) of sucrose (if each weighed 70 kg). It is difficult to imagine that anyone, much less a Parkinsonian patient, would willingly ingest this amount of APM all at once.

Issue 3: Aspartame and Brain Catecholamine Synthesis

The effects of APM on brain catecholamine synthesis have not been very well studied. In one series of experiments, we observed APM gavage in doses up to 200 mg/kg to have no effects on brain levels of dopamine, norepinephrine, or their metabolites (Fernstrom et al. 1983). We also observed no effects on in vivo tyrosine hydroxylation rate in the corpus striatum, suggesting the absence of an effect on DA synthesis. Yokogoshi and Wurtman (1986) have also studied effects of APM administration (200 mg/kg, largely by the intraperitoneal route) on brain catecholamines. Their results show scattered effects in different brain regions for a variety of catecholamines and catecholamine metabolites, giving the impression that no clear effects are occurring. The study can be faulted for administering a food additive by the intraperitoneal route. Notwithstanding, the study shows very little, and the data are thus generally consistent with those of Fernstrom et al. (1983).

It is well known that the synthesis of catecholamines is not responsive to precursor supply when the neurons that synthesize them are not rapidly firing (Wurtman et al. 1980, Fernstrom 1983). The nigrostriatal DA neurons normally fire quite slowly (Bannon and Roth 1983), and thus are not typically responsive to tyrosine administration (Scally et al. 1977, Sved and Fernstrom 1981). Another set of DA neurons should therefore be studied, which does show higher firing activity and therefore presumably greater tyrosine sensitivity. Only a single, preliminary study has appeared to date that attempts to do this. In this study, the investigators looked for effects of tyrosine and APM on in vivo tyrosine hydroxylation rate in the projection field of the mesocortical DA neurons, the prefrontal and cingulate cortices (Bannon and Roth 1983, Tam et al. 1986). They reported that tyrosine injection stimulates hydroxylation rate in these neurons and that APM does so too, but only at doses approaching 400 mg/kg p.o. This result suggests that DA synthesis may not be remarkably sensitive to APM loading. However, it should be noted that these studies were performed in rats, in which the major effect of APM administration is to raise serum and brain Tyr levels. In humans, an increase in blood (and probably brain) Phe would be the major effect. The effect of selectively incrementing brain Phe levels on DA synthesis has not been studied in vivo in rats or humans, though a recent study using rat brain slices explores the relationship of Phe level to DA release (Milner et al. 1986). Available data suggest the situation may be enzymatically complicated, since Phe may be both a substrate for and an inhibitor of tyrosine hydroxylase (Kaufman 1974), a view supported by the findings of Milner et al. (1986). As a consequence, the effect of APM (or Phe) administration on DA synthesis in the human (or rat) brain remains to be clarified.

Summary and Conclusions

Aspartame ingestion will block the carbohydrate-induced rise in brain Trp levels and 5-HT synthesis in rats, but only at very large doses (>500 mg/kg p.o.). The sweetener can probably also antagonize the transport of LNAA drugs into brain, but the threshold dose is again likely to be very high (in the range of that which influences brain Trp uptake: >500 mg/kg p.o.). Effects of APM ingestion on brain catecholamine synthesis are only poorly studied. However, available data suggest that even if catecholamine synthesis is shown to respond to APM ingestion, the dose required for such effects will probably be very large. Thus, the studies purporting to connect APM administration to effects on brain LNAA transport and monoamine synthesis in brain center on pharmacologically high doses of the sweetener. Their relevance to normal APM consumption is, at best, unclear. It should be kept in mind that for all of these demonstrated and postulated effects of APM on brain neurotransmitters, the method of treatment has been a single oral dose (or injection) of the dipeptide. The chronic intake of large amounts of APM will probably have small effects in comparison, owing to metabolic adaptation of amino acid catabolic enzymes. Finally, in future studies, APM administration should be limited to the *oral route*. The sweetener is a dipeptide food additive; the normal entry of its amino acid constituents into the circulation is via the gut. The peptide does not normally enter the circulation in appreciable amounts, and certainly not via the peritoneum or the nape of the neck. Further experimentation using nonoral routes should therefore be strongly discouraged.

Acknowledgments. Some of the studies described were supported by the National Institute of Mental Health (MH38178; MH00254).

References

Bannon, M.J., and Roth, R.H. (1983). Pharmacology of mesocortical dopamine neurons. Pharmacol. Rev. **35:**53–68.

Bianchine, J.R. (1980). Drugs for Parkinson's disease. *In* Gilman, A.G., Goodman, L.S., and Gilman, A. (eds.), The pharmacologic basis of therapeutics, 6th ed. New York: Macmillan, pp. 481–482.

Carlsson, A., and Lindqvist, M. (1978). Dependence of 5-HT and catecholamine synthesis on concentrations of precursor amino acids in rat brain. Naunyn Schmiedeberg's Arch. Pharmacol. **303:**157–164.

Fernstrom, J.D. (1983). Role of precursor availability in the control of monoamine biosynthesis in brain. Physiol. Rev. **63:**484–546

Fernstrom, J.D. (1984). Effects of acute aspartame ingestion on large neutral amino acids and monoamines in rat brain. *In* Stegink, L.D., and Filer, L.J. (eds.), Aspartame: physiology and biochemistry. New York: Marcel Dekker, pp. 641–653.

Fernstrom, J.D., and Faller, D.V. (1978). Neutral amino acids in the brain: changes in response to food ingestion. J. Neurochem. **30:**1531–1538.

Fernstrom, J.D., and Wurtman, R.J. (1971). Brain serotonin content: increase following ingestion of carbohydrate diet. Science **174:**1023–1025.

Fernstrom, J.D., Fernstrom, M.H., and Gillis, M.A. (1983). Acute effects of aspartame on large neutral amino acids and monoamines in rat brain. Life Sci. **32:**1651–1658.

Fernstrom, J.D., Fernstrom, M.H., and Grubb, P.E. (1986). Effects of aspartame ingestion on the carbohydrate-induced rise in tryptophan hydroxylation rate in rat brain. Am. J. Clin. Nutr. **44:**195–205.

Kaufman, S. (1974). Properties of pterin-dependent aromatic amino acid hydroxylases. *In* Wolstenholme, G.E.W., and FitzSimons, D.W. (eds.), Aromatic amino acids in the brain. Amsterdam: Elsevier, pp. 85–108.

Markovitz, D.C., and Fernstrom, J.D. (1977). Diet and uptake of aldomet by the brain: competition with natural large neutral amino acids. Science **197:**1014–1015.

Mazur, R.H. (1984). Discovery of aspartame. *In* Stegink, L.D., and Filer, L.J. (eds.), Aspartame: physiology and biochemistry. New York: Marcel Dekker, pp. 3–9.

Milner, J.D., Irie, K., and Wurtman, R.J. (1986). Effects of phenylalanine on the release of endogenous dopamine from rat striatal slices. J. Neurochem. **47:**1444–1448.

Milsom, J.P., Morgan, M.Y., and Sherlock, S. (1979). Factors affecting plasma amino acid concentrations in control subjects. Metabolism **28:**313–319.

Munro, H.N. (1969). Evolution of protein metabolism in mammals. *In* Munro, H.N. (ed.), Mammalian protein metabolism, Vol III. New York: Academic Press, pp. 133–183.

Nutt, J.G., Woodward, W.R., Hammerstad, J.P., Carter, J.H., and Anderson, J.L. (1984). The "on–off" phenomenon in Parkinson's disease: Relation to levodopa absorption and transport. New Engl J Med. **310:**483–488.

Pardridge, W.M. (1977). Kinetics of competitive inhibition of neutral amino acid transport across the blood-brain barrier. J. Neurochem. **28:**103–108.

Pardridge, W.M., and Oldendorf, W.H. (1975). Kinetic analysis of blood brain barrier transport of amino acids. Biochim. Biophys. Acta **401**:128–136.

Scally, M.C., Ulus, I., and Wurtman, R.J. (1977). Brain tyrosine level controls striatal dopamine synthesis in haloperidol-treated rats. J. Neural Transm. **41**: 1–6.

Stegink, L.D., Filer, L.J., Baker, G.L., and McDonnell, J.E. (1979). Effect of aspartame loading upon plasma and erythrocyte amino acid levels in phenylketonuric heterozygotes and normal adult subjects. J. Nutr. **109**:708–717.

Sved, A.F., and Fernstrom, J.D. (1981). Tyrosine availability and dopamine synthesis in the striatum: studies with gamma-butyrolactone. Life Sci. **29**:743–748.

Tam, S.Y., Ono, N., and Roth, R.H. (1986). Precursor control and influence of aspartame on midbrain dopamine. J. Cell. Biochem. **31**(2)[Suppl]:18.

Wurtman, R.J. (1983). Neurochemical changes following high-dose aspartame with dietary carbohydrates. N. Engl. J. Med. **309**:429–430.

Wurtman, R.J., Hefti, F., and Melamed, E. (1980). Precursor control of neurotransmitter synthesis. Pharmacol. Rev. **32**:315–335.

Yokogoshi, H., Roberts, C.H., Caballero, B., and Wurtman, R.J. (1984). Effects of aspartame and glucose administration on brain and plasma levels of large neutral amino acids and brain 5-hydroxyindoles. Am. J. Clin. Nutr. **40**:1–7.

Yokogoshi, H., and Wurtman, R.J. (1986). Acute effects of oral or parenteral aspartame on catecholamine metabolism in various regions of rat brain. J. Nutr. **116**:356–364.

CHAPTER 10

Aspartame, Phenylalanine, and Seizures in Experimental Animals

Judith M.B. Pinto and Timothy J. Maher*

An association has recently been proposed between seizure promotion and consumption of the phenylalanine-containing artificial sweetener aspartame. Since consumption of aspartame, unlike dietary protein, can elevate brain phenylalanine, and thus potentially inhibit the synthesis and release of neurotransmitters known to protect against seizure activity, we have studied the effects of aspartame and of its metabolic breakdown products (aspartic acid, methanol, and phenylalanine) on seizure activity, using two well-established experimental models, i.e., pentylenetetrazole and fluorothyl-induced seizures. The doses employed were chosen to produce changes in the plasma phenylalanine ratio which would mimic, in the mouse, the changes seen when humans consume moderate amounts of the sweetener. Mice receiving aspartame in doses of 1000 mg/kg or greater exhibited a significant *lowering* in the threshold to seizures induced by pentylenetetrazole. This response was mimicked by equimolar phenylalanine, but not by equimolar aspartic acid or methanol. Additionally, we tested the ability of tyrosine, a precursor of seizure-protecting catecholamines, to influence pentylenetetrazole-induced seizures. Tyrosine doses of 500 mg/kg or greater significantly *elevated* the seizure threshold in this model. *Coadministration* of tyrosine with either phenylalanine or aspartame *prevented* the deleterious effects of these compounds on the seizure threshold. Similarly, coadministration of the large neutral amino acid valine, which competes with phenylalanine for entry into the brain, also prevented the phenylalanine-induced changes. When phenylalanine or aspartame were administered prior to fluorothyl exposure, a significant decrease in the seizure threshold was also noted. Doses as low as 500 mg/kg were able to potentiate these seizures in immature mice. We conclude that in these commonly employed seizure models, aspartame, via its metabolic breakdown product phenylalanine, is capable of lowering the threshold to seizures. This may bear on the reported association between aspartame and seizures in humans.

A report by the Centers for Disease Control (1984) documented the adverse effects on the central nervous system associated with the consump-

*Department of Pharmacology, Massachusetts College of Pharmacy, Boston, MA 02115, USA.

tion of the artificial sweetener aspartame (L-aspartyl-L-phenylalanine methyl ester, APM, NutraSweet), which included seizures, insomnia, and mood changes. Since that time, additional reports have appeared in the literature describing the association between APM consumption and headaches (Johns 1986), inappropriate behavioral responses (Ferguson 1985, Drake 1986), and seizures (Wurtman 1985, Walton 1986). Since the consumption of APM in humans produces selective increases in the plasma phenylalanine ratio (phenylalanine concentration divided by the summed concentrations of competing large neutral amino acids), and presumably in brain phenylalanine, and since this amino acid is known to interfere with the synthesis and release of seizure-protective catecholamines, we investigated the ability of APM to potentiate seizures in two experimental animal models.

In numerous animal models of epilepsy (those used by most pharmaceutical companies for anticonvulsant screening), treatments that alter catecholaminergic neutrotransmission are known to influence seizure thresholds. For instance, the catecholamine precursor L-DOPA is known to exert anticonvulsant effects against maximal electroshock and pentylenetetrazole-induced seizures in mice and rats (Kilian and Frey 1973, Chen et al. 1968, Rudzik and Johnson 1970, McKenzie and Soroko 1972, Kleinrok et al. 1978). Additionally, nonselective dopamine agonists such as apomorphine or the selective D-2 receptor agonists lisuride and (+)-4-propyl-9-hydroxynaphthoxazine are effective anticonvulsants. Some investigators have demonstrated that alterations in noradrenergic neurotransmission may be involved in the anticonvulsant effects of L-DOPA, since pretreatment with a dopamine β-hydroxylase inhibitor can attenuate the anticonvulsant effects of L-DOPA against maximal electroshock seizures (McKenzie and Soroko 1973). Catecholamines may also be involved in the development of kindling in rats, since depletion of central norepinephrine or destruction of noradrenergic fibers (by lesions or by injection of 6-hydroxydopamine) results in a dramatic acceleration in the rate of kindling (Altman and Corcoran 1983, Araki et al. 1983, Arnold et al. 1973, Corcoran and Mason 1980, McIntyre and Edson 1981, Lewis et al. 1987).

The seizure susceptibility of genetically bred animal models of epilepsy is also altered by catecholaminergic modifications. The DBA/2J mouse, which is prone to audiogenic seizures, exhibits an anticonvulsant effect following L-DOPA administration which correlates well with increments in brain dopamine levels (Dailey and Jobe 1984). Similarly, other treatments that increase or decrease central catecholamine levels attenuate or intensify, respectively, seizure activity (Schlesinger et al. 1968, Boggan and Seiden 1971). In the genetically epilepsy-prone rat, where norepinephrine levels and turnover rates are decreased in numerous brain areas (Jobe et al. 1984), treatments that enhance noradrenergic neurotransmission decrease seizure severity (Jobe et al. 1986), while treatments that reduce noradrenergic neurotransmission increase seizure severity (Jobe et al. 1973).

Finally, the seizures induced by intermittent light stimulation in the Senegalese baboon (*Papio papio*) are attenuated by intracerebroventricular injections of norepinephrine, while reserpine intensifies these seizures (Altshuler et al. 1976), suggesting a noradrenergic involvement in seizure activity in this animal model.

Previous studies have shown that increasing the availability of tyrosine to rapidly firing neurons can lead to enhanced catecholamine synthesis and release. Thus, when dopamine-containing neurons that run from the substantia nigra to the corpus striatum are unilaterally injected with the neurotoxin 6-hydroxydopamine, the surviving neurons increase their firing frequency, and the administration of tyrosine produces increases in striatal concentrations of dopamine metabolites (Melamed et al. 1980). Similarly, tyrosine can, by enhancing norepinephrine production, lower blood pressure in hypertensive rats (Sved et al. 1979), increase blood pressure in rats made hypotensive by hemorrhage (Conlay et al. 1981), and alter the behavioral and neurochemical consequences of acute stress in rats (Lehnert et al. 1984); it can also increase tyrosine hydroxylation in light-stimulated, but not dark-adapted, retina (Fernstrom et al. 1984).

While the administration of APM to rats has been shown to enhance catecholamine synthesis and release [i.e., lowering blood pressure in hypertensive rats (Kiritsy and Maher 1986, Thakore and Crane 1987) and increasing norepinephrine and its metabolite in various rat brain regions (Yokogoshi and Wurtman 1986)], these studies have employed doses of APM which preferentially elevate brain *tyrosine* and *not phenylalanine*. This increase in tyrosine reflects the extremely active hepatic phenylalanine hydroxylase of the rodent. The enzyme is much less active in humans; hence, all phenylalanine or APM doses in humans preferentially elevate plasma phenylalanine. Since phenylalanine can interfere with catecholamine synthesis and release (Milner et al. 1986, Nagatsu et al. 1964), and since tyrosine protects against this effect, we have given rodents APM doses which elevate the plasma phenylalanine ratio as much as or more than that of tyrosine's—a situation similar to that observed when humans consume moderate amounts of APM (Caballero et al. 1986).

The first experiments were designed to determine the appropriate APM dose to mimic biochemically the changes that occur in brain phenylalanine in humans consuming this sweetener. Male CD-1 mice received, via oral intubation, an APM dose of 0, 200, 500, 1000, or 2000 mg/kg in a volume of 20 ml/kg body weight. One hour later the animals were sacrificed by decapitation and blood was collected from the cervical wound for amino acid analysis. As shown in Table 10.1, APM doses of 500 mg/kg or greater produced significant elevations in the plasma phenylalanine ratio. Doses of 2000 mg/kg were required to produce elevations in the plasma phenylalanine ratio that were greater than those in the plasma tyrosine ratio. Since phenylalanine can interfere with catecholamine synthesis and release, and since tyrosine may serve as the antidote in this situation, we used these

TABLE 10.1. Plasma phenylalanine and tyrosine concentrations and ratios in mice administered aspartame.[a]

Aspartame dose (mg/kg)	Plasma concentration (nmol/ml)		Plasma ratio[b]	
	Phe	Tyr	Phe/LNAA	Tyr/LNAA
0	78 ± 9	92 ± 6	0.11 ± 0.01	0.12 ± 0.01
200	99 ± 5[c]	136 ± 17[c]	0.11 ± 0.01	0.17 ± 0.02[c]
500	149 ± 22[c]	214 ± 17[c]	0.17 ± 0.07[c]	0.27 ± 0.05[c]
1000	213 ± 28[c]	251 ± 11[c]	0.24 ± 0.03[c]	0.30 ± 0.01[c]
2000	635 ± 35[c]	406 ± 24[c]	0.77 ± 0.06[c]	0.38 ± 0.02[c]

[a] Male CD-1 mice received via oral intubation 0 to 2000 mg/kg of aspartame and were killed one hour later. Bloods were collected and plasma was separated and analyzed for amino acid content.

[b] Phenylalanine (Phe) and tyrosine (Tyr) ratios were determined as the concentration of Phe and Tyr divided by the summed concentrations of the other large neutral amino acids (LNAA).

[c] $p < 0.05$ significantly different from the 0-mg/kg group.

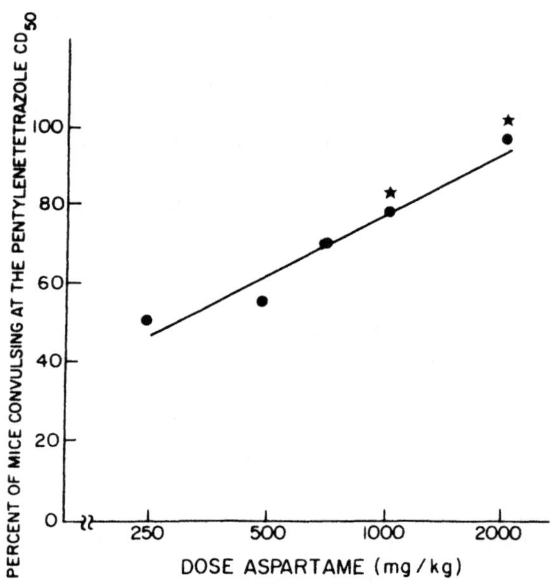

FIGURE 10.1. Effect of aspartame pretreatment on the percentage of mice convulsing following the administration of the CD_{50} of pentylenetetrazole. Groups of male CD-1 mice (average $N = 24$) received 0 to 2000 mg/kg of aspartame via oral intubation followed by a subcutaneous injection of 65 mg/kg of pentylenetetrazole one hour later. The number of mice convulsing was determined. $^*p < 0.05$ significantly different from the 0-mg/kg group as determined by the chi-square test (Pinto and Maher 1986).

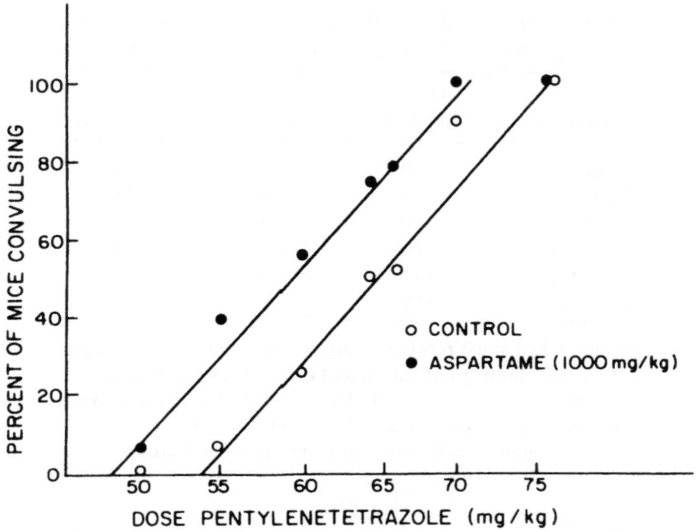

FIGURE 10.2. Effect of aspartame (1000 mg/kg) on the percentage of mice convulsing at various doses of pentylenetetrazole. Groups (average $N = 24$) of male CD-1 mice received water or 1000 mg/kg of aspartame via oral intubation followed by various doses of pentylenetetrazole, an hour later. The number of animals convulsing was determined. Aspartame pretreatment significantly ($p < 0.05$) shifted the dose-response curve to the left, as determined by the method of Litchfield and Wilcoxon.

seemingly high doses of APM in the subsequent studies, because our overriding goal was to use the animals to mimic what actually occurs when people consume APM.

Aspartame administration produced a dose-dependent increase in seizure frequency among animals subsequently receiving the CD_{50} dose of pentylenetetrazole (65 mg/kg, subcutaneously, Figure 10.1). At the 1000- and 2000-mg/kg doses, 78% and 100% of the animals experienced seizures, compared with 50% in the water-pretreated group.

Other mice pretreated with a fixed dose (1000 mg/kg) of APM, or with water, and given various doses (50 to 75 mg/kg) of pentylenetetrazole an hour later exhibited a significant leftward shift of the dose-response curve (Figure 10.2). The calculated CD_{50}'s (95% confidence intervals) for the APM- and water-pretreated mice were 59 (56–63) and 66 (64–68) mg/kg, respectively ($p < 0.05$). Enhanced susceptibility to pentylenetetrazole was also observed among mice pretreated with phenylalanine (in doses equimolar to effective doses of APM) but not among animals pretreated with aspartic acid or methanol.

Coadministration of APM with the large neutral amino acid valine, which competes with phenylalanine for passage across the blood-brain bar-

TABLE 10.2. Effect of tyrosine pretreatment on pentylenetetrazole-induced seizures in the mourse.[a]

Treatment	Dose (mg/kg)	Fraction convulsing	Percent convulsing
Control	0	38/77	49
Tyrosine	100	18/40	45
Tyrosine	250	15/40	38
Tyrosine	500	11/40	28[b]
Tyrosine	615	12/40	30[b]
Tyrosine	1000	13/40	32[b]

[a] Male CD-1 mice received water or tyrosine via oral intubation one hour prior to subcutaneously administered pentylenetetrazole (65 mg/kg). The number of animals convulsing during the next two hours was determined.
[b] $p < 0.05$ significantly different from the control group.

rier, protected mice from the seizure-promoting effects of APM, while alanine, an amino acid that does not compete with phenylalanine for brain uptake, failed to attenuate the sweetener's effect on pentylenetetrazole-induced seizures. The seizure-enhancing activity of APM has also been recently confirmed by Garattini (1986).

While a previous report in the literature had documented the seizure-enhancing activity of phenylalanine in rats (Gallagher 1969), this report had also suggested that tyrosine might possess seizurogenic potential. Therefore, we determined the ability of acutely administered tyrosine to potentiate pentylenetetrazole-induced seizures. Animals received a tyrosine dose of 0, 100, 250, 500, 615, or 1000 mg/kg orally one hour prior to pentylenetetrazole (65 mg/kg). As shown in Table 10.2, tyrosine failed to potentiate seizures; in fact, tyrosine actually had an *anticonvulsant* action. Additionally, coadministration of a tyrosine dose of 615 mg/kg with the previously determined seizurogenic dose of phenylalanine (560 mg/kg) or APM (1000 mg/kg) *prevented* the previously observed potentiation of seizures.

A seizure-promoting effect of APM was also observed in mice exposed to the inhalation convulsant fluorothyl. As shown in Figure 10.3, APM doses of 1000 mg/kg or greater significantly reduced the time to clonic seizure ($p < 0.05$). This response could be mimicked with equimolar phenylalanine, but not aspartic acid or methanol. Additionally, valine coadministration prevented the APM-induced lowering of seizure threshold. When immature mice were tested, a dose of only 500 mg/kg was required for potentiation. A recent report using lidocaine-induced seizures has found a similar seizure-potentiating effect of APM (Kim and Kim 1987).

The data presented above clearly demonstrate the seizure-promoting

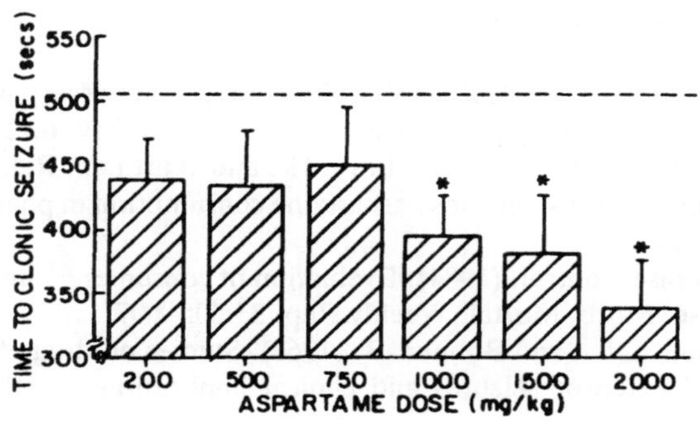

FIGURE 10.3. Effect of aspartame pretreatment on the time to clonic seizure in mice challenged with fluorothyl. Aspartame (0 to 2000 mg/kg) was administered orally to groups of mice (average $N = 14$) one hour prior to exposure to inhaled fluorothyl. $^*p < 0.05$, significantly different from the 0-mg/kg group (— —) as determined by one-way ANOVA and Dunnett's test.

activity of APM in animal models that are widely used to identify compounds affecting (i.e., usually protecting against) seizure activity. Increased brain phenylalanine is most likely involved in this response since equimolar phenylalanine mimics the APM-induced response in each case, and since the coadministration of valine attenuates this response. The evidence does not indicate that APM itself *causes* seizures; rather it *promotes* seizures in animals already at risk (that is animals treated with pentylenetetrazole or fluorothyl). Similarly, it is possible that doses of the sweetener that lead to significant elevations in brain phenylalanine might increase or intensify seizure activity among susceptible humans. Whether or not APM actually does promote seizures in susceptible humans will have to be explored in appropriately designed controlled clinical trials.

Acknowledgment. These studies were supported in part by a grant from the National Institute of Neurological and Communicative Diseases and Stroke (NS21231).

References

Altman, I.M., and Corcoran, M.E. (1983). Facilitation of neocortical kindling by depletion of forebrain noradrenaline. Brain Res. **270:**174–177.

Altshuler, H.J., Killam, E.K., and Killam, K.F. (1976). Biogenic amines and the photomyoclonic syndrome in the baboon *Papio papio*. J. Pharmacol. Exp. Ther. **196:**156–166.

Araki, H., Aihara, H., Watanabe, S., Ohta, H., Yamamoto, T., and Ueki, S. (1983). The role of noradrenergic and serotonergic systems in the hippocampal kindling effect. Jpn. J. Pharmacol. **33:**57–64

Arnold, P., Racine, R., and Wise, R. (1973). Effect of atropine, reserpine, 6-

hydroxydopamine, and handling on seizure development in the rat. Exp. Neurol. **40:**457–460.

Boggan, W.O., and Seiden, L.S. (1971). Dopa reversal of reserpine enhancement of audiogenic seizure susceptibility in mice. Physiol. Behav. **6:**215–217.

Caballero, B., Mahon, B., Rohr, F., Levy, H., and Wurtman, R.J. (1986). Plasma amino acid levels after single dose aspartame consumption in phenylketonuria. J. Pediatr. **109:**668–671.

Centers for Disease Control. (1984). Evaluation of consumer complaints related to aspartame use. Morb. Mortal. Weekly Rep. **33:**605–607.

Chen, G., Ensor, C.R., and Bohner, B. (1968). Studies on drug effects on electrically induced extensor seizures and clinical implications. Arch. Int. Pharmacodyn. **172:**183–218

Conlay, L.C., Maher, T.J. and Wurtman, R.J. (1981). Tyrosine increases blood pressure in hypotensive rats. Science **212:**559–560.

Corcoran, M.E., and Mason, S.T. (1980). Role of forebrain catecholamines in amygdaloid kindling. Brain Res. **190:**473–484.

Dailey, J.W., and Jobe, P.C. (1984). Effects of increments in the concentration of dopamine in the central nervous system on audiogenic seizures in DBA/2J mice. Neuropharmacology **23:**1019–1024.

Drake, M.E. (1986). Panic attacks and excessive aspartame ingestion. Lancet **ii:**631.

Ferguson, J.M. (1985). Interaction of aspartame and carbohydrates in an eating disorder patient. Am. J. Psychiatr. **142:**271.

Fernstrom, M.H., Volk, E.A., and Fernstrom, J.D. (1984). In vivo tyrosine hydroxylation in the diabetic rat retina: effect of tyrosine administration. Brain Res. **298:**167–170.

Gallagher, B.B. (1969). Amino acids and cerebral excitability. J. Neurochem. **16:**701–706.

Garattini, S. (1986). Aspartame, brain monoamines and neurochemical mediators. International Aspartame Workshop, November, Marbella, Spain.

Jobe, P.C., Picchioni, A.L., and Chin, L. (1973). Effect of lithium carbonate and alpha-methyl-p-tyrosine on audiogenic seizure intensity. J. Pharm. Pharmacol. **25:**830–831.

Jobe, P.C., Ko, K.H., and Dailey, J.W. (1984). Abnormalities in norepinephrine turnover rate in the central nervous system of the genetically epilepsy-prone rat. Brain Res. **290:**357–360.

Jobe, P.C., Dailey, J.W., and Reigel, C.E. (1986). Noradrenergic and serotonergic determinants of seizure susceptibility and severity in genetically epilepsy-prone rats. Life Sci. **39:**775–782.

Johns, D.R. (1986): Migraine provoked by aspartame. N. Engl. J. Med. **315:**456.

Kilian, M., and Frey, H.H. (1973). Central monoamines and convulsive thresholds in mice and rats. Neuropharmacology **12:**681–692.

Kim, K.C., and Kim, S.H. (1987). Studies on the effect of aspartame and lidocaine interaction in central nervous system of mice. Fed. Proc. **46:**705.

Kiritsy, P.J., and Maher, T.J. (1986). Acute effects of aspartame on systolic blood pressure in spontaneously hypertensive rats. J. Neural Transm. **66:**121–128.

Kleinrok, Z., Czuczwar, S., Wojcik, A., and Przegalinski, E. (1978). Brain dopamine and seizure susceptibility in mice. Pol. J. Pharmacol. Pharm. **30:**513–519.

Lehnert, H., Reinstein, D.K., Trowbridge, B.W., and Wurtman, R.J. (1984).

Neurochemical and behavior consequences of acute, uncontrollable stress: effect of dietary tyrosine. Brain Res. **303:**215–223.

Lewis, J., Westerberg, V., and Corcoran, M.E. (1987). Monoaminergic correlates of kindling. Brain Res. **403:**205–212.

Litchfield, J.T. Jr., and Wilcoxon, F. (1949). A simplified method of evaluating dose-effect experiments. J. Pharmacol. Exp. Ther. **96:**99–113.

McIntyre, D.C., and Edson, N. (1981). Facilitation of amygdaloid kindling after norepinephrine depletion with 6-hydroxydopamine in rats. Exp. Neurol. **74:**748–757.

McKenzie, G.M., and Soroko, F.E. (1972). The effects of apomorphine, (+)-amphetamine and L-dopa on maximal electroshock convulsions—a comparative study in the rat and mouse. J. Pharm. Pharmacol. **24:**696–701.

McKenzie, G.M., and Soroko, F.E. (1973). Inhibition of the anticonvulsant activity of L-dopa by FLA-63, a dopamine-beta-hydroxylase inhibitor. J. Pharm. Pharmacol. **25:**76–77.

Melamed, E., Hefti, F., and Wurtman, R.J. (1980). Tyrosine administration increases striatal dopamine release in rats with partial nigrostriatal lesions. Proc. Natl. Acad. Sci. USA **77:**4305–4309.

Milner, J.D., Irie, K., and Wurtman, R.J. (1986). Effects of phenylalanine on the release of endogenous dopamine from rat striatal slices. J. Neurochem. **47:**1444–1448.

Nagatsu, T., Levitt, M., and Udenfriend, S. (1964). Tyrosine hydroxylase, the initial step in norepinephrine biosynthesis. J. Biol. Chem. **239:**2910–2917.

Pinto, J.M.B. and Maher, T.J. (1988). Aspartame administration potentiates fluorothyl- and pentylenetetrazole-induced seizure in mice. *Neuropharmacol.* **27:**51–55.

Rudzik, A.D., and Johnson, G.A. (1970). Effect of amphetamine and amphetamine analogs on convulsive thresholds. *In* Costa, E., and Garratini, S. (eds.), Amphetamines and related compounds. New York: Raven Press, p. 715.

Schlesinger, K. Boggan, W., and Freedman, D.X. (1968). Genetics of audiogenic seizures. II. Effects of pharmacological manipulation of brain serotonin, norepinephrine and gamma-aminobutyric acid. Life Sci. **7:**437–447

Sved, A.F., Fernstrom, J.D., and Wurtman, R.J. (1979). Tyrosine administration reduces blood pressure and enhances brain norepinephrine release in spontaneously-hypertensive rats. Proc. Natl. Acad. Sci. *USA* **76:**3511–3514.

Thakore, K., and Crane, S.C. (1987). Blood pressure and regional brain tyrosine, norepinephrine and dopamine in spontaneously hypertensive rats fed aspartame or sucrose. Fed. Proc. **46:**904.

Walton, R.G. (1986). Seizure and mania after high intake of aspartame. Psychosomatics **27:**218–220.

Wurtman, R.J. (1985). Aspartame: possible effect on seizure susceptibility. Lancet **ii:**1060.

Yokogoshi, H., and Wurtman, R.J. (1986). Acute effects of oral or parenteral aspartame on catecholamine metabolism in various regions of rat brain. J. Nutr. **116:**356–364.

CHAPTER 11

Behavioral and Neurological Effects of Aspartame

Hugh A. Tilson, Dayao Zhao, N. John Peterson, Kevin Nanry, and J.-S. Hong*

The effects of high doses of aspartame on sensorimotor function were studied in rats. Intragastric administration of 250 to 1000 mg/kg of aspartame had no effect on the magnitude of the acoustic startle reflex, prepulse inhibition of the startle reflex, or motor activity, all of which are sensitive to the effects of convulsant agents, including picrotoxin, strychnine, and lindane. Aspartame also had no effects on the acquisition or retention of a passive avoidance task or a two-way shuttle box response. These data indicate that aspartame has little, if any, effect on sensorimotor functioning of rats, at least in single doses up to 1000 mg/kg.

Introduction

Dietary ingestion of aspartame (L-aspartyl-L-phenylalanine methyl ester), an artificial sweetener, is known to increase plasma levels of phenylalanine, aspartic acid, and methanol, all of which are metabolic by-products of aspartame. Aspartate is an excitatory amino acid neurotransmitter, an excess of which may cause neuronal death (Olney and Ho 1970), while the toxicity of methanol is well known (Potts et al. 1955). In the rat, phenylalanine is rapidly metabolized to tyrosine; phenylalanine and tyrosine are taken up into the brain by a macromolecule that functions as a transport system, and these amino acids may affect the biosynthesis and utilization of adrenergic neurotransmitters such as norepinephrine and dopamine (Wurtman 1983). Recent experiments have shown that oral administration of aspartame can produce large increases in phenylalanine and tyrosine in the brain (Fernstrom et al. 1983, Wurtman 1983).

Extensive laboratory and clinical studies have indicated the safety of aspartame, at least in normal populations (Aspinall et al. 1980, Potts et al. 1980, Ranney and Oppermann 1979, Stegink et al. 1979). However, some

*Laboratory of Behavioral and Neurological Toxicology, National Institute of Environmental Health Sciences, P.O. Box 12233, Research Triangle Park, NC 27709, USA.

investigators have used phenylalanine to induce phenylketonuria, an inborn disorder of phenylalanine metabolism (Polidora et al. 1966a,b, Waisman and Harlow 1965, Brunner et al. 1979). Reports using this animal model of phenylketonuria have found deficits in learning associated with high dietary levels of phenylalanine (Loo et al. 1962, Polidora et al. 1963, 1966b). Schalock and Kloper (1967a,b) reported that when elevated levels of phenylalanine were introduced during prenatal or early postnatal development, long-lasting behavioral abnormalities could be observed.

Since phenylalanine is a major constituent of aspartame, it is possible that aspartame could cause an amino acid imbalance resulting in behavioral alterations. Brunner et al. (1979) investigated this possibility by exposing rats continuously from conception through 90 days of age. High concentrations of aspartame caused delays in several developmental milestones (eye opening, surface righting, and swimming development); hypoactivity; delayed auditory startle; and delayed forward quadripedal locomotor development prior to weaning. Subsequent assessment indicated persistent changes in the open field maze, decreased brain weights, and decreased cerebellar granule and olfactory bulb granule cell counts. Exposure to phenylalanine also produced behavioral alterations; morphological measurements (brain cell counts) were not made on animals receiving phenylalanine. These investigators concluded that a diet of 6% aspartame was approximately equivalent to one of 3% phenylalanine. Since aspartame consists of approximately equal amounts on phenylalanine and aspartic acid, these data suggest that any behavioral effects associated with aspartame ingestion may be associated with the phenylalanine moiety.

The neurobehavioral effects of aspartame or phenylalanine in the adult have not been well studied. Shortly after approval for usage in humans, several reports of aspartame-associated neurological effects were reported, including headaches, dizziness, and mood alterations (Massachusetts Medical Society 1984). Wurtman (1985) recently pointed out that diminished brain levels of catecholamines are associated with lower seizure thresholds in genetically epilepsy-prone rats (Jobe et al. 1984) and suggested that exposure to large amounts of aspartame may ultimately decrease brain levels of norepinephrine and adversely affect some people predisposed to seizures. In spite of these observations, few data have been reported concerning the effects of aspartame on behavioral and neurological functioning in experimental animals. Potts et al. (1980) reported that acute exposure to aspartame did not affect the motor coordination of mice nor did it alter hexobarbital-induced hypnosis. Aspartame showed no significant anticonvulsant activity in the maximal electroshock seizure procedure nor did it alter ptosis induced by Ro 4-1284, which is a test for antidepressant activity. Analgesic activity was determined using two different tests, but no consistent effects were observed. Potts et al. (1980) also reported that aspartame or L-phenylalanine in doses up to 200 mg/kg was without effect on the acquisition of conditioned avoidance by rats in a shuttle box procedure.

Potts et al. (1980) also gave aspartame or L-phenylalanine to rats via the diet for 90 to 92 days beginning at 21 days of age. Behavioral testing was done between day 66 and day 86 of exposure. Significant increases in motor activity were noted in males and females given the 9% aspartame and 2.5% phenylalanine diet, respectively. Male rats fed a diet containing 5% phenylalanine or 9% aspartame made significantly fewer conditioned avoidance responses in a shuttle box task. No consistent effects were observed in females receiving aspartame or phenylalanine. In addition, the acquisition and performance of a noncued continuous avoidance response in a two-way shuttle box was studied. Male rats receiving the 5% L-phenylalanine, 4.5% aspartame, or 9% aspartame diets received more shocks in this paradigm, indicating impaired performance. Female rats did not appear to be affected by aspartame.

Potts et al. (1980) gave rodents a single administration of aspartame in a dose (200 mg/kg) estimated to be approximately 10 times the normal total daily dietary ingestion of 20 mg/kg for a human; this calculation was based on the daily consumption of 1 g of aspartame per day by a 50-kg adult. However, these calculations were based on the assumption that the route of metabolism for aspartame in humans is similar to that in rodents. Wurtman (personal communication) has pointed out that rodents metabolize phenylalanine to tyrosine more efficiently than humans. Therefore, to achieve equivalent levels of phenylalanine from aspartame, higher doses of aspartame might be appropriate for rodent studies.

Neurobehavioral Evaluation of Aspartame

Recent research in our laboratory has examined the possible effects of higher doses (250, 500, and 1000 mg/kg) of aspartame on several indicators of central nervous system (CNS) function of rats. Because it has been speculated that aspartame may alter CNS excitability or the seizure susceptibility of certain individuals, we have examined the effects of aspartame in several tests used routinely in our laboratory to evaluate the effects of environmental agents on reactivity or CNS arousal. In many of these experiments, we have compared the effects of aspartame to several agents known to produce seizures, including picrotoxin, lindane, and strychnine. Picrotoxin is a GABA-activated chloride channel antagonist (Olsen 1981), while lindane may be a competitive inhibitor at the picrotoxin binding site within the GABA receptor-ionophore complex (Lawrence and Casida 1984). Strychnine is a specific glycine antagonist (Curtis et al. 1971).

Startle Response

The acoustic startle reflex, which is a response mediated by a five-synapse pathway beginning at the auditory nerve and ending in the lower motor neurons, frequently has been used to measure the effects of drugs and chemicals on behavioral reactivity to sensory stimulation (Davis 1980, Tilson and Mitchell 1984). Rats were dosed intragastrically with 250, 500, or 1000 mg/kg of L-*d*-aspartyl-L-phenylalanine methyl ester suspended in isotonic saline and Tween 80 and, two hours later, tested for reactivity to the repeated presentation of a 8-kHz, 110-dB tone for 20 trials. Aspartame was without effect on startle reactivity. Picrotoxin (0.75 or 1.5 mg/kg, i.p., 30 min prior to testing) produced a significant decrease in startle reactivity, while lindane (20 or 40 mg/kg, p.o., 3 h prior to testing) increased startle reactivity. Strychnine (0.75 or 1.5 mg/kg, i.p., 5 min prior to testing) also increased the acoustic startle reflex. These observations are in accord with those reported elsewhere (Gallager et al. 1983, Kehne et al. 1981, Tilson et al. 1985).

Prepulse Inhibition

The magnitude of the startle reflex can be modified by the presentation of a brief, relatively weak stimulus prior to another stimulus capable of eliciting the startle reflex. This phenomenon, known as prepulse inhibition, has been studied as a model of reflex modification (Hoffman 1984) and may be mediated by an inhibitory pathway parallel to the startle reflex arc (Hoffman and Ison 1980). Prepulse inhibition was tested in rats using a 20-ms, 70-dB white noise presented at various intervals (0, 10, 40, 80, 100, or 4000 ms) prior to a 8-kHz, 110-dB eliciting stimulus. Oral administration of aspartame (250 to 1000 mg/kg) had no significant effect on prepulse inhibition. Picrotoxin (1.5 mg/kg) decreased, while strychnine (1.5 mg/kg) increased, the response to the tone without affecting the ability of the white noise to inhibit the startle response.

Motor Activity

Spontaneous motor activity has been used extensively in rodents to study the pharmacological and toxicological effects of chemicals (Reiter 1978, Reiter and MacPhail 1979). Rats were dosed intragastrically with aspartame (0, 500, or 1000 mg/kg), and two hours later, their activity was measured for 60 min in rectangular chambers equipped with photocells to record vertical and horizontal movements. Aspartame had no significant effect on either measure of motor activity. Picrotoxin (1.5 mg/kg) and strychnine (1.5 mg/kg) decreased motor activity.

Learning and Memory

Failing to observe any significant effects of aspartame on routine measures of sensorimotor function relative to those of picrotoxin or strychnine, we turned our attention to the study of possible effects of aspartame on learning and memory. Rats were given aspartame (500 or 1000 mg/kg) two hours prior to being placed into a chamber separated by a guillotine door. After 5 s, the door was raised, and contingent upon the rats crossing into the opposite chamber, the door was closed and a mild footshock was administered for 1 s. Retention was assessed 48 hours later. Aspartame had no significant effect on retention. In another experiment, rats were dosed with aspartame (500 or 1000 mg/kg) immediately after training and then retested 48 hours later. Again, aspartame had no significant effect on retention of passive avoidance.

In another experiment, rats were dosed with aspartame (500 or 1000 mg/kg), and two hours later, acquisition of discriminated avoidance responding was assessed in a two-way shuttle box during a 60-trial acquisition session. Aspartame had no significant effect on the number of avoidances, responses made during the intertrial interval, or avoidance or escape latencies.

Summary and Conclusions

Acute intragastric administration of aspartame in doses of up to 1000 mg/kg had no significant effect on routine measures of sensorimotor function, including the acoustic startle response, prepulse inhibition of startle, or spontaneous motor activity. These tests have been used frequently to assess chemical-induced alterations in CNS arousal and behavioral reactivity to environmental stimulation. In addition, aspartame doses of 500 or 1000 mg/kg had no significant effect in two different tests of learning and memory. Future studies will determine the possible effects of aspartame on seizure thresholds in rats. Additional studies are also needed to assess the effects of repeated exposure to aspartame on sensorimotor function and to determine more precisely the adverse effects of developmental exposure to aspartame.

References

Aspinall, R.L., Saunders, R.N., Pautsch, W.F., and Nutting, E.F. (1980). The biological properties of aspartame. J. Environ. Pathol. Toxicol. **3**:383–391

Brunner, R.L., Vorhees, C.V., Kinney, L., and Butcher, R.E. (1979). Aspartame: assessment of developmental psychotoxicity of a new artificial sweetener. Neurobehav. Toxicol. **1**:79–86.

Curtis, D.R., Duggan, A.W., and Johnston, G.A.R. (1971). The specificity of strychnine as a glycine antagonist in the mammalian spinal cord. Brain Res. **12**:547–552.

Davis, M. (1980). Neurochemical modulation of sensory motor reactivity: acoustic and tactile startle reflexes. Neurosci. Biobehav. Rev. **4**:241–263.

Fernstrom, J.D., Fernstrom, M.H., and Gillis, M.A. (1983). Acute effects of aspartame on large neutral amino acids and monoamines in rat brain, Life Sci. **14**:425–436.

Gallager, D.W., Kehne, J.H., Wakeman, E.A., and Davis, M. (1983). Developmental changes in pharmacological responsivity of the acoustic startle reflex: effects of picrotoxin. Psychopharmacology **79**:87–93.

Hoffman, H.S. (1984). Methodological factors in the behavioral analysis of startle: the use of reflex modification procedures and the assessment of threshold. *In* Eaton, R.C. (ed.), Neural mechanisms of startle behavior. New York: Plenum, pp. 267–285.

Hoffman, H.S., and Ison, J.R. (1980). Reflex modification in the domain of startle. I. Some empirical findings and their implications for how the nervous system processes sensory input. Psychol. Rev. **87:**175–189.

Jobe, P.C., Ko, K.H., and Dailey, J.W. (1984). Abnormalities in norepinephrine turnover rate in the central nervous system of the genetically epilepsy-prone rat. Brain Res. **290:**357–360.

Kehne, J.H., Gallager, D.W., and Davis, M. (1981). Strychnine: brainstem and spinal mediation of excitatory effects on acoustic startle. Eur. J. Pharmacol. **76:**177–186.

Lawrence, L.J., and Casida, J.E. (1984). Interactions of lindane, toxaphene and cyclodienes with brain-specific *t*-butyl bicyclophosphorothionate receptor. Life Sci. **35:**171–178.

Loo, Y.H., Diller, E., and Owen, J.E. (1962). Effect of phenylalanine diet of learning in the rat. Nature **194:**1286–1287.

Massachusetts Medical Society (1984). Morb. Mortal. Weekly Rep. **33:**605–607.

Olney, J.W., and Ho, C.L. (1970). Brain damage in infant mice following oral intake of glutamate, aspartate or cysteine. Nature **227:**609–611.

Olsen, R.W. (1981). GABA-benzodiazepine-barbiturate receptor interactions. J. Neurochem. **37:**1–13

Polidora, V.J., Boggs, D.E., and Waisman, H.A. (1963). A behavioral deficit associated with phenylketonuria in rats. Proc. Soc. Exp. Biol. Med. **113:**817–820.

Polidora, V.J., Boggs, D.E., and Waisman, H.A. (1966a). Phenylketonuria in rats, reversibility of behavioral deficit. Science **151:**219

Polidora, V.J., Cunningham, R.F., and Waisman, H.A. (1966b). Dosage parameters of a behavioral deficit associated with phenylketonuria in rats. J. Comp. Physiol. Psychol. **61:**436–441.

Potts, A.M., Praglin, J., Farkas, I., Orbison, L., and Chickering, D. (1955). Studies on the visual toxicity of methanol. VIII. Additional observations on methanol poisoning in the primate test object. Am. J. Ophthalmol. **40:**76–82.

Potts, W.J., Bloss, J.L., and Nutting, E.F. (1980). Biological properties of aspartame. I. Evaluation of central nervous system effect. J. Environ. Pathol. Toxicol. **3:**341–353.

Ranney, R.E., and Oppermann, J.A. (1979). A review of the metabolism of the aspartyl moiety of aspartame in experimental animals and man. J. Environ. Pathol. Toxicol. **2:**979–985.

Reiter, L. (1978). Use of activity measures in behavioral toxicology. Environ. Health Perspect. **26:**9–20.

Reiter, L., and MacPhail, R. (1979). Motor activity: a survey of methods with potential use in toxicity testing. Neurobehav. Toxicol. Teratol. **1:**53–66.

Schalock, R.L., and Kloper, F.D. (1967a). Phenylketonuria: enduring behavioral deficits in phenylketonuric rats. Science **168:**147–151.

Schalock, R.L., and Kloper, F.D. (1967b). Induced phenylketonuria in rats: behavioral effects. J. Ment. Defic. Res. **11:**282–287.

Stegink, L.D., Filer, L.J., and Baker, G.L. (1979). Effect of aspartame loading upon plasma and erythrocyte amino acid levels in phenylketonuric heterozygotes and normal adult subjects. J. Nutr. **109:**2173–2181.

Tilson, H.A., and Mitchell, C.L. (1984). Neurobehavioral techniques to assess the effects of chemicals on the nervous system. Annu. Rev. Pharmacol. Toxicol. **24:**425–450.

Tilson, H.A., Hong, J.S., and Mactutus, C.F., (1985). Effects of 5, 5-diphenylhydantoin (phenytoin) on neurobehavioral toxicity of organochlorine insecticides and permethrin. J. Pharmacol. Exp. Ther. **223:**285–289.

Waisman, H.A., and Harlow, H.F. (1965). Experimental phenylketonuria in infant monkeys. Science **147:**685.

Wurtman, R.J. (1983). Neurochemical changes following high-dose aspartame with dietary carbohydrates. N. Engl. J. Med. **309:**429–430.

Wurtman, R.J. (1985). Aspartame: possible effect on seizure susceptibility. Lancet **ii:**1060.

Chapter 12

Role of Monoamines in Seizure Predisposition in the Genetically Epilepsy-Prone Rat

Phillip C. Jobe and John W. Dailey*

The genetically epilepsy-prone rat (GEPR) is proving to be a useful model of the human epilepsies. Seizure predisposition in these animals is partially determined by widespread noradrenergic and serotonergic deficits in the brain. Drug-induced perturbations of these neurotransmitter systems cause predictable changes in seizure severity and/or susceptibility. Anticonvulsant effects occur in response to drugs or procedures which increase noradrenergic and/or serotonergic activity in the brain. In contrast, proconvulsant effects occur in response to interventions which substantially diminish either or both types of neurotransmission. Inhibition of norepinephrine synthesis alone does not exacerbate seizure activity in the GEPR. However, inhibition of synthesis either at the tyrosine hydroxylase step or the dopamine-beta-hydroxylase step prolongs seizure facilitating effects of drugs which inactivate norepinephrine storage vesicles. Synthesis inhibition also interacts with norepinephrine release produced by cold-induced stress or false transmitter-induced displacement to cause seizure facilitation. Some precursors to norepinephrine and serotonin are among the pharmacologic agents which produce anticonvulsant effects in the GEPR. The norepinephrine precursors, L-DOPA and dopamine, lose anticonvulsant effectiveness if conversion to norepinephrine is prohibited by synthesis inhibitors. Aspartame, an agent which can increase plasma phenylalanine and tyrosine concentrations, fails to facilitate or suppress sound-induced seizures in GEPRs even when doses are as high as 2000 mg/kg. Since in some circumstances phenylalanine can act as a tyrosine hydroxylase inhibitor and because tyrosine serves as the major dietary precursor to norepinephrine, an effect of large doses of aspartame on noradrenergic transmission might be hypothesized. However, our observations using the GEPR suggest that any possible effects of large aspartame doses on brain noradrenergic and serotonergic activity are functionally insignificant as determinants of seizure predisposition. These initial tests suggest that dietary aspartame has little or no risk of facilitating convulsions in subjects with seizure predisposition dependent upon innate noradrenergic and serotonergic deficits.

* Department of Basic Sciences, University of Illinois College of Medicine at Peoria, Peoria, IL 61656, USA.

Introduction

A substantial body of evidence shows that monoamines (norepinephrine, dopamine, and serotonin) have the potential to modulate seizure threshold, seizure severity, and other manifestations of seizure activity in normal animals (Jobe 1987, Burley and Ferrendelli 1984, Browning 1987, Craig and Colasanti 1987, Chapman and Meldrum 1987). In addition, noradrenergic activity appears to be an important determinant of epilepsy in certain animal models (Jobe and Laird 1981, 1987, McNamara et al. 1987, Meldrum et al. 1975). For example, noradrenergic decrements appear to be a partial and important correlate of seizure predisposition in the genetically epilepsy-prone rat (Jobe and Laird 1987, Jobe et al. 1986a, b, Laird and Jobe 1987). On the other hand, dopaminergic decrements or noradrenergic increments may be determinants of seizure predisposition in the DBA/2J (Dailey and Jobe 1984, Jobe and Laird 1987, Chapman and Meldrum 1987) and tottering mouse (Jobe and Laird 1987, Nobels 1984).

Models in which noradrenergic influences appear to be anticonvulsant include the genetically epilepsy-prone rat (Jobe and Laird 1987), electroshock seizures in nonepileptic rats (Browning 1987), cobalt-induced seizures in rat (Craig and Colasanti 1987), and the genetically epileptic baboon (Killam and Killam 1984). In addition, emerging evidence supports the concept that deficits in noradrenergic and/or dopaminergic activity may contribute to the epileptic state of humans as they do in some animal models of epilepsy (e.g., genetically epilepsy-prone rats and probably the genetically epileptic baboon).

The purpose of this chapter is to review briefly the pharmacologic and pathophysiologic evidence obtained in genetically epilepsy-prone rats. This animal model is particularly well suited to the study of the effects of compounds, like phenylalanine, that may alter brain levels of norepinephrine and thus change seizure susceptibility and/or severity.

Characteristics of the Genetically Epilepsy-Prone Rat Model

Numerous characteristics of the genetically epilepsy-prone rat (GEPR) support its use as a model for certain forms of human epilepsy:

1. The GEPR has a lower seizure threshold for an electrical or convulsant drug stimulus than nonepileptic controls (Jobe 1981, Jobe and Laird 1981, Reigel et al. 1986).
2. The adult GEPR is susceptible to audiogenic and hyperthermic seizures (Jobe and Laird 1981, Jobe et al. 1982b).
3. As is typically the case in human epilepsy, the GEPR exhibits a protracted period of susceptibility to seizures.

4. The cortical EEG pattern of the GEPR during audiogenic seizure is characterized by a spike and wave pattern (Tacke et al. 1984).
5. Spontaneous seizures occur in the severe-seizure GEPR (i.e., the GEPR-9).
6. The GEPR is remarkably sensitive to kindling (Savage et al. 1986).
7. Established antiepileptic drugs effectively attenuate seizures in the GEPR, and distinctions can be made between those that are clinically effective in generalized tonic-clonic seizures and those that are useful in absence seizures (Consroe et al. 1979, Dailey and Jobe 1985).
8. Numerous psychoactive drugs which either enhance or suppress seizures in the GEPR produce similar effects in humans (Jobe and Laird 1981, Jobe et al. 1984b).
9. Epileptic seizures in both humans and the GEPR are linked to genetic factors (Jobe et al. 1981).

Evidence for Role of Central Monoaminergic Systems in Seizure Predisposition in Genetically Epilepsy-Prone Rats

Evidence that a noradrenergic deficit is a primary determinant of seizure severity in the GEPR has been obtained from a number of detailed pharmacologic studies. Pharmacologic evidence indicates that seizure intensity in the GEPR is inversely related to the magnitude of brain norepinephrine concentration (see Jobe et al. 1986a). Accordingly, excessive formation of norepinephrine by precursor loading decreases the severity of seizures. Activation of β-receptors and α_1-receptors also attenuates seizure severity. In contrast, drug treatments which reduce norepinephrine concentrations also enhance seizure severity in the GEPR.

Although dopaminergic neurons were affected by some treatments, the time course and the magnitude of the changes in this neurotransmitter system did not correspond to changes in seizure severity. Even large increases or decreases in dopamine concentrations failed to alter seizure patterns in the GEPR. Thus, dopaminergic systems appear not to be determinants of seizure intensity in the GEPR.

Serotonergic neurons have also been shown to exert an attenuating effect on sound-induced seizures in the GEPR (see Jobe et al. 1986a). Whereas drug-induced increments in serotonin reduce seizure severity, drug-induced depletion of serotonin intensifies the convulsive response to sound.

These pharmacologic data indicate that noradrenergic and/or serotonergic activity is inversely related to seizure *severity*, but no evidence regarding the regulation of seizure *susceptibility* in the GEPR is provided. The question of seizure susceptibility was addressed by administering the monoamine depleting agent Ro 4-1284 to two types of rats: (1)

non-audiogenic-seizure-susceptible progeny of nonepileptic parents; and (2) non-audiogenic-seizure-susceptible progeny of epileptic parents. Only in the progeny of epileptic parents was monoamine depletion associated with an increased susceptibility to seizures (Jobe et al. 1981). Thus, it is likely that monoaminergic decrements represent only a part of the total set of neurochemical defects which underlie the seizure-prone state in the GEPR. Nonsusceptible progeny of nonepileptic parents do not become susceptible to audiogenic seizures in response to decreased brain levels of norepinephrine because they do not carry the obligatory nonmonoaminergic determinants. Conversely, nonsusceptible progeny of GEPR parents do experience seizures when brain norepinephrine levels are decreased because these animals do carry the hypothetical nonmonoaminergic determinants.

Pathophysiology of Seizure Severity and Susceptibility in Genetically Epilepsy-Prone Rats

Pathophysiological evidence provides additional support for the concept that noradrenergic and/or serotonergic deficits within the brain are primary determinants of the epileptic state of the GEPR. Based on studies performed in nonepileptic control rats and in both seizure-experienced and non-seizure-experienced GEPRs, it is now clearly evident that widespread deficits in norepinephrine concentration and turnover rate characterize the GEPR brains. In addition, deficits in tyrosine hydroxylase activity are present in these GEPR brains (Dailey and Jobe 1986).

Four types of noradrenergic conditions are known to characterize different parts of the adult GEPR brain (Jobe et al. 1986a). These conditions have been identified by comparing noradrenergic indices in discrete areas of the brains of nonepileptic control rats with the same areas of moderate-seizure (GEPR-3) and severe-seizure (GEPR-9) GEPRs. The four neuroanatomically dependent conditions can be summarized as: (1) areas with equal noradrenergic deficits in GEPR-3's and GEPR-9's; (2) an area with opposing abnormalities such that noradrenergic activity in GEPR-3's is greater than in controls which, in turn, is greater than in GEPR-9's; (3) areas with forward-order graded deficits in which noradrenergic indices are highest in controls, intermediate in GEPR-3's, and lowest in GEPR-9's; and (4) an area characterized by normality in noradrenergic indices.

Areas in which noradrenergic deficits appear to be of equal magnitude in GEPR-3's and GEPR-9's include the hippocampus, hypothalamus, inferior colliculus, and the remainder of the midbrain. The area with opposing abnormalities in the two types of GEPRs is the cerebellum. Forward-order graded deficits exist in the telencephalon minus the hippocampus and striatum as well as in the thalamus and the pons/medulla. Noradrenergic normality appears to characterize the striatum.

The different noradrenergic conditions in the various brain areas have led to a series of hypotheses regarding the roles of this neurotransmitter system in seizure regulation (Dailey and Jobe 1986). As pointed out above, several areas of the brain are characterized by noradrenergic deficits of similar or equal magnitudes in both GEPR-3's and GEPR-9's. Behaviorally, both GEPR-3's and GEPR-9's are susceptible to seizures precipitated by stimuli that do not trigger seizures in nonepileptic control rats. Simultaneous consideration of the similar magnitude of concentration deficits and the common abnormality of seizure susceptibility shared between the two types of epileptic animals suggests a functional relationship, namely, that these noradrenergic deficits may be determinants of susceptibility.

In contrast, the equal noradrenergic deficit areas do not appear to regulate seizure severity. If any noradrenergic deficits were directly and primarily responsible for allowing severe seizures in the GEPR-9 and for preventing severe seizures in the GEPR-3, the deficits in the severe-seizure subjects would be significantly greater than those in the moderate-seizure subjects.

The brain area with opposing noradrenergic abnormality in GEPR-3's and GEPR-9's suggests a role for noradrenergic terminals in this structure in the regulation of seizure severity but not susceptibility. Perhaps the noradrenergic increment in GEPR-3's prevents the moderate-seizure animals from experiencing the more severe tonic extensor convulsion characteristic of GEPR-9's. This particular noradrenergic increment clearly does not negate susceptibility to seizures that occur in response to an acoustic stimulus.

Areas with forward-order graded deficits may regulate seizure severity as well as seizure susceptibility. Deficits in these areas could determine susceptibility because both GEPR-3's and GEPR-9's exhibit such deficits. Moreover, these deficits could determine seizure severity because they are greater in GEPR-9's than in GEPR-3's. Consequently, the moderate deficits of the GEPR-3 may allow the occurrence of seizures but prevent the appearance of the severe tonic extensor convulsions. In contrast, severe noradrenergic deficits in the GEPR-9 may not only allow the occurrence of the seizures but also facilitate the development of maximal degrees of seizure severity.

Noradrenergic factors in the one area of the brain characterized by normal indices for this transmitter system are not presently candidates for regulation of the seizure-prone state of the GEPR. Both GEPR-3's and GEPR-9's differ from controls in that they are susceptible to seizures which do not occur in nonepileptic controls. Moreover, they differ from each other in that the GEPR-9 has more severe seizures than the GEPR-3. Yet, they fail to differ in terms of noradrenergic activity in this one area of the brain.

Innate serotonergic conditions which have been detected in GEPR brains are in some instances similar and in other instances different from those identified for the noradrenergic systems. Based on our studies of serotonin (Jobe et al. 1986a), only one area—the striatum—appears to be a candidate for serotonergic regulation of seizure severity. Susceptibility but not severity may be determined by serotonergic abnormalities in the telencephalon minus hippocampus, thalamus, hypothalamus, midbrain, and the pons/medulla. Serotonergic abnormalities in the cerebellum do not appear to be determinants of either seizure susceptibility or severity or in the GEPR.

These pathophysiological hypotheses represent an interim interpretation of data. Further tests of the concepts are essential. Experimentally induced alterations in the monoaminergic indices in numerous anatomic sites must be evaluated to determine whether they will cause the predicted alterations in seizure susceptibility and/or severity in the GEPR. Appropriately designed breeding paradigms must also be tested to evaluate whether noradrenergic and/or serotonergic abnormalities in specific brain regions are inherited together with susceptibility or the postulated levels of seizure severity.

Effects of Phenylalanine on Seizure Susceptibility and Severity

Given the role of phenylalanine in the production of tyrosine, a precursor in monoamine synthesis, the influence of phenylalanine on seizure susceptibility and severity has been examined for several years. Schlesinger et al. (1969) reported the effects of phenylalanine given to DBA/2J mice in an intraperitoneal dose of 1.2 mg/kg daily, beginning on day 14 of age and terminating on day 29. When given according to this paradigm, phenylalanine failed to cause an increase in the mean score for seizure severity or in the incidence of lethal convulsions. In contrast, a slight increase in the incidence of clonic and tonic seizures was noted. Similarly, when large amounts of phenylalanine were administered in the diet (1% or approximately 2500 mg/kg), an increase in the incidence of tonic seizures or in seizure severity was noted (Schlesinger et al. 1969, Truscott 1975). Interestingly, feeding a diet *devoid* of phenylalanine or tyrosine to DBA/2J mice from 14 to 28 days of age *also* increased seizure susceptibility.

Gallagher (1970, 1971) and colleagues (1968) studied the effects of phenylalanine on seizures induced by hexafluorodiethyl ether in male albino rats. Extremely high dietary dosages (3000 to 5000 mg/kg per day) of phenylalanine or tyrosine caused a reduced latency for onset of seizures. Additional insight into the effects of phenylalanine on seizure susceptibility has been obtained in more recent studies by Haigler and coworkers (1986), Pinto and Maher (1986), and Garattini et al. (1986), who used aspartame as a "vehicle" for phenylalanine delivery since phenylalanine comprises approximately 56% of the aspartame molecule.

Haigler and coworkers (1986) failed to observe an effect of aspartame (50 and 500 mg/kg by gavage) on seizure threshold in mice given the convulsant drug pentylenetetrazol or electroconvulsive shock. Pinto and Maher (1986) extended the no-effect dose of aspartame in mice to 750 mg/kg but found that an oral aspartame dose of 1000 mg/kg or higher increased the incidence of pentylenetetrazol-induced seizures in mice. In contrast to findings by Gallagher et al. (1968), Pinto and Maher (1986) failed to observe an increase in seizure indices after administration of tyrosine at a dose equivalent to 1000 mg/kg of phenylalanine.

Garattini et al. (1986) described an increase in susceptibility to pentylenetetrazol-induced seizures after aspartame doses of 750 and 1000 mg/kg. These researchers found that dividing the 1000-mg/kg dose of aspartame into two equal doses reversed the altered seizure susceptibility. Furthermore, Garattini and coworkers suggest that the interaction between high-dose aspartame and pentylenetetrazol-induced seizure susceptibility may be specific to this chemical seizure model in the mouse since 1000 mg/kg of aspartame did not alter seizure susceptibility of mice to quinolinic acid.

Thus, extremely high doses of phenylalanine (or aspartame) have been shown to influence seizure susceptibility in the DBA/2J mouse (audiogenic seizures), the CD-1 mouse (pentylenetetrazol-induced seizures), and the albino rat (hexafluorodiethyl ether-induced seizures). In contrast, no effects were observed in quinolinic acid-induced seizures (rats) and in some electroshock-induced seizures (mice). The large dosages of phenylalanine (given alone or derived from aspartame administration) required to produce any effects on drug- or sound-induced seizures in rodents suggest that phenylalanine/aspartame may not be acting through monoaminergic systems but may, in fact, be acting via nonspecific mechanisms such as changes in blood osmolarity. Clearly, the phenylalanine and/or aspartame doses required to elicit significant effects on seizure susceptibility greatly exceed those that can be realistically ingested in the human diet.

Use of Genetically Epilepsy-Prone Rats to Evaluate the Effects of Phenylalanine on Seizures

The above overview has presented several important attributes of the GEPR that make it an appropriate seizure model to test the possible effects of phenylalanine on seizure susceptibility and severity. To digress briefly, it is important to restate the hypothesized mechanism whereby phenylalanine may influence brain monoaminergic activity. Due to the presence of a common transport system for large neutral amino acids in brain, elevations of plasma phenylalanine may compete with tyrosine and/or 5-hydroxytrytophan for entry into the central nervous system, thereby reducing precursors for norepinephrine and/or serotonin. Moreover, because tyrosine is the primary substrate for catecholamine synthesis in brain, an increase in the quantity of phenylalanine relative to tyrosine may result in reduced synthesis of norepinephrine. Such reductions have been hypothesized to be at least partly responsible for the alleged lowering of seizure thresholds in humans ingesting large quantities of aspartame (Wurtman 1985, Maher 1986).

Because noradrenergic and serotonergic deficits are two of the primary determinants of the epileptic state of the GEPR, this model is particularly well suited to the study of the effects of large doses of phenylalanine on seizure susceptibility and/or severity. Accordingly, we have recently begun a comprehensive research program to determine whether aspartame-derived phenylalanine alters the seizure state of GEPRs. *Preliminary* data from our laboratory indicate that over a large range of *acutely* administered (oral) aspartame doses (50 to 2000 mg/kg; equivalent to approximately 28 to 1120 mg of phenylalanine per kg), no change in either seizure susceptibility or severity occurs in moderate-seizure (GEPR-3) or severe-seizure (GEPR-9) rats.

Acknowledgments. The neurochemical studies reviewed in this paper were partially supported by grants from the Veterans' Administration (MRIS4725) and the National Institutes of Health (NS 16829). Recent work with aspartame has been supported by a grant from the NutraSweet Company.

References

Bourn, W.M., Chin, L., and Picchioni, A.L. (1972). Enhancement of audiogenic seizure by 6-hydroxydopamine. J. Pharm. Pharmacol. **24**:913–914.

Browning, R.A. (1987). The role of neurotransmitters in electroshock seizure models. *In* Jobe, P.C., and Laird, H.E., II (eds.), Neurotransmitter and epilepsy. Clifton, New Jersey: Humana Press, pp. 277–320.

Burley, E.S., and Ferrendelli, J.A. (1984). Regulatory effects of neurotransmitters on electroshock and pentylenetetrazol seizures. Fed. Proc. **43**:2521–2524.

Chapman, A.G., and Meldrum, B.S. (1987). Epilepsy prone mice: genetically-determined sound-induced seizures. *In* Jobe, P.C., and Laird, H.E., II (eds.), Neurotransmitters and epilepsy. Clifton, New Jersey: Humana Press, pp. 9–40.

Consroe, P., Picchioni, A., and Chin, L. (1979). Audiogenic seizure susceptible rats. Fed. Proc. **38**:2411–2416.

Craig, C.R., and Colasanti, B.K. (1987). Experimental epilepsy induced by direct topical placement of chemical agents on the cerebral cortex. *In* Jobe, P.C., and Laird, H.E., II (eds.), Neurotransmitters and epilepsy. Clifton, New Jersey: Humana Press, pp. 191–214.

Dailey, J.W., and Jobe, P.C. (1983). Noradrenergic abnormalities in the genetically epilepsy-prone rat: do they cause or result from seizures? Soc. Neurosci. Abs. **9** (part 1):399.

Dailey, J.W., and Jobe, P.C. (1984). Effect of increments in the concentration

dopamine in the central nervous system on audiogenic seizures in DBA/2J mice. Neuropharmacology **23:**1019–1024.
Dailey, J.W., and Jobe, P.C. (1985). Anticonvulsant drugs and the genetically epilepsy-prone rat. Fed. Proc. **44:**2640–2644.
Dailey, J.W., and Jobe, P.C. (1986). Indices of noradrenergic function in the central nervous system of seizure-naive genetically epilepsy-prone rats. Epilepsia **27:**665–670.
Dailey, J.W., Lasley, S.M., Frasca, J., and Jobe, P.C. (1987). Aspartame (ASM) is not pro-convulsant in the Genetically Epilepsy Prone Rat (GEPR). The Pharmacologist, pp. 142.
Gallagher, B.B. (1970). The relationship between phenylalanine, phenylpyruvate, vitamin B_6 and seizure threshold. Neurology **20:**412 (Abstract).
Gallagher, B.B. (1971). The influence of tyrosine, phenylpyruvate, vitamin B_6 upon seizure thresholds. J. Neurochem. **18:**799–808.
Gallager, B.B., Prichard, J.W., and Glaser, G.H. (1968). Seizure threshold and excess dietary amino acids. Neurology **18:**208–212.
Garattini, S., Caccia, S., and Salmona, M. (1986). Aspartame, brain amino acids and neurochemical mediators. International Aspartame Workshop Proceedings, Marbella, Spain.
Haigler, H.J., Nevins, M.E., and Arnolde, S.M. (1986). Aspartame: lack of effect on convulsant thresholds in mice. J. Cell. Biochem. **31:**R31.
Jobe, P.C. (1970). Relationship of brain amine metabolism to audiogenic seizure in the rat. Dissertation, The University of Arizona.
Jobe, P.C. (1981). Pharmacology of audiogenic seizures. *In* Brown, R.D., and Daigneault, E.A. (eds.), Pharmacology of hearing: experimental and clinical bases. New York: John Wiley and Sons, pp. 271–304.
Jobe, P.C. (1987). Spinal seizures induced by electrical stimulation. *In* Burnham, W.M., Browning, R., Faingold, C., and Fromm, G. (eds.), Epilepsy and the reticular formation. New York: Alan R. Liss, pp. 81–91.
Jobe, P.C., and Laird, H.E. (1981). Neurotransmitter abnormalities as determinants of seizure susceptibility and intensity in the genetic models of epilepsy. Biochem. Pharmacol. **30:**3137–3144.
Jobe, P.C., and Laird, H.E. (1987). Neurotransmitter systems and the epilepsy models: distinguishing features and unifying principles. *In* Jobe, P.C., and Laird, H.E., II (eds.), Neurotransmittors and epilepsy. Clifton, New Jersey: Humana Press, pp. 339–366.
Jobe, P.C., Picchioni, A.L., and Chin, L. (1969). Enhancement of audiogenic seizure by modification of brain amine metabolism. Fed. Proc. **28:**795.
Jobe, P.C., Picchioni, A.L., and Chin, L. (1973). Effect of the dopamine receptor stimulation and blockade on Ro 4-1284-induced enhancement of electroshock seizure. J. Pharm. Pharmacol. **25:**830–831.
Jobe, P.C., Dailey, J.W., and Brown, R.D. (1981). Effect of Ro 4-1284 on audiogenic seizure susceptibility and intensity in epilepsy-prone rats. Life Sci. **28:**2031–2038.
Jobe, P.C., Laird, H.E., Ko, K.H., Ray, T., and Dailey, J.W. (1982a). Abnormalities in monoamine levels in the central nervous system of the genetically epilepsy-prone rat. Epilepsia **23:**359–366.
Jobe, P.C., Woods, T.W., McNatt, L.E., Kearns, G.L., Wilson, J.T., and Dailey,

J.W. (1982b). Genetically epilepsy prone rats (GEPR), a model for febrile seizures? Fed. Proc. **41**:1560.

Jobe, P.C., Ko, K.H., and Dailey, J.W. (1984a). Abnormalities in norepinephrine turnover rate in the central nervous system of the genetically epilepsy-prone rat. Brain Res. **290**:357–360.

Jobe, P.C., Woods, T.W., and Dailey, J.W. (1984b). Proconvulsant and anticonvulsant effects of tricyclic antidepressants in genetically epilepsy-prone rats. *In* Porter, R.J., Mattson, R.H., Ward, A.A., and Dam, M. (eds.), Advances in epileptology: XVth epilepsy international symposium. New York: Raven Press, pp. 187–191.

Jobe, P.C., Reigel, C.E., Woods, T.W., Penney, J.E., and Dailey, J.W. (1985). Do noradrenergic deficits in the genetically epilepsy-prone rat (GEPR) correlate with seizure susceptibility or severity? Fed. Proc., **44**:1107.

Jobe, P.C., Dailey, J.W., and Reigel, C.E. (1986a). Noradrenergic and serotonergic determinants of seizure susceptibility and severity in genetically epilepsy-prone rats. Life Sci. **39**:775–782.

Jobe, P.C., Reigel, C.E., Mishra, P.K., and Dailey, J.W. (1986b): Neurotransmitter abnormalities as determinants of seizure predisposition in the genetically epilepsy-prone rat. *In* Engel, J., Farriello, R.G., Lloyd, K.G., Morselli, P.L., and Nistico, G. (eds.), Neurotransmitters, seizures and epilepsy III. New York: Raven Press, pp. 387–397.

Killam, E.K., and Killam, K.F. (1984). Evidence for neurotransmitter abnormalities related to seizure activity in the epileptic baboon. Fed. Proc., **43**:2510–2515.

Kim, K.C., and Kim, S.H. (1987). Studies on the effect of aspartame and lidocaine interaction in the central nervous system in mice. Fed. Proc., **46**:705 (abstract #2275).

Ko, K.H., Dailey, J.W., and Jobe, P.C. (1982). Effect of increments in norepinephrine concentrations on seizure intensity in the genetically epilepsy-prone rat. J. Pharm. Exp. Ther. **222**:662–669.

Ko, K.H., Dailey, J.W., and Jobe, P.C. (1984). Evaluation of monoaminergic receptors in the genetically epilepsy-prone rat. Experiential **40**:70–73.

Laird, H.E., II (1974). 5-Hydroxytryptamine as an inhibiting modulator of audiogenic seizures. Doctoral dissertation, The University of Arizona.

Laird, H.E., and Jobe, P.C. (1987). The genetically epilepsy-prone rat. *In* Jobe, P.C., and Laird, H.E., III (eds.), Neurotransmitters and epilepsy. Clifton, New Jersey: Humana Press, pp. 57–87.

Laird, H.E., II, Chin, L., and Picchioni, A.L. (1974). Enhancement of audiogenic seizure intensity of p-chloromephamphetamine. Proc. West. Pharmacol. Soc. **17**:46–50.

Laird, H.E., II, Dailey, J.W., and Jobe, P.C. (1983). Innate abnormalities in norepinephrine levels in genetically epilepsy-prone rats (GEPRs). Fed. Proc. **42**:363.

Maher, T.J. (1986). Neurotoxicology of food additives. Neurotoxicology **7**:183–196.

McNamara, J.O., Bonhaus, D.W., Crain, B.J., Gellman, R.L., and Shin, C. (1987). Biochemical and pharmacologic studies of neurotransmitters in the kindling model. *In* Jobe, P.C., and Laird, H.E., II (eds.), Neurotransmitters and epilepsy. Clifton, New Jersey: Humana Press, pp. 115–160.

Meldrum, B.S., Anlezark, G.F., and Trimble, M. (1975). Drugs modifying dopaminergic activity and behavior, the EEG and epilepsy in *Papio papio*. Eur. J. Pharmacol. **32**:203–215.

Nobels, J.L. (1984). A single gene error of noradrenergic axon growth synchronizes central neurons. *Nature* **310**:409–411.

Pinto, J.M.B., and Maher, T. (1986). High dose aspartame lowers the seizure threshold to subcutaneous pentylenetetrazol in mice. The Pharmacologist **28**:155.

Reigel, C.E., Dailey, J.W., and Jobe, P.C. (1986). The genetically epilepsy-prone rat: an overview of seizure-prone characteristics and responsiveness to anticonvulsant drugs. Life Sci., **39**:763–774.

Savage, D.D., Reigel, C.E., and Jobe, P.C. (1986). Angular bundle kindling is accelerated in rats with a genetic predisposition to acoustic stimulus-induced seizures. Brain Res. **376**:412–415.

Schlesinger, K., Schreiber, R.A., Griek, B.J., and Henry, K.R. (1969). Effects of experimentally induced phenylketonuria on seizure susceptibility in mice. J. Comp. Physiol. Psychol. **67**:149–155.

Tacke, U., Tuomisto, L., and Danner, R. (1984). Cortical spike wave discharges during audiogenic convulsions in rats. Exp. Neurol. **85**:233–238.

Truscott, T.C. (1975). Effects of phenylalanine and 5-hydroxytryptophan on seizure severity in mice. Pharmacol. Biochem. Behav. **3**:939–941.

Wurtman, R.J. (1985). Aspartame: Possible effect on seizure susceptibility (letter to the editor). Lancet **ii**:1060.

CHAPTER 13

Aspartame Administration Decreases the Entry of α-Methyldopa into the Brain of Rats

Timothy J. Maher and Paul J. Kiritsy*

We have previously shown that the administration of aspartame (APM) to spontaneously hypertensive rats (SHR) produces a dose-dependent decrease in blood pressure (BP), a response that is attenuated by the coadministration of the large neutral amino acid (LNAA) valine. α-Methyldopa (αMD) is a clinically useful, LNAA antihypertensive agent which appears to act (via its metabolite, α-methylnorepinephrine) on central α_2 receptors in the brain. While studies have shown that APM administration does not alter the hypotensive effect of αMD in SHR, we have investigated the ability of APM administration to attenuate the entry of αMD into the brain of rats. Male rats randomly received a 200-mg/kg dose of APM or water, orally after a 12-hour fast. All animals received a 50-mg/kg dose of αMD, i.p., 15 minutes following oral intubation and were sacrificed 1 hour later. Animals receiving APM exhibited smaller increases in brainstem αMD concentrations (898 ± 171 vs. 2421 ± 215 ng/g, $p < 0.001$). Since APM itself would *not* be expected to lower BP in humans, due to the sluggish conversion of phenylalanine to tyrosine when compared to rodents, the coadministration of APM may lead to a decreased hypotensive effectiveness of αMD in humans.

Introduction

The oral administration of aspartame (L-aspartyl-L-phenylalanine methyl ester; APM) to laboratory animals causes disproportionate increases in the blood and brain levels of the amino acids phenylalanine (Phe) and tyrosine (Tyr) (Yokogoshi et al. 1984). While the administration of Tyr to normotensive animals produces no significant effects on blood pressure (BP), its administration to spontaneously hypertensive rats (SHR) produces a dose-dependent decrease in BP, which is associated with enhanced catecholamine synthesis in, and release from, brainstem (Sved et al. 1979). We have previously shown that the administration of APM to SHR similarly produces a dose-dependent decrease in BP (Kiritsy and Maher 1986), a

*Department of Pharmacology, Massachusetts College of Pharmacy, Boston, MA 02115, USA.

response that is attenuated by the coadministration of the large neutral amino acid (LNAA) valine (which competes for brain uptake with the Tyr derived from APM metabolism).

α-Methyldopa (αMD) is a clinically useful, centrally active antihypertensive agent which appears to act on brainstem postsynaptic α_2 adrenoceptors via its metabolite, α-methylnorepinephrine (Henning and van Zwieten 1968). Thus, the ability of αMD to lower BP is dependent upon its delivery to the brain. Since αMD is a LNAA which enters the brain via a specific transport macromolecule located in the blood-brain barrier, competition among other LNAAs and αMD would be expected. Previous studies have shown that rats allowed to ingest a protein-containing meal exhibit smaller increases in brain levels of αMD following its administration, while rats allowed to ingest an insulin-releasing carbohydrate-containing meal exhibit larger increases in brain levels of αMD when compared to fasting rats (Markovitz and Fernstrom 1977). A recent study has shown that APM administration does not alter the *hypotensive* effect of αMD in SHR (Fernstrom 1984), although one would have predicted that the LNAA derived from APM would retard the entry of αMD into the brain and, thus, interfere with the hypotensive response. The observation that Tyr derived from APM can *itself* lower BP could help to explain this apparent discrepancy.

The hypotensive activity of APM in the rat most likely *cannot* be extrapolated to humans. This anticipated species difference can be attributed to significant differences in the activity of hepatic phenylalanine hydroxylase, since following the administration of APM to rats, larger increments in blood Tyr rather than in blood Phe are observed. In humans, activity of phenylalanine hydroxylase is much lower than in rats, so that APM causes much larger increases in blood Phe than in blood Tyr; thus, APM would not be expected to decrease BP. On the other hand, APM could, by increasing blood LNAA, *interfere* with the hypotensive activity of αMD in humans.

Thus, we investigated the ability of APM administration to alter the entry of αMD into the brain of rats by measuring *actual levels of αMD* in addition to measuring the hypotensive effect.

Materials and Methods

Male Sprague-Dawley rats, 100 to 150 g, were housed individually in our animal facilities and given free access to food (Rat Chow 5001, Purina; 23.1% protein) and water for 1 week prior to experimentation. On the evening before an experiment, food was removed but animals had free access to water. The following morning the animals were divided into two equal groups; one group received a 200-mg/kg dose of APM orally (p.o.), while the second group received an equal volume of water. Both groups

TABLE 13.1. Effects of aspartame pretreatment on brainstem levels of α-methyldopa (αMD).[a]

Treatment	α-Methyldopa (ng/g)
Saline	N.D.
Water + α-methyldopa	2421 ± 215
Aspartame + α-methyldopa	898 ± 171[b]

[a] Animals ($N = 8$ or more) were fasted overnight and divided into two groups on the day of an experiment. Aspartame (200 mg/kg) and αMD (50 mg/kg) were prepared fresh and administered orally and i.p., respectively. Values represent ng αMD/g brainstem (±SEM).
[b] $p < 0.001$ different from water-αMD group as determined by Student's *t* test. N.D. = Not detectable.

then received a 50-mg/kg dose of αMD, interperitoneally (i.p.), 15 minutes after oral intubation. Animals were sacrificed by decapitation 1 hour following αMD, and brainstem was quickly removed. Brainstem tissue was analyzed for αMD content using high-pressure liquid chromatography with electrochemical detection with carbidopa as internal standard. Concentrations of αMD are expressed as ng/g brain tissue. Statistical differences were determined by ANOVA.

Results

While all animals receiving αMD intraperitoneally exhibited increases in its concentration in brainstem 1 hour following i.p. injection, those receiving prior treatment with APM exhibited a significantly smaller increase in brainstem αMD concentrations when compared to the water control group (898 ± 171 vs. 2421 ± 215 ng/g, Table 13.1). Previous experiments failed to detect any endogenous αMD in brainstem of saline- or APM-pretreated rats.

In other experiments, the coadministration of αMD and APM to SHR produced a significant decrease in BP compared to saline, as did αMD or APM alone (−68.8 ± 6.6 vs. −3.9 ± 5.8 mm Hg, Table 13.2). There was no significant difference between the APM and APM/αMD groups.

Discussion

These data affirm that coadministration of APM with the antihypertensive αMD attenuates the entry of this antihypertensive into rat brainstem. Aspartame is an artificial sweetener composed of the simple amino acids aspartic acid and Phe which are rapidly produced upon gut hydrolysis and are quickly absorbed. Upon passage through the liver, Phe is hydrolyzed to Tyr. It has been shown previously that the administration of APM to fasted

TABLE 13.2. Effects of α-methyldopa, aspartame, or aspartame + α-methyldopa on systolic blood pressure in SHR measured indirectly.[a]

Treatment	Blood pressure change (mm Hg)
Saline	-3.9 ± 5.8
α-Methyldopa	-78.4 ± 5.1[b]
Aspartame	-21.5 ± 6.7[b]
Aspartame + α-methyldopa	-68.8 ± 6.6[b]

[a] Animals ($N = 7$ or more) were acclimated to the indirect tail-cuff method of blood pressure measurement prior to experiment and fasted overnight. Aspartame and α-methyldopa (200 and 50 mg/kg, respectively) were prepared on the day of an experiment and administered i.p. Values represent mean change in blood pressure after two hours (mm Hg ± SEM).

[b] $p < 0.001$ compared to saline as determined by ANOVA and Newman-Keuls tests.

rats will increase both the plasma and brain levels of the LNAA, Phe, and Tyr (Fernstrom et al. 1983). This effect is potentiated by the coadministration of a carbohydrate. α-Methyldopa is a LNAA quite similar in chemical structure to the naturally occurring amino acids Phe and Tyr. For this reason, it would appear reasonable that competition would exist between αMD and other LNAAs at the specific transport uptake site(s) in the blood-brain barrier (Pardridge 1977). Indeed, studies have shown that competition among αMD and other amino acids for transport exists in the kidneys and intestine (Young and Edwards 1964, 1966). Moreover, the concurrent injection of LNAAs can decrease the proportion of an αMD dose that is subsequently detected in the brain of rats. Further studies are required to determine the importance of this food additive–drug interaction.

References

Fernstrom, J.D. (1984). Effects of acute aspartame ingestion on large neutral amino acids and monoamines in rat brain. *In* Stegink, L.D., and Filer, L.J. (eds.), Aspartame: physiology and biochemistry. New York: Marcel Dekker, pp. 641–653.

Fernstrom, J.D., Fernstrom, M.H., and Gillis, M.A. (1983). Acute effects of aspartame on large neutral amino acids and monoamines in rat brain. Life Sci. **32:**1651–1658.

Henning, M., and van Zwieten, P.A. (1968). Central hypotensive action of α-methyldopa. J. Pharm. Pharmacol. **20:**409–417.

Kiritsy, P.J., and Maher, T.J. (1986). Acute effects of aspartame on systolic blood pressure in spontaneously hypertensive rats. J. Neural Transm. **66:**121–128.

Markovitz, D.C., and Fernstrom, J.D. (1977). Diet and uptake of aldomet by the brain: competition with natural large neutral amino acids. Science **197:**1014–1015.

Pardridge, W. (1977). *In* Wurtman, R.J., and Wurtman, J.J. (eds.), Nutrition and the brain, Vol. 1. New York: Raven Press, pp. 141–204.

Sved, A.F., Fernstrom, J.D., and Wurtman, R.J. (1979). Tyrosine administration reduces blood pressure and enhances brain norepinephrine release in SHR rats. Proc. Natl. Acad. Sci. USA **76:**3511–3514.

Yokogoshi, H., Roberts, C.H., Caballero, B., and Wurtman, R.J. (1984). Effects of aspartame and glucose administration on brain and plasma levels of large neutral amino acids and brain 5-hydroxyindoles. Am. J. Clin. Nutr. **40:**1–7.

Young, J.A., and Edwards, K.D.G. (1964). Studies on the absorption, metabolism, and excretion of methyldopa and other catechols and their influence on amino acid transport in rats. J. Pharmacol. Exp. Ther. **145:**102–112.

Young, J.A., and Edwards, K.D.G. (1966). Competition for transport between methyldopa and other amino acids in rat gut loops. Am. J. Physiol. **210:**1130–1136.

CHAPTER 14

The Effect of Aspartame on 50% Convulsion Doses of Lidocaine

K.C. Kim,* M.D. Tasch,* and S.H. Kim†

Aspartame doses of 50 mg/kg and 100 mg/kg, given intraperitoneally, reduced 50% convulsion doses of lidocaine by 29.5% and 37.8%, respectively. These values are statistically significant. Latency time and recovery time were not significantly changed. Orally administered doses of aspartame up to 600 mg/kg did not change the 50% convulsion dose of lidocaine.

Our study may suggest that phenylketonuric patients as well as asymptomatic heterozygous individuals may be especially sensitive to the toxic effect of local anesthetics on the central nervous system.

Introduction

Aspartame (NutraSweet), L-aspartyl-L-phenylalanine methyl ester, is a currently popular sugar substitute. Aspartame intake is associated with significantly increased phenylalanine and tyrosine levels in the brain and the blood (Stegink 1984a,b). Inconsistent and modest decreases in brain tryptophan concentrations are also observed in some models (Stegink 1984a). In patients with phenylketonuria, an increase of phenylalanine concentration is thought to be responsible for seizures (Baird 1983). Elevated phenylalanine and tyrosine levels alter plasma tryptophan levels and the synthesis and release of serotonin; these compounds may be involved in the genesis of lidocaine-induced convulsions. Seizures have been sporadically reported after aspartame intake. This study was designed to investigate a possible drug interaction between aspartame and lidocaine in the central nervous system (CNS).

* Department of Anesthesia, Indiana University Medical Center, Indianapolis, IN 46227, USA.
† Ball State University, Muncie, IN 47306, USA.

Materials and Methods

Female white mice of the CD-1 strain, weighing 20 ± 1 g, were used for this study. All animals were allowed free access to both food and water until the time of experimental procedures. Drugs were administered intraperitoneally in a volume of 0.005 ml/g body weight. Doses used in this study were aspartame, 20 mg/kg to 200 mg/kg, followed by lidocaine, 20 mg/kg to 80 mg/kg. Convulsion was defined by the following three criteria: (a) loss of righting reflex, (b) clonic and tonic seizures, and (c) opishotonus. Latency time (from lidocaine injection to the first clonic seizure), duration of convulsion, and recovery time (to regaining normal gait after cessation of convulsion) were observed. To determine the mean 50% convulsion dose (CD_{50}) of lidocaine, groups of 10 mice were tested with various drug doses until at least four points were established between doses at which 0% of mice and 100% of mice convulsed. The CD_{50} was computed from the probit–log dose regression line using the method of Litchfield and Wilcoxon. Parallelism and significance were tested by the same method (Litchfield and Wilcoxon 1949).

Results

Control CD_{50} of lidocaine was 61 mg/kg. This was reduced to 50 mg and 49 mg/kg when aspartame was given at 5 mg/kg and 20 mg/kg, respectively. The differences in these values were not statistically significant. When the dose of aspartame was increased to 50 mg/kg and 100 mg/kg, CD_{50} of lidocaine was reduced to 43 mg/kg and 38 mg/kg, respectively. These differences were statistically significant (see Table 14.1).

TABLE 14.1. Effect of intraperitoneal administration of aspartame on CD_{50} of lidocaine.

Aspartame dose (mg/kg)	CO_{50} of lidocaine	Percent decrease from control
0 (control)	61	—
5	50	18.1
20	49	19.7
50	43	29.5[a]
100	38	37.8[a]

[a] Statistically significant.

TABLE 14.2. Latency and recovery time of convulsion.

	Latency time (min)	Recovery time (min)
Lidocaine	2.8 ± 0.4	4.3 ± 1.2
Lidocaine + aspartame	3.4 ± 0.7	5.2 ± 1.9

[a] Differences are not statistically significant.

Latency and recovery time were not changed significantly (see Table 14.2). Orally administered doses of aspartame up to 600 mg/kg had no effect on CD_{50} of lidocaine.

Discussion

Circulating lidocaine is cleared by the liver, with metabolism influenced by hepatic blood flow and by the activity of hepatic microsomal monoxygenase enzymes (Geddes 1967). There are no available data to indicate that aspartame impairs either hepatic perfusion or drug-metabolizing enzymes. No effect of aspartame on the volume of distribution of lidocaine has been documented either. In the absence of such pharmacokinetic phenomena, the reduction of lidocaine CD_{50} by aspartame seems likely to represent a pharmacodynamic drug interaction in the brain.

Aspartame is metabolized to yield phenylalanine. It does not yield any of the other large neutral amino acids which, when released from proteins, compete with phenylalanine for transit across the blood-brain barrier. Aspartame administration has been shown to raise serum and brain concentrations of phenylalanine and tyrosine above normal values. Experiments have, in contrast, shown that aspartame induced decreases in brain levels of tryptophan without altering the concentrations of serotonin, dopamine, norepinephrine, or a number of their metabolites. A study in rats demonstrated that aspartame at 200 mg/kg, given by gastric tube, caused striking increases in blood and brain concentrations of phenylalanine and tyrosine and that these alterations were significantly exaggerated when glucose was also administered enterally (Yokogoshi et al. 1984). The authors indicated that the mechanism of this glucose-aspartame interaction was that insulin secretion, elicited by glucose, lowered the plasma levels of other LNAA that would competitively inhibit phenylalanine entry into the brain.

The results of the above studies, taken with ours, suggest that alterations in brain amino acid concentrations may in some way be involved in the lowering of lidocaine CD_{50} by aspartame. Some theoretical clinical implications arise. One is that phenylketonuric patients, as well as asymptomatic heterozygous individuals with subclinical reductions in phenylalanine hydroxylase activity, may manifest increased susceptibility to lidocaine-induced seizures, related to increased CNS phenylalanine levels. In addition, the recent consumption of aspartame along with carbohydrate could conceivably render the normal patient more sensitive to the toxic effects of local anesthetics on the CNS. Drawing clinical implications from these animal experiments is, of course, made speculative by differences in species and dosages. However, an argument has been made that notable elevations in CNS phenylalanine levels can be achieved by a person (especially a child) consuming relatively large but plausible amounts of aspartame-containing liquids with a high-carbohydrate meal or snack. In such a setting, the safe maximal dose of local anesthetic subject to systemic uptake could possibly be diminished.

Acknowledgment. Authors are greatly appreciative of Dr. B. Harris who helped us to initiate and complete this study.

References

Baird, H.W. (1983). Convulsive disorders. *In* Behrman, R.E., and Vaughan, V.C. (eds.), Textbook of pediatrics. 12th ed. Philadelphia: W.B. Saunders, pp. 1234.

Geddes, I.C. (1967). Metabolism of local anesthetic agents. Anesthesiol. Clin. **5**:525–541.

Litchfield, J.T., and Wilcoxon, F. (1949). A simplified method of evaluating dose-effect experiments. J. Exp. Pharm. Ther. **96**:90–113.

Stegink, L.D. (1984a). Aspartame and glutamate metabolism. *In* Stegink, L.D., and Filer, L.J. (eds.), Aspartame: physiology and biochemistry. New York: Marcel Dekker, pp. 47–75.

Stegink, L.D. (1984b). Aspartame metabolism in humans. *In* Stegink, L.D., and Filer, L.J. (eds.), Aspartame: physiology and biochemistry. New York: Marcel Dekker, pp. 509–553.

Yokogoshi, H., Roberts, C.H., Caballero, B., and Wurtman, R.J. (1984). Effects of aspartame and glucose administration on brain and plasma levels of large neutral amino acids and brain 5-hydroxyindoles. Am. J. Clin. Nutr. **40**:1–7.

CHAPTER 15

Studies on the Susceptibility to Convulsions in Animals Receiving Abuse Doses of Aspartame

Silvio Garattini, Silvio Caccia, Maria Romano, Luisa Diomede, Giovanna Guiso, Annamaria Vezzani, and Mario Salmona*

Aspartame and its major metabolite phenylalanine were studied for their capacity to potentiate metrazol, electroshock, and quinolinic-acid-induced seizures in rats. Pretreatment with 0.75 and 1 g/kg aspartame or 0.250 and 0.500 g/kg phenylalanine administered by gavage significantly increased the number of animals undergoing convulsions after a dose of 65.9 mg/kg of metrazol. This effect was no longer significant when 1 g/kg aspartame was given as three divided doses over two hours, after a meal or overnight with food and drinking water. However, a dose of 1 g/kg aspartame or 0.500 g/kg phenylalanine did not influence the threshold electrical current necessary to elicit convulsions in 50% of animals or the limbic seizures induced by convulsant doses of quinolic acid. After doses of 1 g/kg aspartame or 0.500 g/kg phenylalanine, striatal and hippocampal levels of 5-HT, NA, DA and their metabolites were not significantly modified. Also, in vivo release of striatal DA measured by brain microdialysis was not changed by treatment with these compounds. Under different treatment schedules, significant increases in brain phenylalanine and tyrosine were observed in rats treated acutely with doses of 250 mg/kg or more.

The brain ratio of phenylalanine to tyrosine was calculated for all the experiments performed. This value was slightly different from controls after a dose of 1 g/kg aspartame (0.88 versus 0.81). A dose of 0.5 g/kg of phenylalanine lowered this ratio to 0.60. It thus appears that the brain ratio of phenylalanine to tyrosine played no role in the control of seizure susceptibility. In our experimental conditions, this event occurs only after "abuse doses" of aspartame or phenylalanine and the mechanism is still unknown.

Aspartame, used as a sweetener, is a dipeptide which is rapidly metabolized in the body before entering the systemic circulation. Three main metabolites are formed—phenylalanine, aspartic acid, and methanol (Oppermann 1984).

Despite rather extensive studies, knowledge about this agent is still not complete or even satisfactory, considering the continuous advances in biomedicine and the increasing per capita intake of aspartame.

* Istituto di Ricerche Farmacologiche "Mario Negri," Via Eritrea 62, 20157 Milan, Italy.

TABLE 15.1. Influence of aspartame and phenylalanine on an effective convulsive dose (ED_{50}) of metrazol in rats.[a]

Phenylalanine (g/kg p.o.)	Metrazol convulsions	Aspartame (g/kg p.o.)	Metrazol convulsions
0	5/10	0	35/60
0.125	6/10	0.50	15/20
0.250	8/10[b]	0.75	10/10[b]
0.500	9/10[b]	1.00	18/20[b]
Leucine 1.000	5/10	1.00[c]	16/20
		1.00[d]	13/20
		2.00[e]	14/20

[a] Phenylalanine, leucine and aspartame were administered orally to fasted animals 60 min before the convulsive ED_{50} of metrazol (65.91 mg/kg i.p.). Data represent the number of animals showing tonic/clonic convulsions and the total number of animals.

[b] $p < 0.05$; significantly different from vehicle group.

[c] Divided in 3 doses during a period of 2 h.

[d] Administered after a meal.

[e] Ingested overnight with food and drinking water.

Because one of aspartame's metabolites, phenylalanine, has a role as a possible precursor of important chemical mediators (dopamine, noradrenaline, and adrenaline) and because of the pathological findings in subjects with an excessive level of phenylalanine in a genetically determined disease such as phenylketonuria, several investigators have searched for possible effects of aspartame on the central nervous system. This paper reviews the effect of aspartame on convulsions elicited experimentally in animals with the aim of assessing its significance for man.

Effect of Aspartame on Experimental Convulsions

Previous investigations have shown that aspartame potentiates the convulsions elicited by metrazol, an agent used in pharmacology to screen for anticonvulsant drugs, in mice (Pinto and Maher 1987) and in rats (Garattini 1986). Aspartame's potentiating effect on metrazol requires an oral dose of 1.0 g/kg in mice (Pinto and Maher 1987) and 750 mg/kg in rats (Garattini 1986). As summarized in Table 15.1, phenylalanine has a similar effect, at doses which are compatible with the view that aspartame acts through its "active" metabolite. This effect exerted by high doses of phenylalanine is not common to all amino acids, as shown by the lack of effect of leucine under similar experimental conditions (1.0 g/kg p.o.) (Table 15.1).

Aspartame lowers the dose of metrazol necessary to elicit convulsions in 50% of the animals, but the effect is quite weak. In fact, from the data of Pinto and Maher (1987), it can be calculated that in mice the ED_{50} of metrazol is shifted from about 66 mg/kg s.c. in controls to 59 mg/kg when given 60 min after aspartame (1.0 g/kg by gavage). Similarly, in rats the

TABLE 15.2. Effect of aspartame and phenylalanine on electroshock-induced tonic convulsions.

Electroshock (CC_{50})[a]			
Control (vehicle)	Aspartame (1 g/kg p.o.)	Control (vehicle)	Phenylalanine (0.5 g/kg p.o.)
14.0 (12.7–15.2)	12.0 (11.4–12.5)	18.6 (15.8–21.9)	13.7 (9.0–20.8)
17.4 (15.1–19.5)	15.0 (14.1–15.8)	19.5 (17.0–21.8)	14.8 (11.0–20.0)
20.4 (19.5–21.4)	15.8 (13.8–17.8)	22.9 (21.6–24.3)	15.1 (12.6–17.8)

[a] The values represent the electroconvulsive threshold expressed as mA. CC_{50} is the threshold defined as the amperage necessary to cause the hindleg tonic extensor component of the seizure in 50% of animals according to the method of Kimball et al. (1957) ($n = 15$ for each value).

ED_{50} of metrazol was 65.9 mg/kg i.p. (95% confidence limits, 55.5 to 75.9 mg/kg) in untreated animals and 50.7 mg/kg i.p. (44.0 to 58.2) when animals were treated 1 h before with aspartame (1.0 g/kg by gavage) (Garattini et al. 1987). The metabolite aspartic acid is not effective (Pinto and Maher 1987), and methanol is formed at concentrations too low to have any pharmacologic effects (Tephly and Mc Martin 1984).

Less clear is the effect of aspartame on other convulsant agents. For instance, Wurtman and Maher (1987) found an increase of seizures with a convulsant anesthetic, fluorothyl (10% delivered in a sealed chamber at a rate of 0.05 ml/min). Aspartame shortens the lag time for convulsions elicited by maximal electroshock in mice (1.0 g/kg p.o.) (Wurtman and Maher 1987), a proconvulsant effect not observed in rats (1.0 g/kg p.o.), when the threshold of electrical current necessary to elicit convulsions in 50% of the animals is measured. Phenylalanine (0.5 g/kg p.o.) lowers this parameter more consistently (Table 15.2).

Aspartame (1.0 g/kg p.o.) does not affect the seizures (Garattini 1986, Garattini et al. 1987) induced by an endogenous convulsant agent such as quinolinic acid (Wolfensberger et al. 1983), showing that this artificial sweetener should not be considered capable of increasing all types of convulsions. This statement is reinforced by observations in animals with a genetic propensity to convulsions where aspartame or phenylalanine showed either proconvulsant or anticonvulsant effects, depending on the experimental conditions (Woodbury 1986, Jobe 1986). Other pertinent data are reported by other authors in this book.

Mechanisms Involved in Aspartame's Effect on Metrazol Convulsions

Various hypotheses can be generated to explain the effect of aspartame:

(i) Effect on brain metrazol concentrations: The high doses of aspartame necessary to potentiate the effect of metrazol might increase the entry

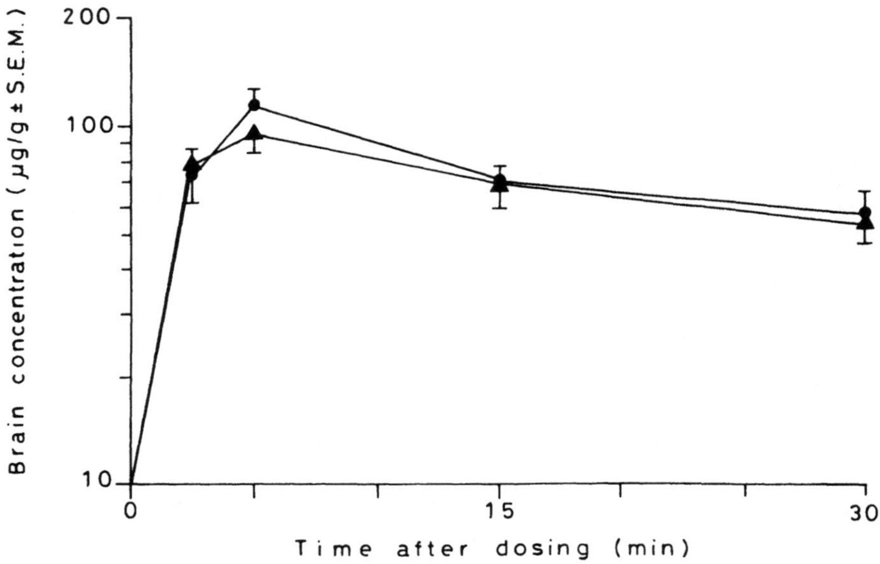

FIGURE 15.1. Brain concentrations of metrazol at different times after administration of metrazol (65.9 mg/kg i.p.) alone (▲) or preceded 1 h before by aspartame (1.0 g/kg p.o.) (●). Brain metrazol was measured by a gas-chromatographic method (Ramzan and Levy 1985).

of metrazol into the brain. That this is not the case is shown by the fact that aspartame (1.0 g/kg p.o.) does not change the kinetics of brain metrazol measured at different times (2, 5, 15, and 30 min) after administration of metrazol (Figure 15.1). However, since the decrease of metrazol threshold induced by aspartame is about 20%, these negative data must be critically evaluated.

(ii) Effect on benzodiazepine-GABA receptors: Metrazol convulsions are highly sensitive to the inhibitory effect of benzodiazepine (BDZ) tranquilizers (Garattini et al. 1973), and this has led to the hypothesis that metrazol may interact with the GABA receptor involved in BDZ action (Gee and Yamamura 1983). It is therefore possible that aspartame and phenylalanine increase metrazol convulsions by interacting with BDZ-GABA receptors. In vitro studies have shown that phenylalanine does not inhibit specific binding of [^{3}H]-flunitrazepam to specific brain structures up to a concentration of 100 μmol/ml (Karobath and Sperk 1979), which is quite low in relation to the phenylalanine concentrations found in rat brain after an active dose of aspartame (Garattini 1986). Studies were therefore made in vivo to establish whether [^{3}H]-flunitrazepam binding was altered after aspartame pretreatment (1.0 g/kg p.o. 1 h before). The results show that aspartame pretreatment significantly decreased total [^{3}H]-flunitrazepam concentration in brainstem (treated, 0.23 ± 0.04; controls, 0.31 ± 0.02

fmol/mg protein; $p < 0.02$), thus resulting in about a 30% decrease of [^{3}H]-flunitrazepam, which is either free or specifically bound. Similar, although not significant, trends were seen in cerebellum and cortex of treated rats (Mennini, T., personal communication).

Further experiments are needed to ascertain whether this effect of aspartame occurs at the receptor level or involves other mechanisms regulating the free tracer distribution into the brain.

(iii) Effect on brain catecholamines: Brain catecholamines have an inhibitory role on convulsions (Jobe et al. 1986, Lovell 1971, Maynert 1969). Results are conflicting (Jobe et al. 1974, Meldrum et al. 1975) regarding the relative roles of noradrenaline (NA) and dopamine (DA), although the most recent findings indicate that NA has a more important role than DA (Mason and Corcoran 1979). Phenylalanine, by raising the level of tyrosine, could increase catecholamine synthesis, although tyrosine hydroxylase already appears to be saturated by brain tyrosine at physiological levels (Mc Geer et al. 1971, Ikeda et al. 1967). However, at high concentrations, phenylalanine is a competitive inhibitor of tyrosine for tyrosine hydroxylase, thus causing inhibition of catecholamine synthesis (Levitt et al. 1965, Ikeda et al. 1967). The facilitation of convulsions by phenylalanine currently has been interpreted as being related to an inhibitory effect on brain catecholamines; it is therefore not surprising that aspartame at high doses, by producing phenylalanine, potentiates the action of convulsants such as metrazol. Indeed, Wurtman's group has shown that aspartame does affect brain catecholamines, thus explaining its interaction with convulsants (Yokogoshi and Wurtman 1986). However, a closer look at the data indicates that aspartame has this effect in spontaneously hypertensive rats (SHR) but not in Sprague-Dawley rats, where an aspartame dose of 200 mg/kg raises NA rather than lowering it (Yokogoshi and Wurtman 1986). In addition, the effects are very small and limited to given brain areas, and they are certainly very different from those exerted by 6-hydroxy-DOPA (Mason and Corcoran 1979) or reserpine (Jobe et al. 1986) (at least an 80% decrease), both known to increase metrazol convulsions.

Studies in this laboratory do not support the view that aspartame markedly affects brain catecholamines or indoleamines. For instance, the administration of 250 mg of aspartame per kg by gavage did not change the levels of serotonin (5-HT) or its metabolite 5-HIAA in the hippocampus of rats pretreated with water or a glucose load (3 g/kg orally) (Garattini 1986, Garattini et al. 1987). Under the same experimental conditions, no changes could be detected in cortical NA or its metabolite MHPG-SO_4, or striatal DA and its metabolites DOPAC and HVA. Similarly, rats receiving 1.0 g of aspartame per kg by gavage did not show changes in hippocampal 5-HT and 5-HIAA or in striatal 5-HT, 5-HIAA, NA, DOPAC, and HVA (Garattini 1986, Garattini et al. 1987) (Table 15.3).

TABLE 15.3. Monoamines and acid metabolite levels in the striatum and hippocampus of rats after an acute oral dose of aspartame (1.0 g/kg) or phenylalanine (0.5 g/kg).

Amine or metabolite	Level ± SE (ng/g tissue)[a] Controls	Aspartame-treated[b]	Controls	Phenylalanine-treated[b]
Striatum				
5-HT	365 ± 27	329 ± 9	258 ± 11	236 ± 14
5-HIAA	395 ± 26	373 ± 14	322 ± 14	314 ± 11
NA	189 ± 11	225 ± 14	140 ± 9	122 ± 14
DA	8688 ± 197	8701 ± 442	6200 ± 48	5290 ± 130
DOPAC	762 ± 12	810 ± 25	960 ± 77	768 ± 27
HVA	484 ± 18	524 ± 30	707 ± 42	677 ± 30
Hippocampus				
5-HT	414 ± 32	415 ± 24	230 ± 38	203 ± 9
5-HIAA	351 ± 8	347 ± 24	204 ± 31	192 ± 6

[a] Brain amines and their metabolites were measured according to Achilli et al. (1985).
[b] Animals were killed 1 h after aspartame or phenylalanine.

In order to improve the sensitivity of the system, measurements were made on extracellular levels, rather than on total DOPAC or 5-HIAA, by pulse voltammetry (De Simoni et al. 1985). No effects were observed in striatum or nucleus accumbens DOPAC or 5-HIAA up to 3 h after administration of an aspartame dose of 1.0 g/kg by gavage.

Another method used for detecting changes of chemical mediators was trans-striatal dialysis (Imperato and Di Chiara 1985). Aspartame given at the dose of 1.0 g/kg by gavage did not modify the release of DA, DOPAC, or HVA in rat striata for a period of 3 h (Figure 15.2). These results are in agreement with the findings obtained with a phenylalanine dose of 500 mg/kg by gavage, indicating that NA and DA release is not significantly affected by the treatment (Table 15.3). This obviously does not exclude the possibility that catecholamine release may be altered in other specific brain areas.

Furthermore, our data agree with findings by Brass and Greengard (1982) that brain catecholamines did not decrease even with brain phenylalanine levels about 10 times those necessary under our conditions to potentiate metrazol convulsions.

Significance of the Effect of Aspartame on Convulsions for Man

In attempting extrapolations of experimental findings to man, a number of limitations and problems must be borne in mind.

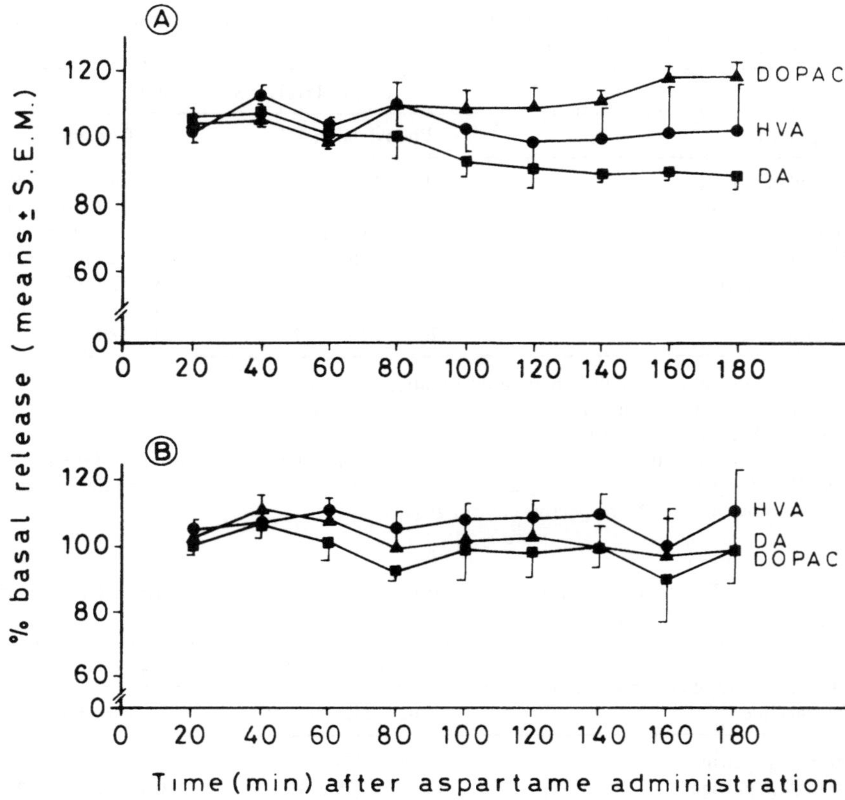

FIGURE 15.2. Effect of aspartame (1.0 g/kg p.o.) (panel A) or phenylalanine (500 mg/kg p.o.) (panel B) on the release of striatal dopamine (DA) (■), dioxyphenylacetic acid (DOPAC) (▲), and homovanillic acid (HVA) (●) measured by transstriatal dialysis (Imperato and Di Chiara 1985). Mean ± SE ($n = 4$).

First of all, man is not exposed to metrazol since this compound—formerly used as an analeptic—is no longer available as a drug.

Second, the findings refer to aspartame given under well-defined experimental conditions. The sweetener, administered by gavage, was always given for only a short time in order to maximize absorption and overcome the metabolic capacity of the liver. In contrast, dietary use of aspartame implies its intake even at high doses during relatively long periods over a large part of the day. The implications of this difference are illustrated by the fact that phenylalanine, for example, measured in rats after a single aspartame dose of 250 mg/kg by gavage, doubles plasma levels while the same dose given in the diet for a period of 14 days induces no changes. Similar results are obtained with higher doses; for instance, an aspartame dose of 1.0 g/kg by gavage raises plasma phenylalanine by more than 3

TABLE 15.4. Plasma levels of phenylalanine and tyrosine after administration of aspartame under different conditions.

Dose and route of administration of aspartame	Plasma levels ± SE (nmol/ml)	
	Phenylalanine	Tyrosine
100 mg/kg, gavage	90 ± 13 (71 ± 3)	102 ± 14 (92 ± 3)
250 mg/kg, gavage	132 ± 6 (72 ± 2)	158 ± 16 (86 ± 5)
250 mg/kg, gavage[a]	134 ± 7 (73 ± 3)	170 ± 14 (75 ± 4)
250 mg/kg, diet[b]	57 ± 10 (56 ± 4)	113 ± 12 (101 ± 13)
250 mg/kg, diet[c]	61 ± 6 (58 ± 4)	106 ± 15 (103 ± 17)
1.0 g/kg, gavage	237 ± 59 (67 ± 9)	290 ± 58 (76 ± 9)
2.0 g/kg, diet[d]	76 ± 17 (65 ± 17)	150 ± 18 (92 ± 5)

[a] 1h after 3 g/kg glucose orally, then killed 1 h after aspartame.
[b] Given with a normocaloric diet for 14 days (Romano et al. 1987a).
[c] Given with a carbohydrate-rich diet for 14 days (Romano et al. 1987a).
[d] Given overnight with the food and drinking water, then killed in the morning (9.00 A.M.).
Note: When aspartame was given by gavage (10% solution), rats were killed 1 h later. Values in parentheses are for the respective control groups.

TABLE 15.5. Brain levels of phenylalanine and tyrosine after administration of aspartame under different conditions.

Dose and route of administration of aspartame	Brain levels ± SE (nmol/ml)	
	Phenylalanine	Tyrosine
100 mg/kg, gavage	71 ± 4 (71 ± 5)	91 ± 4 (104 ± 7)
250 mg/kg, gavage[a]	105 ± 7 (75 ± 3)	130 ± 10 (77 ± 2)
250 mg/kg, diet[b]	47 ± 5 (42 ± 5)	87 ± 12 (83 ± 12)
250 mg/kg, diet[c]	42 ± 3 (38 ± 4)	92 ± 8 (87 ± 19)
250 mg/kg, gavage	101 ± 3 (65 ± 1)	118 ± 8 (66 ± 3)
1.0 g/kg, gavage	202 ± 23 (71 ± 5)	206 ± 34 (104 ± 7)
2.0 g/kg, diet[d]	66 ± 28 (70 ± 1)	144 ± 16 (108 ± 12)

[a] 1h after 3 g/kg glucose orally, then killed 1 h after aspartame (Romano et al. 1987a).
[b] Given with a normocaloric diet for a period of 14 days (Romano et al. 1987a).
[c] Given with a carbohydrate-rich diet for a period of 14 days (Romano et al. 1987a).
[d] Given overnight with the food and drinking water, then killed in the morning (9.00 A.M.).
Note: When aspartame was given by gavage (10% solution), rats were killed 1 h later. Values in parentheses are for the respective control groups.

times while 2 g/kg given overnight with the food and drinking water causes no significant changes in plasma phenylalanine (Table 15.4) (Garattini et al. 1987, Romano et al. 1987a). The brain levels were in good agreement with those measured in plasma (Table 15.5). However, under our experimental conditions, we could not confirm that the administration of aspartame with a glucose load raised brain phenylalanine more than under normal conditions (Yokogoshi et al. 1984, Wurtman 1983) (Table 15.4).

These findings also are of significance for the effect on metrazol convulsions. When 1.0 g/kg of aspartame is given divided in three doses over a period of 2 h rather than as a single dose, it has no effect on metrazol convulsions; similarly, 2 g/kg of aspartame given with the diet and the drinking water overnight do not potentiate metrazol convulsions (Table 15.1).

Third, the results on convulsions were obtained in fasted animals. If animals are fed before aspartame administration, metrazol convulsions are no longer potentiated (Table 15.1). This may be explained by the fact that neutral amino acids absorbed after a meal compete for the transport of phenylalanine into the brain (Fernstrom and Faller 1978, Yokogoshi et al. 1984). Indeed, valine, which by itself does not affect metrazol convulsions, antagonizes the potentiating effect of aspartame (Pinto and Maher 1987).

However, it has been argued that data obtained in rodents may give an optimistic picture in relation to man as far as aspartame is concerned. In fact, in rodents phenylalanine is rapidly transformed into tyrosine, which may compete with phenylalanine in terms of effects on brain catecholamines, while in man phenylalanine is only slowly hydroxylated to form tyrosine. Accordingly, data obtained in human volunteers indicate that rapid intake of aspartame up to about 8 mg/kg (corresponding to 25 tablets of aspartame) does not affect the level of plasma phenylalanine or tyrosine 30 or 180 min after administration (Garattini et al. 1987, Romano et al. 1987b). At 34 mg/kg aspartame, plasma phenylalanine roughly doubles while plasma tyrosine rises about 25% (Stegink 1984). At 50 mg/kg aspartame (single rapid intake), plasma phenylalanine increases about 3 times, while plasma tyrosine rises about 70% (see Figure 15.3). At even higher doses—200 mg/kg aspartame—plasma phenylalanine rose about 9 times while plasma tyrosine increased by 3 times (Stegink 1984).

However, it has not yet been completely demonstrated that the ratio of phenylalanine to tyrosine is the decisive factor in the increased susceptibility to convulsions and that, therefore, man should be more sensitive to the effect of aspartame. In fact, Table 15.6 sets out the plasma levels of phenylalanine and tyrosine after high oral doses of aspartame or phenylalanine in relation to their capacity to potentiate metrazol convulsions. Aspartame and phenylalanine were equally active in potentiating metrazol despite quite different relative increases of plasma tyrosine and a different ratio of phenylalanine to tyrosine. Therefore, it does not look as though a relatively higher level of tyrosine protects against the action of phenylalanine. Further studies are, however, necessary to reach any firm conclusion. (See Chapter 10 of this volume.)

All these data taken together indicate that the potentiation of metrazol convulsions by aspartame has no relevance to the practical use of aspartame even at "abuse" doses. In addition, it should be pointed out that even 50 mg/kg of aspartame in orange juice ingested over 4 h does not affect plasma phenylalanine or tyrosine levels (see Figure 15.3). These data are in

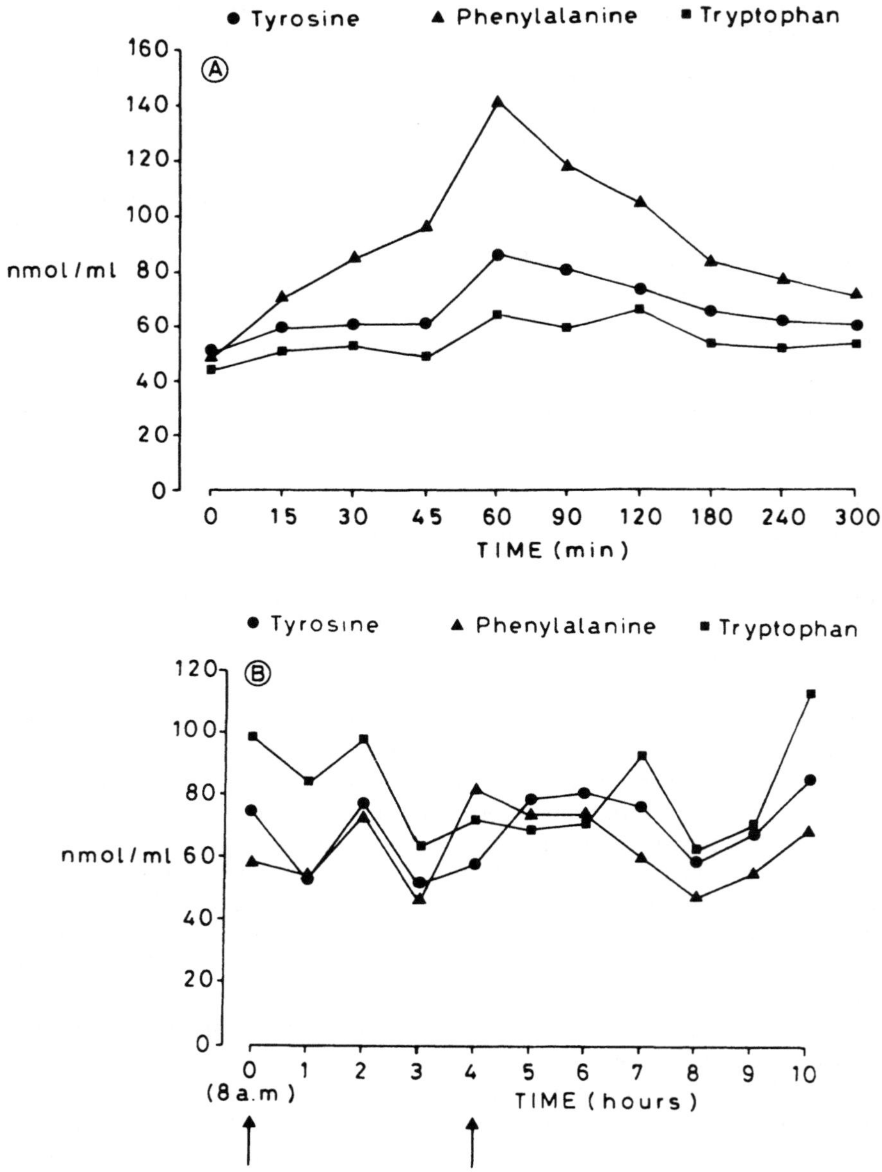

FIGURE 15.3. Plasma tyrosine, phenylalanine, and tryptophan levels in normal adults given 3 g/60 kg body weight of aspartame dissolved in 200 ml of orange juice in a single intake (panel A) or in 500 ml of orange juice over a period of 4 h (panel B).

TABLE 15.6. Ratio of plasma phenylalanine to tyrosine in relation to metrazol convulsions.

Treatment (mg/kg, gavage)	Plasma levels (noml/ml ± SE)		Phenylalanine/ tyrosine ratio	Metrazol convulsions
	Phenylalanine	Tyrosine		
Vehicle	67 ± 9	76 ± 9	0.88	17/30
Phenylalanine (500)	380 ± 2[b]	627 ± 2[b]	0.60	9/10[a]
Aspartame (1000)	237 ± 67[b]	290 ± 58[b]	0.81	18/20[a]

[a] $p < 0.05$ with respect to vehicle. [b] $p < 0.01$ with respect to vehicle.

agreement with previous studies indicating that more than 50 mg/kg of aspartame per day given for 13 weeks in a variety of foods and beverages did not change plasma levels of phenylalanine or tyrosine in normal or heterozygous phenylketonuric adults (Visek 1984).

Concluding Remarks

Evidence from a variety of studies in animals and in man indicate that the plasma and—in animals—brain phenylalanine levels reached after aspartame administration depend on the conditions under which the study is conducted. A severalfold increase of phenylalanine was obtained only when aspartame was given at unrealistic doses in a single rapid intake. The same unrealistic doses given over a period of a few hours, even repeatedly for several weeks, did not affect plasma and—in animals—brain phenylalanine or tyrosine levels. Only when aspartame was given by gavage at high doses was there a potentiation of metrazol-induced, but not of electroshock- or quinolinic-acid-induced, convulsions in rats. This potentiation was mild and related to the rise in brain phenylalanine but not to changes of brain catecholamines or serotonin.

This potentiation of metrazol convulsions by aspartame is unlikely to represent a hazard for man, although obviously any definitive judgment must await full clarification of the mechanism of this potentiation.

Acknowledgment. This work was supported by CNR (National Research Council), Rome, Italy, under an institutional grant for studies on clinical pharmacology.

References

Achilli, G., Perego, C., and Ponzio, F. (1985). Application of the dual-cell coulometric detector: a method for assaying monoamines and their metabolites. Anal. Biochem. **148:**1–9.

Brass, C.A., and Greengard, O. (1982). Modulation of cerebral catecholamine concentrations during hyperphenylalaninaemia. Biochem. J. **208:**765–771.

De Simoni, M.G., Giglio, R., Dal Toso, G., Kostowski, W., and Algeri, S. (1985). Interaction between serotonergic and dopaminergic systems detected in striatum by differential pulse voltammetry. Eur. J. Pharmacol. **110**:289–290.
Fernstrom, J.D., and Faller, D.V. (1978). Neutral amino acids in the brain: changes in response to food ingestion. J. Neurochem. **30**:1531–1538.
Garattini, S. (1986). Aspartame, brain amino acids and neurochemical mediators. *In* International aspartame, workshop proceedings, Nov. 17–21, 1986, Marbella, Spain, Aspartame Technical Committee, International Life Sciences Institute-Nutrition Foundation, Washington, D.C., Session III.
Garattini, S., Mussini, E., Marcucci, A., and Guaitani, A. (1973). Metabolic studies on benzodiazepines in various animal species. *In* Garattini, S., Mussini, E., and Randall, L.O. (eds.), The benzodiazepines. New York: Raven Press, pp. 75–97.
Garattini, S., Perego, C., Caccia, S., Vezzani, A., and Salmona, M. (1987). Aspartame, brain, aminoacids and neurochemical mediators. Presented at: Sweeteners: Health Effects, American Health Foundation, New York, Feb. 18–20 1987 (in press).
Gee, K.W., and Yamamura, H.I. (1983). Photoaffinity labeling of benzodiazepine receptors in rat brain with flunitrazepam alters the affinity of benzodiazepine receptor agonist but not antagonist binding. J. Neurochem. **41**:1407–1413.
Ikeda, M., Levitt, M., and Udenfriend, S. (1967). Phenylalanine as substrate and inhibitor of tyrosine hydroxylase. Arch. Biochem. **120**:420–427.
Imperato, A., and Di Chiara, G. (1985). Dopamine release and metabolism in awake rats after systemic neuroleptics as studied by trans-striatal dialysis. J. Neurosci. **5**:297–306.
Jobe, P.C. (1986). Effects of amino acids on seizure thresholds in animal models. *In* International aspartame workshop proceedings, Nov. 17–21, 1986, Marbella, Spain, Aspartame Technical Committee, International Life Sciences Institute-Nutrition Foundation, Washington, D.C., Session V.
Jobe, P.C., Stull, R.E., and Geiger, P.F. (1974). The relative significance of norepinephrine, dopamine and 5-hydroxytryptamine in electroshock seizure in the rat. Neuropharmacology **13**:961–968.
Jobe, P.C., Dailey, J.W., and Reigel, C.E. (1986). Noradrenergic and serotonergic determinants of seizure susceptibility and severity in genetically epilepsy-prone rats. Life Sci. **39**:775–782.
Karobath, M., and Sperk, G. (1979). Stimulation of benzodiazepine receptor binding by γ-aminobutyric acid. Proc. Natl. Acad. Sci. USA **76**:1004–1006.
Kimball, A.W., Burnett, W.T., Jr., and Doherty, D.G. (1957). Chemical protection against ionizing radiation. I. Sampling methods for screening compounds in radiation-protection studies with mice. *Radiat. Res.* **7**:1–12.
Levitt, M., Spector, S., Sjoerdsma, A., and Udenfriend, S. (1965). Elucidation of the rate-limiting step in norepinephrine biosynthesis in the perfused guinea-pig heart. J. Pharmacol. Exp. Ther. **148**:1–8.
Lovell, R.A.(1971). Some neurochemical aspects of convulsions. *In* Lajtha, A. (ed.), Handbook of neurochemistry, Vol. 6. New York: Plenum Press, pp. 63–102.
Mason, S.T., and Corcoran, M.E. (1979). Catecholamines and convulsions. Brain Res. **170**:497–507.
Maynert, E.W. (1969). The role of biochemical and neurohumoral factors in the

laboratory evaluation of antiepileptic drugs. Epilepsia (Amsterdam) **10:**145–162.

McGeer, E.G., McGeer, P.L., and Wada, J.A. (1971). Distribution of tyrosine hydroxylase in human and animal brain. J. Neurochem. **18:**1647–1658.

Meldrum, B., Anlezark, G., and Trimble, M. (1975). Drugs modifying dopaminergic activity and behavior, the EEG and epilepsy in *Papio papio*. Eur. J. Pharmacol. **32:**203–213.

Oppermann, J.A. (1984). Aspartame metabolism in animals. *In* Stegink, L.D., and Filer, L.J., Jr. (eds.), Aspartame: physiology and biochemistry. New York: Marcel Dekker, pp. 141–159.

Pinto, J.M.B., and Maher, T.J. (1987). Aspartame decreases the threshold for pentylenetetrazole-induced seizures in mice. Neuropharmacology (in press).

Ramzan, I.M., and Levy, G. (1985). Kinetics of drug action in disease states. XIV. Effect of infusion rate on pentylenetetrazol concentrations. J. Pharmacol. Exp. Ther. **234:**624–628.

Romano, M., Casacci, F., De Marchi, F., Esteve, A., Pacei, T., Lomuscio, G., Mennini, T., and Salmona, M. (1987a). Effect of aspartame and carbohydrate-rich diet on plasma and brain aminoacid levels and brain monoamines in rat. J. Nutr. (submitted for publication).

Romano, M., Casacci, F., Pacei, E., De Marchi, F., Esteve, A., and Salmona, M. (1987b). Effect of aspartame associated with a carbohydrate-rich meal on plasma levels of large neutral aminoacids in man. J. Nutr. (submitted for publication).

Stegink, L.D. (1984). Aspartame metabolism in humans: acute dosing studies. *In* Stegink, L.D., and Filer, L.J., Jr. (eds.), Aspartame: physiology and biochemistry. New York: Marcel Dekker, pp. 509–553.

Tephly, T.R., and McMartin, K.E. (1984). Methanol metabolism and toxicity. *In* Stegink, L.D., and Filer, L.J., Jr. (eds.), Aspartame: physiology and biochemistry. New York: Marcel Dekker, pp. 111–140.

Visek, W.J. (1984). Chronic ingestion of aspartame in humans. *In* Stegink, L.D., and Filer, L.J., Jr. (eds.), Aspartame: physiology and biochemistry. New York: Marcel Dekker, pp. 495–508.

Wolfensberger, M., Amsler, U., Cuénod, M., Foster, A.C., Whetsell, W.O., Jr., and Schwarcz, R. (1983). Identification of quinolinic acid in rat and human brain tissue. Neurosci. Lett. **41:**247–252.

Woodbury, D.M. (1986). Experimentally-induced seizures in mice as genetic models of epilepsy for evaluation of the effects of aspartame and its metabolites. *In* International aspartame workshop proceedings, Nov. 17–21, 1986, Marbella, Spain, Aspartame Technical Committee, International Life Sciences Institute-Nutrition Foundation, Washington, D.C., Session V.

Wurtman, R.J. (1983). Neurochemical changes following high-dose aspartame with dietary carbohydrates. N. Engl. J. Med. **309:**429–430.

Wurtman, R.J., and Maher, T.J. (1987). Effects of aspartame on the brain. Prev. Med. (in press).

Yokogoshi, H., Roberts, C.H., Caballero, B., and Wurtman, R.J. (1984). Effects of aspartame and glucose administration on brain and plasma levels of large neutral amino acids and brain 5-hydroxyindoles. Am. J. Clin. Nutr. **40:**1–7.

Yokogoshi, H., and Wurtman, R.J. (1986). Acute effects of oral or parenteral aspartame on catecholamine metabolism in various regions of rat brain. J. Nutr. **116:**356–364.

CHAPTER 16

General Discussion: Calculation of the Aspartame Dose for Rodents that Produces Neurochemical Effects Comparable to Those Occurring in People

Richard J. Wurtman* and Timothy J. Maher†

Doses of aspartame or phenylalanine that raise brain phenylalanine levels more than those of tyrosine diminish the release of brain (striatal) dopamine and reportedly enhance the test rodent's susceptibility to seizure production (by pentylenetetrazole, fluorothyl, or electroconvulsive shock). Exogenous tyrosine has opposite effects on dopamine release and, reportedly, on sensitivity to pentylenetetrazole. This paper attempts to identify the quantitative relationships between the doses needed to elevate plasma phenylalanine, relative to tyrosine, by a certain amount in people and the doses needed to produce a comparable effect in experimental rodents.

All aspartame doses that have been tested in *humans* selectively elevate plasma *phenylalanine* relative to tyrosine; however, in *rodents*, aspartame doses of *500 mg/kg, or more,* are required to raise blood phenylalanine by proportionately as much as, or more than, tyrosine. (This reflects the very high activity of hepatic phenylalanine hydroxylase in rodents.) Analysis of published data on plasma amino acid levels after various aspartame doses suggests that the rat or mouse must receive about 60 times as much aspartame (mg/kg) as the human to obtain comparable changes in its plasma phenylalanine and tyrosine.

Consumption of the artificial sweetener aspartame is anecdotally associated with various neurological and behavioral side effects (Bradstock et al. 1986, Department of Health and Human Services 1986, John 1986, Drake 1986, Wurtman 1985). Attempts to develop animal models for investigating these effects are complicated by major differences in the rates at which human and rodent livers convert the phenylalanine in aspartame to tyrosine. Phenylalanine hydroxylase activity is much greater in rodents than in humans; thus, customary aspartame doses for rodents (e.g., 100 to 300 mg/kg) elevate plasma tyrosine levels more than those of phenylalanine (Wurtman, 1983). Hence, most of the reported effects of aspartame in rodents have been mediated by *tyrosine* [e.g., increased brain levels of

*Department of Brain and Cognitive Sciences, Massachusetts Institute of Technology, Cambridge, MA 02139, USA.
†Massachusetts College of Pharmacy, Boston, MA 02115, USA.

catecholamines and their metabolites (Yokogoshi and Wurtman 1986); reduced blood pressure in spontaneously hypertensive rats (Kiritsy and Maher 1986)], and not—as would occur in people—by *phenylalanine*.

We have attempted to identify aspartame doses which, by surpassing the rodent liver's ability to hydroxylate phenylalanine, can transiently elevate blood and brain phenylalanine by more than tyrosine. Use of such doses would allow rodents to serve as appropriate models for the human's neurochemical responses to the sweetener. The assumption that aspartame's neurological effects in humans are mediated by phenylalanine, not tyrosine, is based on phenylalanine's known neurotoxic activity (Swaiman 1969) and its ability to inhibit catecholamine synthesis and release (Milner et al. 1986). Tyrosine protects against the latter effects (Milner and Wurtman 1985).

Phenylalanine and tyrosine compete with each other and with other large neutral amino acids for facilitated diffusion across the blood-brain barrier; however, the affinity of phenylalanine for the transport carrier ($K_m = 32\ \mu M$) is almost threefold that of tyrosine ($K_m = 86\ \mu M$) (Pardridge 1986). After consumption of a given aspartame dose, the relative fluxes of phenylalanine and tyrosine from blood to brain will depend on these affinity differences, and on the relative increments in their plasma concentrations [i.e., ($Phe_t/Phe_0/Tyr_t/Tyr_0$)]. Using published data (Caballero et al. 1986, Stegink et al. 1979a,b, 1980, Garratini 1986, Stegink 1984, Yokogoshi et al. 1984) on plasma amino acid levels, we have compared the changes that occur in this ratio after humans and rodents consume various aspartame doses.

In humans, all reported doses increased this ratio, reflecting the slow hydroxylation of phenylalanine. In rodents, doses of 500 mg/kg or less raised plasma phenylalanine by *less* than tyrosine, and 1000 mg/kg was needed to increase the phenylalanine by more than tyrosine (Figure 16.1). Beyond that dose, the curve relating the aspartame dose to the increment in plasma phenylalanine relative to tyrosine paralleled the curve for humans. However, about 60 times as much aspartame, per kg, was needed to obtain comparable elevations in the ratio. (It should be noted that, because of the above affinity differences, the increase in plasma phenylalanine probably need not be as great as that in tyrosine to elevate brain phenylalanine by more than tyrosine.)

In studies described in Chapter 8 of this volume, we have found similar dosage relationships for the effects of phenylalanine on the release of dopamine within the rat's brain (the corpus striatum): A dose of 200 mg/kg, given intraperitoneally, *increased* dopamine release, as did tyrosine (100 to 250 mg/kg); a dose of 500 mg/kg neither increased nor decreased dopamine release; while a dose of 1000 mg/kg markedly *inhibited* dopamine release. (The precise effects of parenteral phenylalanine on brain phenylalanine and tyrosine levels probably differ slightly from those of an equimolar oral aspartame dose. In future studies, it will be interesting to compare the

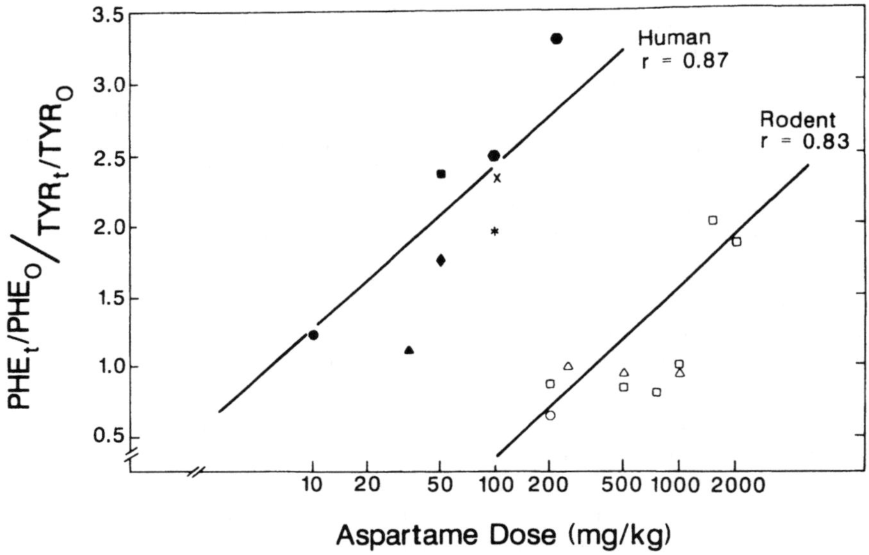

FIGURE 16.1. Effects of oral aspartame doses on plasma phenylalanine (Phe) and tyrosine (Tyr) levels in humans and rodents. Blood samples were obtained one hour after ingestion of the sweetener by humans, or its administration by gavage to rats or mice. Data were taken from the following publications: ● Caballero et al. 1986, ▲ Stegink et al. 1979a, ◆ Garattini 1986, ■ Stegink et al. 1979b, * Stegink et al. 1980, × ● Stegink et al. 1984, ○ Yokogoshi et al. 1984, □ Pinto and Maher 1988, △ Garattini 1986.

effects of intraperitoneal phenylalanine and oral aspartame on phenylalanine and tyrosine levels in the brain's extracellular fluid.) As discussed above, it appears that *any* aspartame or phenylalanine dose given to people raises brain phenylalanine by more than tyrosine, and may therefore have the same inhibitory effects on dopamine release as the 1000-mg/kg phenylalanine dose used in these studies.

When mice receive 1000 mg/kg or more of aspartame, or an equivalent dose of phenylalanine, seizure induction by pentylenetetrazole (PTZ) (Pinto and Maher 1988) or fluorothyl (Maher and Pinto 1987) is potentiated: for example, PTZ doses (55, 60, or 66 mg/kg) which, given alone, induce seizures in 10%, 25%, or 52% of test animals will, after aspartame (1000 mg/kg), induce seizures in 40%, 56%, or 79% (Pinto and Maher 1988). Tyrosine, in contrast, increases the seizure threshold of PTZ-treated mice. These effects are compatible with the known direct relationship between brain catecholamines and seizure thresholds (Jobe et al. 1986): pharmacologic depletion of dopamine and norepinephrine tends to increase seizure susceptibility, while drugs (like monoamine oxidase inhibitors) that increase brain monoamines are protective against seizures. Presumably, the aspartame dose affects seizure susceptibility by decreasing the release of brain monoamines, while the tyrosine has opposite effects. If the dose of aspartame, per kg, required to cause an increase in blood phenylalanine relative to tyrosine in rodents is 60 times the dose needed in humans, our data suggest that as little as 15 to 20 mg of aspartame per kg may diminish catecholamine release in people.

Acknowledgment. These studies were supported by a grant (NS-21231) from the National Institutes of Health.

References

Bradstock, M.K., Serdula, M.K., Marks, J.S., Barnard, R.J., Crane, N.T., Remmington, P.L., and Trowbridge, F.L. (1986). Evaluation of reactions to food additives: the aspartame experience. Am. J. Clin. Nutr. **43**:464–469.

Caballero, B., Mahon, B., Rohr, F., and Wurtman, R.J. (1986). Plasma amino acid levels after single dose aspartame consumption in phenylketonuria. J. Pediatr. **109**: 669–671.

Department of Health and Human Services. Quarterly Report on Adverse Reactions Associated with Aspartame Ingestion. October 1, 1986.

Drake, M.E. (1986). Panic attacks and excessive aspartame ingestion. Lancet **ii**:631.

Garattini, S. (1986). Aspartame, brain monoamines and neurochemical mediators. International Aspartame Workshop. Nov. 17–21, 1986, Marbella, Spain.

Jobe, P.C., Dailey, J.W., and Reigel, C.E. (1986). Noradrenergic and serotonergic determinants of seizure susceptibility and severity in genetically epilepsy-prone rats. Life Sci. **39**:775–782.

John, D.R. (1986). Migraine provoked by aspartame. N. Engl. J. Med. **315**:456.

Kiritsy, P.J., and Maher, T.J. (1986). Acute effects of aspartame on systolic blood pressure in spontaneously hypertensive rats. J. Neural Transm. **66**:121–128.

Maher, T.J., and Pinto, J.M.B. (1987). Aspartame administration potentiates fluorothyl-induced seizures in mice. J. Neurochem. **48**:S52.

Milner, J.D., and Wurtman, R.J. (1985). Tyrosine availability determines stimulus-evoked dopamine release from rat striatal slices. Neurosci. Lett. **59**:215–220.

Milner, J.D., Irie, K., and Wurtman, R.J. (1986). Effects of phenylalanine on the release of endogenous dopamine from rat striatal slices. J. Neurochem. **47**:1444–1448.

Pardridge, W.M. (1986). Potential effects of the dipeptide sweetener aspartame on the brain. *In* Wurtman, R.J., and Wurtman, J.J. (eds.), Nutrition and the brain. New York: Raven Press, Vol. 7, pp. 199–241.

Pinto, J.M.B., and Maher, T.J. (1988). Aspartame administration potentiates fluorothyl- and pentylenetetrazole-induced seizure in mice. Neuropharmacol. **27**:51–55.

Stegink, L.D. (1984). Aspartame metabolism in humans: acute dosing studies. *In* Stegink, L.D., and Filer, L.J., (eds.), Aspartame: physiology and biochemistry. New York: Marcel Dekker, pp. 509–553.

Stegink, L.D., Filer, L.J., Jr., Baker, G.L., and McDonnell, J.E. (1979a). Effect of aspartame loading upon plasma and erythrocyte amino acid levels in phenylketonuric heterozygotes and normal adult subjects. J. Nutr. **109**:708–717.
Stegink, L.D., Filer, L.J., Jr., and Baker, G.L. (1979b). Plasma, erythrocyte and human milk levels of free amino acids in lactating women administered aspartame or lactose. J. Nutr. **109**:2173–2181.
Stegink, L.D., Filer, L.J., Jr., and Baker, G.L., and McDonnell, J.E. (1980). Effect of an abuse dose of aspartame upon plasma and erythrocyte levels of amino acids in phenylketonuric heterozygous and normal adults. J. Nutr. **110**:2216–2224.
Swaiman, K.F. (1969). The effect of phenylalanine and its metabolites on glucose utilization in developing brain. J. Neurochem. **16**:385–388.
Wurtman, R.J. (1983). Neurochemical changes following high-dose aspartame with dietary carbohydrates. N. Engl. J. Med. **309**:429–430.
Wurtman, R.J. (1985). Aspartame: possible effect on seizure susceptibility. Lancet **ii**:1060.
Yokogoshi, H., Roberts, C.H., Caballero, B., and Wurtman, R.J. (1984). Effects of aspartame and glucose administration on brain and plasma levels of large neutral amino acids and brain 5-hydroxyindoles. Am. J. Clin. Nutr. **40**:1–7.
Yokogoshi, H., and Wurtman, R.J. (1986). Acute effects of oral or parenteral aspartame on catecholamine metabolism in various regions of rat brain. J. Nutr. **116**:356–364.

Part IV Behavioral and Electroencephalographic Effects of Aspartame and Phenylalanine in Humans: Possible Involvement in Seizure Thresholds

CHAPTER 17

Monoamines and Seizures in Humans

Donald L. Schomer*

Seizures in humans have many different etiologies. Only when seizures recur is a clinical diagnosis of epilepsy made. The monoaminergic systems of the brain appear to have both diffuse projections and specific projections. Attempts to manipulate the various components, i.e., the norepinephrine, the dopamine, and the serotonin-dependent systems, have resulted in increased or decreased seizure susceptibility in different animal models of epilepsy. Humans appear to have variable responses that may or may not affect their relative susceptibility to seizure.

Introduction

Epilepsy is a clinical diagnosis that, by its very nature, implies susceptibility to recurrent seizures. It affects approximately 0.5% of the general population, i.e., 1.2 million people in the United States. About 8% of the general population will suffer a clinical seizure at some point during their lives. A seizure is defined as an episode of pathological hypersynchronous behavior of large numbers of neurons temporally associated with a change in behavior. The behavioral change or clinical manifestation of the seizure varies considerably.

Types of Human Epilepsy

The current scheme of naming clinical seizures was accepted in 1981 (Anonymous 1981). The epilepsies were divided into two major subdivisions, primary generalized and partial. Primary generalized seizures, in turn, are divided into tonic, clonic, tonic-clonic, myoclonic, atonic, and absence. These types of seizures account for about one-third of the cases of recurrent seizures. These seizures tend to have clinical onset earlier in life and, when diagnosed properly, appear to respond at a higher rate to currently available medications than the partial seizures. There is often a con-

*Department of Neurology, Harvard University, Beth Israel Hospital and Children's Hospital, Boston, MA 02215, USA.

comitant history of other family members affected with similar behaviors, suggesting that in some cases there is a genetic component to the disorder. The patients affected usually have normal physical and neurological examinations and neurocognitive profiles when tested during an asymptomatic period. Some individuals also appear to be sensitive to certain manipulations, most notably intermittent photic stimulation or hyperventilation. In addition, there are exceptions to the above generalizations, and these types of seizures on occasion are associated with significant metabolic or degenerative diseases that effect the central nervous system. The primary generalized seizures probably have disturbances of regulation between neocortical and subcortical regions such as the anterior thalamus, interlaminar nuclei, selective basal ganglia nuclei (Gloor 1979), and mammillary bodies (Mirski and Ferrendelli 1986). The disorder therefore seems to affect both hemispheres in a synchronous fashion when the patient becomes symptomatic.

Partial seizures, on the other hand, are episodes of abnormal hypersynchrony that occur in focally discrete regions of the neocortex. When this occurs, a variety of signs and symptoms may result. This form of epilepsy accounts for between one-half and two-thirds of the cases of recurrent seizures. It tends to have onset later in life than primary generalized seizures and usually is less responsive to medication. Perhaps one-third to one-half of patients with this disorder describe themselves as either poorly controlled or uncontrolled (Rodin 1968). Because there is an area of focal onset to these disorders, there is often an abnormal neurological or cognitive examination (Schomer 1983). This form of epilepsy may follow insults to the brain such as strokes, head injuries, or infections. It may also be associated with progressive lesions like tumors or static lesions like vascular malformations or hamartomas. Like the primary epilepsies, these seizures also may be made more frequent by sleep deprivation or concomitant systemic illnesses.

The clinical manifestations of partial epilepsy depend upon the area of onset and how the electrographic seizure spreads to other structures in the central nervous system. Partial seizures may have sensory, motoric, autonomic, hallucinatory, experiential, or emotional components. If consciousness is spared during the seizure, it is referred to as a simple partial seizure. If consciousness is impaired, it is referred to as a complex partial event. If the patient has a partial seizure that goes on to have secondary major motor involvement, it is referred to as a secondarily generalized seizure.

Animal Models of Human Epilepsy

A variety of animal models exist that more or less represent one of the types of seizures or manifestations of seizures. Perhaps the oldest test model is the maximal electrical shock induced seizure. Sudden alterations of electrolyte balance such as hyponatremia, hypomagnesemia, hyperkalemia, and hypercalcemia may produce seizures. Chronic sleep deprivation and hyperthermia may induce seizures in animal models. The systemic administrations of agents like bicucculine, penicillin, and pentylenetetrazol all reproducibly induce seizures. All of these manipulations tend to produce generalized seizures in the animal models and may serve as more or less reasonable models to study the primary or secondary generalized events in humans. In addition, several genetically based disorders exist in the animal world such as those of the photosensitive baboon, *Papio papio*, the tottering mouse (Noebels 1984), and the ketamine sensitive fruit fly. The intermittent administration of subthreshold electrical or chemical stimulations that lead to recurrent and even spontaneous seizures is referred to as the kindling model (McNamara 1984). It most resembles partial seizures. The application of focal agents like penicillin, cobalt, iron salts, and estrogen into areas of the neocortex also produce focal epileptic disorders and these, in turn, have been extensively studied.

The reasons these models have been studied are many. Most often they are used as screening models to determine whether agents may have anticonvulsant properties. Even so, some successful anticonvulsants have defied predictions based upon these models. These models are also used to help in our understanding of the physiology of the human forms of epilepsy. It has been shown that these models represent multiple aberrations of normal physiology, and it is expected that the human state probably also has multiple different physiologies associated with it. On the other hand, these models may not be predictive of the tendency to the development of seizures by humans. Therefore, showing that certain substances may reduce the degree of protection against seizures in certain animal models is not to be equated with a similar response on the part of the human to similar manipulations. If such relationships in the animal models, however, can be shown, it is certainly cause for reflection. It is entirely possible that the human may be more or less susceptible to the same types of manipulations. As predicted by the animal models, humans also seem to have different inherited susceptibilities to seizure generation, and so there may be considerable variation in susceptibility to exogenous manipulations.

What we have learned about the underlying pathophysiology of the epilepsies from both human and animal studies is that several different factors seem to determine the overall likelihood for seizure development (Prince 1987). These are the intrinsic membrane properties of some cells which lead to burst discharges and to pacemakerlike activities; the effectiveness of inhibitory events; and the potency of excitatory coupling among neurons, whether through synaptic or nonsynaptic mechanisms. If monoamine manipulation does affect relative epileptogenic potential, it will probably do so based upon the second mechanism of altered effectiveness of inhibitory events.

Monoamines and Human Epilepsy

What is known about the three monoaminergic systems in human epilepsy is by all accounts very little and, in many cases, anecdotal (Ferrendelli 1984). Norepinephrine-containing cells are found in the locus coeruleus and project diffusely to the neocortex, to the specific thalamic nuclei, certain hypothalamic regions, to the cerebellar cortex, and to specific spinal regions (Cooper et al. 1982a). Their actions are felt to be primarily inhibitory (Cooper et al. 1982a). Mixed responses to the blockade or destruction of norepinephrine innervation in the animal models probably reflect the fact that the models may differ significantly from each other and/or that selective loss of norepinephrine projection may have a differential effect. It is entirely possible that the loss of certain projections may be equated with a loss of inhibition, while loss of other norepinephrine projection pathways may be associated with a loss of inhibition in an inhibitory system, i.e., disinhibition. Those drugs that have anticonvulsant properties in humans and also affect norepinephrine systems tend to block norepinephrine reuptake at the presynaptic site (Maynert et al. 1975). The principal drug here is carbamazepine. In addition, some of the tricyclic drugs which have similar effects on norepinephrine also tend to have some anticonvulsant properties (Ojemann et al. 1983, Fromm et al. 1972, 1978). Drugs that deplete the brain of norepinephrine, i.e., 6-hydroxydopamine and reserpine (Gross and Ferrendelli 1979 & 1982), seem to reduce the number of trials necessary to kindle animals (Scatton et al. 1981). If kindling exists in human epilepsy, then the use of such drugs may promote the development of seizures but may not necessarily induce them. There is suggestive evidence from animal studies (Scatton et al. 1981) and human studies (Briere et al. 1986) that there is a decrease in norepinephrine concentration in focal epileptogenic regions as well as a decrease in the number of norepinephrine receptors locally and in homotopic contralateral neocortex. In addition, since the locus coeruleus system plays a significant role in the general level of neuronal excitability, manipulations of that system in turn may be associated with significant changes in sleep/wake cycle. Such appears to be the case with the drug reserpine, which is known to reduce norepinephrine concentration in presynaptic vesicles (Gross and Ferrendelli 1979). This drug is definitely associated with excessive drowsiness and recurrent nightmares. Certain manipulations of the sleep/wake cycle may also affect relative degree of seizure susceptibility. Therefore, a possible effect of norepinephrine depletion may be increased epileptogenic potential through a disturbance of the sleep/wake cycle, not by any direct effect on neuronal excitability.

Dopamine, based in the nigrostriatal system, the mesolimbic system, and the tuberoinfundibular system, appears to affect selective areas as well as to have effects indirectly in a diffuse fashion (Cooper et al. 1982a). The photosensitive baboon, *Papio papio*, was felt to have abnormal dopaminergic systems that, when treated with a dopamine agonist like apomorphine or bromocryptine, were associated with marked reduction of the photosensitive response (Anlezark et al. 1981). It is not so clear that the effect of the dopamine agonist was central (Anlezark et al. 1981). In the human primary generalized photosensitive seizure disorders and in selective neurodegenerative diseases associated with lipofuscin storage, the photoinduced seizures also seem to respond to the administration of similar dopamine agonists (Quesney 1980, 1981, Quesney et al. 1981). Other electrical and convulsive events not related to photic stimulation did not seem to respond. Therefore, it is possible that there may be relative dopamine depletion associated with symptoms. It is curious that in the other syndrome associated with dopamine depletion, i.e., Parkinson's disease, there appears to be significantly lower risk for the development of seizures. Drugs that block central dopamine receptors, such as the phenothiazines, seem to be associated with slightly greater risk to the development of seizures and to the development of recurrent seizures in those people known to suffer from epilepsy. Like the sleep/wake cycle aberrations described earlier, it is difficult to ascribe a causal relationship to the administration of such dopamine antagonists and the occurrence of seizures, since the number of variables in that patient population is so great.

Serotonin, located in the median raphe system, also appears to have relatively diffuse projections centrally (Cooper et al. 1982b). As discussed in this volume, serotonin may be decreased centrally in humans and animal models by the selective competitive inhibition of uptake of its precursor, 5-hydroxytryptophan, during periods of increased peripheral phenylalanine concentration. In human models, it has been known for some time that drugs which increase central serotonin such as clonazepam or 5-hydroxytryptophan seem to be associated with a reduction of postanoxic myoclonus and certain other forms of myoclonic epilepsy (Van Woert and Huang 1981). The current speculation on that association is that the system most responsible is the projection system from the raphe to the inferior olivary nucleus (Van Woert and Huang 1981). The myoclonic movements seen with a variety of metabolic disorders may have similar origin to anoxic myoclonus and usually are equally well treated with 5-hydroxytryptophan and clonazepam. They respond best to elimination of the primary offending metabolic disorder. It has been postulated that several of the anticonvulsants also increase 5-HIAA, the metabolic end product of serotonin, and that excessive central serotonin, in turn, may be associated with ataxia and dysarthria (Chadwick et al. 1977). The theory concludes that by blocking central absorption of 5-hydroxytryptophan, one may block the side effects of some of the higher doses of anticonvulsants. More recent studies seem to cast doubt on that clinical theorem (Chadwick 1981).

Conclusion

There is no one animal model that accurately represents all aspects of human epilepsy. There will almost assuredly never be such a model, because human epilepsy is not any one disorder. Sudden changes in electrolyte balance, fluid load, and hormonal balance are all associated with changes that in turn may produce clinical seizures or be associated with increased risk for seizures. The genetic background is also important, and there will probably be many different gene loci found associated with the variable clinical presentations of epilepsy. Structural diseases like injuries, tumors, and scars may lead to the development of areas of intermittent hypersynchronous neuronal behavior that are associated with symptoms. Because the human condition of epilepsy has so many variables, it is not surprising that there are conflicting reports on altering seizure susceptibility with various manipulations of the monoaminergic systems. It is conceivable that lowering central norepinephrine, dopamine, or serotonin may produce increased risk for certain convulsive states in certain subpopulations, while having little or no effect on other subpopulations. The group that comes to mind as perhaps the most interesting to study is that of carriers for phenylketonuria. Another group of major interest would be one with certain types of reflex-dependent seizure disorders, included mostly in the primary generalized category.

If altering central monoamines is associated with increased risk to the development of seizures, it may be for less than obvious reasons. Such possibilities include alterations of the norepinephrine-dependent sleep/wake cycle, alterations of dopamine-dependent hormonal release, diffuse effects of serotonin on cerebellar control, and relative disinhibition. It also does not seem logical that a sudden single significant alteration in the peripheral or central monoamine concentration will be associated with a significant risk for seizures or altered cognitive function. From the metabolic perspective, it would appear more likely that if change occurs, it will be after chronic and prolonged exposure to a monoamine-depleting agent that in turn could be associated with definable changes in plasma or cerebrospinal fluid concentrations of precursors or metabolic by-products.

References

Anlezark, G., Marrosu, F., and Meldrum, B. (1981). Dopamine agonists in reflex epilepsy. *In* Morselli, P.L., Lloyd, K.G., Loscher, W., Meldrum, B., and Reynolds, E.H. (eds.), Neurotransmitters, seizures and epilepsy. New York: Raven Press, pp. 251–262.

Anonymous (1981). Proposal for revised clinical and electroencephalographic classification of epileptic seizures. Epilepsia **22**:489–501.

Briere, R., Sherwin, A., Robitaille, Y., Olivier, A., Quesney, L., and Reader, T. (1986). Alpha-1 adrenoceptors are decreased in human epileptic foci. Ann. Neurol. **19**:26–30.

Chadwick, D. (1981). CSF monoamine metabolites in epileptic patients. *In* Morselli, P.L., Lloyd, K.G., Loscher, W., Meldrum, B., and Reynolds, E.H. (eds.), Neurotransmitters, seizures and epilepsy. New York: Raven Press, pp. 293–300.

Chadwick, D., Jenner, P., and Reynolds, E. (1977). Serotonin metabolism in human epilepsy, the influence of anticonvulsant drugs. Ann. Neurol. **1**:218–224.

Cooper, J.R., Bloom, F., and Roth, R. (1982a). Catecholamine II, CNS aspects, Chapter VII. New York: Oxford Press, pp. 173–221.

Cooper, J.R., Bloom, F., and Roth, R. (1982b). Serotonin, Chapter VIII. New York: Oxford Press, pp. 113–148.

Ferrendelli, J. (1984). Roles of biogenic amines and cyclic nucleotides in seizure mechanisms. Ann. Neurol. **16**:S98–S103.

Fromm, G.H., Amores, C., and Thies, W. (1972). Imipramine in epilepsy. Arch. Neurol. **27**:198–204.

Fromm, G.H., Wessel, H., Glass, J., Alvin, J., and Van Horn, G. (1978). Imipramine in absence and myoclonic-astatic seizures. Neurology **28**:953–957.

Gloor, P. (1979). Generalized epilepsy with spike and wave discharge: a reinterpretation of its electrographic and clinical manifestation. Epilepsia **20**:571–586.

Gross, R.A., and Ferrendelli, J. (1979). Effects of reserpine, propranolol and aminophylline on seizure activity and CNS cyclic nucleotides. Ann. Neurol. **6**:296–301.

Gross, R.A., and Ferrendelli, J. (1982). Relationship between norepinephrine and cyclic nucleotides in brain seizure activity. Neuropharmacology **21**:655–661.

Maynert, E.W., Marczynski, T., and Browning, R. (1975). The role of the neurotransmitters in the epilepsies. Adv. Neurol. **13**:79–147.

McNamara, J.O. (1984). Kindling: an animal model of complex partial epilepsy. Ann. Neurol. **16**:S72–S76.

Mirski, M., and Ferrendelli, J. (1986). Selective metabolic activation of the mammillary bodies and their connections during ethosuximide induced suppression of pentylenetetrazol seizures. Epilepsia **27**:194–203.

Noebels, J. (1984). Isolating single genes of the inherited epilepsies. Ann. Neurol. **16**:S18–S21.

Ojemann, L.M., Friel, P., Trejo, W., and Dudley, D. (1983). Effects of doxepin on seizure frequency in depressed epileptic patients. Neurology **33**:646–648.

Prince, D. (1987). An overview of critical issues in epilepsy. *In* Cotman, C. (ed.), Neural plasticity vs. disease: Focus on the NMDA receptor in epilepsy and mental disorders. New York: Institute for Child Development Research, pp. 1–12.

Quesney, L.F. (1981). Dopamine and generalized photosensitive epilepsy. *In* Morselli, P.L., Lloyd, K.G., Loscher, W., Meldrum, B., and Reynolds, E.H. (eds.), Neurotransmitters, seizures and epilepsy. New York: Raven Press, pp. 263–274.

Quesney, L.F., Andermann, F., Lal, S., and Prelevic, S. (1980). Transient abolition of generalized photosensitive epileptic discharge in humans by apomorphine, a dopamine-receptor agonist. Neurology **30**:1169–1174.

Quesney, L.F., Andermann, F., and Gloor, P. (1981). Dopaminergic mechanism in generalized photosensitive epilepsy. Neurology **31**:1542–1544.

Rodin, E.A. (1968). The prognosis of patients with epilepsy. Springfield, Illinois: Charles C. Thomas.

Scatton, B., Dedek, J., Zivkovic, B., Liegeois, C., Trottier, S., Chauvel, P., and Bancaud, J. (1981). Alterations of catecholamine, serotonin, acetylcholine and

cyclic nucleotides in brain regions after cobalt-induced epilepsy in the rat: influences of anticonvulsant treatment. *In* Morselli, P.L., Lloyd, K.G., Loscher, W., Meldrum, B., and Reynolds, E.H. (eds.), Neurotransmitters, seizures and epilepsy. New York: Raven Press, pp. 215–226.
Schomer, D.L. (1983). Partial epilepsy. N. Engl. J. Med. **309**:536–539.
Van Woert, M.H., and Huang, E.C. (1981). Role of brain serotonin in myoclonus. *In* Morselli, P.L., Lloyd, K.G., Loscher, W., Meldrum, B., and Reynolds, E.H. (eds.), Neurotransmitters, seizures and epilepsy. New York: Raven Press, pp. 239–249.

CHAPTER 18

The Possible Role of Aspartame in Seizure Induction

Ralph G. Walton*

Although aspartame is widely used, and generally considered safe, there is evidence that it has an effect on monoamine metabolism which could theoretically lower seizure threshold. Eight cases are presented in whom it would appear likely that seizures were induced by ingestion of this artificial sweetener. Clinicians are urged to take into account the possible role of aspartame when evaluating patients with seizures, or with disorders in which alterations in catecholamine and indoleamine metabolism have been implicated.

The artificial sweetener aspartame is enjoying ever-increasing markets, and its safety is attested to by both the FDA and numerous clinical studies (Council on Scientific Affairs 1985, Bradstock et al. 1986, Horwitz and Bauer-Nehrling 1983). Currently available data regarding the neurochemical impact of this product, however, would certainly suggest the possibility of adverse clinical effects associated with its use. Wurtman (1983) has demonstrated that aspartame can significantly increase rat brain phenylalanine levels, and aspartame-carbohydrate combinations can raise brain tyrosine levels and suppress the usual increase in tryptophan which follows a carbohydrate-rich meal. An increase in norepinephrine precursors, coupled with a simultaneous decrease in serotonin precursors, could potentially have a significant impact on central nervous system catecholamine-indoleamine balance. The following cases are presented as possible instances of such impact secondary to the use of aspartame.

Case 1

A 91-year-old woman was in excellent health until she began drinking "several glasses a day" of aspartame-containing beverages. After several days, she reported malaise and weakness. Ten days after starting aspartame consumption, she experienced her first grand mal seizure. After re-

*Commissioner, Chautauqua County Mental Health, Hall R. Clothier Building, Mayville, NY 14757, USA.

covery from the post-ictal period, she went to the refrigerator for "another glass of Diet 7-Up to make me feel better." The following day, she experienced her second grand mal seizure. She refused any extensive workup or evaluation of her seizures. On her own, she concluded that her seizures must have been aspartame related, discontinued all use of this product, and has had no further seizures.

Case 2

A 27-year-old woman with a ten-year history of a seizure disorder was seizure-free for nine years, controlled on Dilantin (300 mg per day). She consumed aspartame for the first time when for three days in a row she had "Knox Blocks Sugar-Free Jell-O." On the third day of aspartame consumption, she experienced a grand mal seizure. During the past ten months, she has consumed no more aspartame and has again been seizure-free on the same medication regimen.

Case 3

A 35-year-old woman drank a liter of iced tea per day. Whenever she had iced tea sweetened with aspartame, she would experience epigastric distress. On a particularly hot day, she consumed somewhat more iced tea than usual (between 1 and 2 liters). Unbeknownst to the patient, this was an aspartame-containing iced tea product. She experienced epigastric pain, and, approximately an hour after the onset of pain, a grand mal seizure. Her family physician started her on Dilantin and phenobarbital, but after she convinced him that the seizure could have been aspartame related, he discontinued both anticonvulsants. Over the past year, she has avoided aspartame and has not had further seizures.

Case 4

A 61-year-old woman had been in excellent health until she began consuming an average of half a gallon per day of sugar-free beverages prepared with "Crystal Light" mixes. She experienced the onset of headaches, in the absence of a previous headache history. After three months of daily headaches, she experienced a generalized seizure and was hospitalized. CAT scan and EEG were normal. After discontinuing the use of all aspartame-containing products, she has been headache- and seizure-free.

Case 5

A 65-year-old man experienced his first grand mal seizure while reading in bed. Preceding the seizure, there had been a six-month history of episodic involuntary smacking of the lips, chewing movements, and twitching of the right thumb. The patient calculated that he had been consuming an average of 210 mg of aspartame per day in the form of "Crystal Light" iced tea mix. After discontinuing all aspartame-containing products, there have been no further involuntary movements or seizures.

Case 6

A 19-year-old woman complained of malaise and headaches for three days. She developed garbled speech, poor coordination, lost consciousness, and experienced a grand mal seizure. Her mother found her in a post-ictal state. A CAT scan revealed a venous malformation in the corpus callosum. It was felt that she did not experience any intracranial bleeding. Of possible significance is the fact that for the month prior to her seizures she had been consuming three to four cans per day of aspartame-containing "Diet Coke." The patient has been treated with Dilantin (400 mg per day) and has avoided all use of aspartame. Over the past four months, she has had no further seizure activity or neurological symptoms.

Case 7

A 24-year-old woman experienced four grand mal seizures during the third to the sixth months of pregnancy. Of possible relevance is the fact that for the two-year period prior to her seizures, the patient had been drinking approximately two quarts of aspartame-containing "Diet Pepsi" per day. She discontinued all aspartame-containing products and remained seizure-free for her final trimester.

Case 8

A 41-year-old woman experienced frequent headaches and three grand mal seizures during the last two of her five pregnancies. EEG and CAT scan were interpreted as "normal." During the last two pregnancies, but not the first three, she had been consuming approximately three quarts of aspartame-containing "Diet Pepsi" per day. There has been no recurrence of headache or seizure since she discontinued all aspartame-containing products.

Discussion

The above cases lend further suggestive evidence to the argument that aspartame may lower seizure threshold and in certain vulnerable individuals trigger seizures (Wurtman 1985, Walton 1986). Two of the above patients were pregnant, one had a known seizure disorder (but was well controlled for nine years prior to aspartame use), one was very advanced in age, and one had an intracranial venous malformation; these individuals may well have been at risk and with, perhaps, the burden of a threshold lowering agent such as aspartame, developed overt clinical difficulties.

When one looks at reports of mania (Walton 1986), panic attacks (Wurtman 1983), and weight gain (Blundell and Hill 1986) attributed to aspartame, then a lowering of seizure threshold fits logically into an overall pattern of clinical effects secondary to an alteration in monoamine metabolism induced by this artificial sweetener. In view of the very large number of individuals consuming this product, the vast majority of whom apparently do so with impunity, the effect would seem to be a fairly subtle one. Nevertheless, for individuals at risk for seizures, or for those with affective disorder, panic disorder, or clinical syndromes in which monoamine metabolism is felt to be potentially disordered, the use of aspartame could perhaps be hazardous. Appropriate double-blind studies, utilizing populations at risk, clearly need to be done. In the interim, clinicians are urged to take into account the possible impact of aspartame when evaluating patients with seizures or with disorders in which alterations in catecholamine and indoleamine metabolism have been implicated.

References

Blundell, J.E., and Hill, A.J. (1986). Paradoxical effects of an intense sweetener (aspartame) on appetite. Lancet **i:**1092–1093.

Bradstock, M.K., Serdula, M.K., Marks, J.S., Barnard, R.J., Crane, N.T., Remington, P.L., and Trowbridge, F. (1986). Evaluation of reactions to food additives: the aspartame experience. Am. J. Clin. Nutr. **43:**464–469.

Council on Scientific Affairs (1985). Aspartame: review of safety issues. J. Am. Med. Assoc. **254:**400–402.

Drake, M.E. (1986). Panic attacks and excessive aspartame ingestion. Lancet **ii:**631.

Horwitz, D.L., and Bauer-Nehrling, J.K. (1983). Can aspartame meet our expectations? J. Am. Diet Assoc. **83:**142–146.

Walton, R.G. (1986). Seizure and mania after high intake of aspartame. Psychosomatics **27:**218–220.

Wurtman, R.J. (1983). Neurochemical changes following high-dose aspartame with dietary carbohydrates. N. Engl. J. Med. **309:**429–430.

Wurtman, R.J. (1985): Aspartame: possible effects on seizure susceptibility. Lancet **ii:**1060.

CHAPTER 19

Effects of Aspartame on Seizures in Children

Bennett A. Shaywitz, Sheila M. Gillespie, and Sally E. Shaywitz*

This chapter describes the methods used in two double-blind, placebo-controlled, crossover studies examining the effects of large-dose aspartame ingestion on children with seizures. In addition, some of the descriptive statistics on the demographic data are presented. Eight children with well-documented seizure disorders have been studied on a one-month protocol. Two children have completed a one-week in-hospital protocol designed for children who have allegedly experienced a seizure within 24 hours of ingesting a product containing aspartame. To date (6 April 1987), there have been no documented clinical seizures in either study. However, the studies are ongoing and the codes have not been broken; thus, it is not possible to fully evaluate the effects of aspartame until these studies are completed.

Introduction

Aspartame is the first low-calorie sweetener approved by the Food and Drug Administration (FDA) in over 25 years. In 1981, it was approved for use in some dry products, and in 1983, for use in carbonated beverages. It is currently used in many products under the patented trade name NutraSweet. Since the time of the FDA approval, numerous complaints have arisen regarding the safety of aspartame as a food additive (Stegink and Filer 1984). In reviewing the data, it appears that complaints about the central nervous system are relatively more common than allegations regarding other organ systems. The most commonly reported symptoms which are related to the central nervous system include headache, mood alterations, insomnia, dizziness, fatigue, and seizures. The Center for Disease Control investigated these allegations and found no conclusive evidence that aspartame was responsible (Center for Disease Control, 1984). However, since many of the products which contain aspartame are consumed by children, and since many of the concerns raised pertain to the central nervous system, we believed it necessary to design and carry out a

* Yale University School of Medicine, New Haven, CT 06510, USA.

well-controlled study to assess the effects of aspartame administration on children with seizure disorders.

This paper will describe a randomized, double-blind, placebo-controlled, crossover study designed to investigate the effects of large doses of aspartame on the neurological and electrophysiological status of children who have already experienced well-documented seizures. A second study involves children who have had a documented seizure within 24 hours of ingestion of a product containing aspartame.

Experimental Design and Methods

Subject Selection

In the first study, children investigated are 4 to 15 years of age and of normal IQ (full-scale IQ 80 and above). All children have clinical evidence of a seizure disorder, and many are taking prescribed anticonvulsant medications including phenytoin, phenobarbital, carbamazepine, valproic acid, ethosuximide, and primidone. Children are selected from the Pediatric Neurology Clinic at Yale-New Haven Hospital and are referred from other pediatric neurologists in this area.

In the second study, subjects are referred from all parts of the country. They range in age from 4 to 17 years of age. Children are screened by a pediatric neurologist not affiliated with this institution. If they meet the criteria for the study, they are referred to Yale.

If subjects in either study protocol have been previously prescribed anticonvulsant medications, they must be on a stable regimen for 30 days prior to entry into the studies.

Experimental Protocol

Study Design #1

The study, which will include at least 30 subjects, is designed so that each child is tested on both conditions, i.e., aspartame (APM) and placebo, randomized so that each condition is given first an equal number of times within the experiment. APM (or placebo) is administered for a 2-week period. The compounds are administered by the parents at home in capsule form as a single morning dose; APM is given at a dose of 34 mg/kg. Capsules are coded so that the parents, children, and investigators are "blind" to the particular substance the child is receiving.

Prior to initiation into the study, the parents and children meet with study personnel at our Children's Clinical Research Center (CCRC) outpatient facility. Initially, the study is explained and an informed consent is obtained from the parents and the child (if over 10 years of age). Individualized intelligence (WISC-R) and achievement (Woodcock-Johnson) tests are administered, and parents complete the Yale Children's Inventory (YCI), a comprehensive, computer-indexed form encompassing demographic information, genetic background, pre- and perinatal events, developmental and social history, educational experiences, recent life stresses, and current areas of difficulty (Shaywitz et al. 1986). A complete physical exam including a neurological, neuromaturational, and structured clinical interview is performed. In addition, parents are instructed on how to keep two complete 3-day dietary records. Parents are also given a list of all the products which contain APM, so that an APM-free diet can be assured and maintained throughout the 4 weeks of the study. Parents are also given forms to be filled out for each week of the study to document any seizure activity or behavioral changes. Conners Behavior Rating Scales are used to assess behavior. Children are asked to fill out a Subjects Treatment Emergent Symptoms Scale (STESS) weekly to document any side effects. In addition, the children's teachers are sent copies of the "Multigrade Inventory for Teachers," to be completed by all teachers each week of the study. This is a precoded, computerized form from which scales of activity, attention, behavioral difficulties, and academic evaluation are derived.

On the day a child enters the study, his parents are given a bottle of capsules containing either APM or placebo. Parents are instructed to give the child the appropriate number of capsules every morning before school. The patient continues daily ingestion of this substance for an entire week.

During the second week of the study (on the eighth day of APM or placebo ingestion), the children are admitted to our Children's Clinical Research Center for in-hospital testing. Their diet is APM-free with a low catecholamine content. They were on no monoamine oxidase inhibitors. Complete dietary records are maintained. A standard breakfast is served for day 2 of hospitalization. A 24-hour EEG monitor is placed, which the child will wear during his hospital stay; 12-hour urine collections begin at 9:00 P.M. on the evening of day 1.

On the second day, an intravenous line is placed in the arm to facilitate the research blood studies. At 8:30 A.M. blood is drawn for an amino acid profile, determination of methanol and formate, glucose, CBC with differential, and platelets, liver function tests, and anticonvulsant levels (if applicable). Test article #1 is given at 9:00 A.M. At 10:00 A.M., blood studies include analysis for serotonin and monoamine metabolites (HVA and MHPG), as well as an amino acid profile and determination of methanol and formate. At 11:00 A.M. blood is drawn for amino acid profiles and determination of methanol and formate. In addition, urine is collected from 9:00 A.M. to 12 noon. Urine is analyzed for monoamines and metabolites (HVA, dopamine, MHPG, VMA, norepinephrine, normetanephrine, epinephrine, metanephrine, and 5-HIAA) and formate. In the afternoon, the 24-hour EEG is removed and a standard 21-lead EEG is performed. The child is sent home with the remaining capsules and instructed to continue taking them, as prescribed, for the rest of the week. The parents and child return the following week to receive test article #2 (either APM or placebo, the substance not given the first two weeks). The entire protocol is repeated, including a second hospitalization.

At the end of the 4-week study, each child will have a folder containing a physical exam report, a medical history, a YCI (see above description), intelligence and achievement testing results, (2) standard EEG reports, (2) 24-hour medilog reports, (4) parent rating forms, (4 or more) teacher rating forms, (4) STESS questionnaires, seizure records, and (2) 3-day diet histories filled out by the parents, and (2) 2-day diet histories for the inpatient days. The diets are evaluated for relative intake of aspartate, phenylalanine, tyrosine, and tryptophan. This is accomplished with the use of the program Nutritionist III (N-Squared Computer, Silverton, Oregon). This program analyzes 11+ essential amino acids. Each of the above measures will be analyzed and compared as to APM versus placebo when the study is completed and the "blinding" codes are broken.

Study Design #2

The second protocol, which will include as many subjects as we can identify (ideally at least 30), is also designed as a double-blind, placebo-controlled, crossover study, with APM and placebo being given in a randomized order on two treatment days. Children are admitted to our Children's Clinical Research Center for a 6-day period. On the first day of admission, an informed consent is obtained. The children receive a series of prescreening tests including a medical history, psychological history, drug history, physical exam, neurological exam, CT scan, and standard EEG. Laboratory testing performed includes determination of anticonvulsant medication levels (if applicable), routine hematology, urinalysis, plasma amino acid profile, and serum concentrations of total protein, albumin, calcium, inorganic phosphorus, cholesterol, glucose, urea nitrogen, uric acid, alkaline phosphatase, LDH, total bilirubin, AST, ALT, sodium, potassium, chloride, CO_2, and creatinine. In addition, a questionnaire relating to the previous alleged adverse effects from APM ingestion is completed.

Continuous 24-hour EEG monitoring is performed for the remaining 5 days of the study. Serum anticonvulsant levels and plasma amino acid profiles are drawn every morning. In addition, repeat plasma amino acid profiles are drawn on the two treatment days.

APM (in 300-mg capsules) is given in a randomized order on one of the two treatment days (placebo, on the other day). The dosage is approximately 50 mg/kg/day, given in three equally divided doses at 8:00 A.M., 10:00 A.M., 12 noon. Identical capsules containing placebo are given on the other treatment day. Patients are maintained on an APM-free diet and, in addition, receive standardized meals on the nights prior to and on the two treatment days. These diets have been nutritionally analyzed including an amino acid profile. Patients are monitored by nursing and research staff throughout their hospitalization for any adverse effects.

Results

Study Design #1

As of this date (6 April 1987), nine children have participated, with eight completing both arms of the study. Patient #10 has just entered the study. The subjects have ranged in age from 5 to 13. Five males and four females participated. The majority are caucasian children living in suburban areas, with one Hispanic child living in a large urban setting. Six children had a history of generalized seizures, with two also having absence episodes. Two children solely experienced absence seizures, and one patient had a history of partial complex seizures. Duration of seizure histories ranged from 2 to 5 years, with eight of the nine children currently on anticonvulsant regimens. Anticonvulsants varied from phenytoin, ethosuximide, phenobarbitol, carbemazepine, valproic acid, and primidone. The patient's weights ranged from 17 to 73 kg. APM doses ranged from 600 to 2400 mg, depending on the weight of the child.

No clinical seizures were documented in any of the children during the study protocol.

Study Design #2

Two females 10 and 15 years of age have been evaluated on this protocol. They are both from the midwestern region of the United States. One child had never been prescribed any anticonvulsant medication; the other is on valproic acid. Again, no clinical seizures were witnessed while these children were hospitalized for the study protocol.

Conclusion

Two methodologies have been described to evaluate the effect of large-dose APM ingestion on the neurological and electrophysiological status of children who experience seizures. To date, only descriptive statistics have been performed on these two patient groups. The studies are ongoing, and therefore the codes have not been broken. Thus, although there were no documented clinical seizures, it is not possible to fully evaluate the effects of APM on seizure activity until these studies are completed.

Acknowledgments. These studies were funded by the NutraSweet Company.

References

Center for Disease Control. (1984). Evaluation of consumer complaints related to aspartame use. Morb. Mortal. Weekly Rep. **33**:605–607.

Shaywitz, S.E., Schnell, C., Shaywitz, B.A., and Towle, V.R. (1986). Yale children's inventory (YCI): an instrument to assess children with attentional deficits and learning disabilities I. Scale development and psychometric properties. J. Abnor. Child Psychol. **14**:347–364, 1986.

Stegink, I.D., and Filer I.J. (eds.). (1984). Aspartame: physiology and biochemistry. New York: Marcel Dekker.

CHAPTER 20

Aspartame and Human Behavior: Cognitive and Behavioral Observations

Paul Spiers,* Donald Schomer,* LuAnn Sabounjian,* Harris Lieberman,† Richard Wurtman,† John Duguid,† Riley McCarten,† and Michele Lyden†

Little is known regarding the possible cognitive and behavioral effects of the artificial sweetener aspartame in humans. Few, if any, studies have addressed the issues raised by the potential elevation of brain phenylalanine levels which may be induced by aspartame ingestion and the concomitant depletion in specific neurotransmitter supplies which may result. In this chapter, the relevant literature addressing this topic is reviewed, the results of a new double-blind, acute study are briefly summarized, and findings from the pilot phase of a new chronic exposure study are reviewed.

Formal reports concerning possible behavioral alterations in association with ingestion of the artificial sweetener aspartame are most notable either for their paucity or for the complete absence of careful measurement either of behavioral changes or cognitive functioning. This is surprising given the potentially powerful effects which aspartame may be capable of producing in human brain. These alterations occur via the intermediary effects of phenylalanine, a large neutral amino acid, on tyrosine and tryptophan, which have been discussed in detail in other chapters of this book. The conclusion to be drawn from the neurobehavioral perspective is that aspartame ingestion may alter neurotransmission as a result of its potential effect on serotonin, dopamine, and norepinephrine. Specific experimental studies have demonstrated that the administration of pure amino acids, like tryptophan or tyrosine, or of foods that change plasma amino acid levels results in behavioral effects in humans such as decreased alertness, altered sleep, depressive mood, and increased aggressivity. Clinically, meanwhile, these neurotransmitters have been shown to be important for the mediation of numerous functions and have been associated with disorders of motor control, of abstraction, of inhibition and problem-solving,

*Behavioral Neurology Unit and the Comprehensive Epilepsy Center, Harvard University Medical School, Beth Israel Hospital, Boston, MA 02215, USA.
†Department of Brain and Cognitive Sciences and Clinical Research Center, Massachusetts Institute of Technology, Cambridge, MA 02139, USA.

of sleep, of mood, and, in particular, of depression and may play a role in various psychiatric disorders.

These findings are not particularly surprising considering, for example, that dopaminergic pathways have significant temporolimbic and orbitofrontal projections and that these areas have been independently identified to have important behavioral specializations in the control of emotion, social behavior, arousal, attention, and mental flexibility. Furthermore, these brain regions contain neurons sensitive to other important behavioral neurotransmitters. Given recent findings that a single neuron may respond to as many as three different neurotransmitters, it is possible to conceive of a situation where neuronal populations that react to aspartame may contain cells with overlapping sensitivity, for example, to hormones, which have been shown to be highly represented in these regions, and which clearly can alter behavior.

While metabolic studies have generally indicated that the aspartic acid component of aspartame does not cross the blood-brain barrier, it is important to recognize that aspartate and glutamate are excitatory amino acids. The recent identification of binding sites for these substances, such as the *N*-methyl-D-aspartate (NMDA) receptor, has led to research suggesting that these excitatory amino acids are probably critical to the brain's ability to learn and remember new information. In addition, these excitatory amino acids probably act as neuromodulators by opening slow potential intracellular channels which may, in turn, contribute significantly to the potentiation of excitatory states that result in epilepsy or produce morphological alterations resulting in neuronal loss after cerebral insult. Even though it may be unlikely that aspartame alters aspartic acid levels in normal human brain, this issue has not been adequately addressed, and little is known regarding the possible effect of raising brain concentrations of excitatory amino acids in potentially vulnerable individuals. The influence which such substances may exert on human behavior is obviously substantial, particularly given that the anatomic distribution of NMDA receptors identified by autoradiographic techniques involves most major limbic structures and certain hypothalamic nuclei, linking these substances to hormonal regulation as well. The question, therefore, appears to be not how aspartame might influence human behavior, as the mechanisms for such effects are clearly available, but rather whether it, in fact, does, and what research needs to be done before it can be concluded that it does not.

Before proceeding to consider the literature on aspartame and human behavior, it is worth considering what the structure of an adequate clinical investigation into this question might entail. Precedent certainly exists for such studies, and the literature which has evolved on the relative effects of the various anticonvulsant medications is most instructive. Ideally, such an investigation should begin with some theoretical understanding of the neural mechanisms potentially affected by the substance in question, and a double-blind, placebo-controlled design should be used with administration of the substance for an equivalent period of time and in a comparable dose to the manner in which it is expected to be consumed when freely available. Plasma levels of the substance should be monitored and cor related with outcome variables which, at the very least, should include specific measures of cognitive functioning administered before, during, and after the trial and independent measures of cognitive functioning, such as school or job performance where available. Additionally, specific measures of mood or psychological state administered at regular intervals, as well as independent blind behavioral observations or ratings of mood or psychological state, and monitoring of the subjective complaints reported by participants should be included in the investigation. No study even approaching this fundamental design has ever been conducted to investigate the relationship between aspartame and human behavior. Studies reported in the late 1970s evaluated plasma concentrations of phenylalanine and aspartate and their relation to various physiological parameters in acute dosing, acute dosing at abuse levels, and chronic dosing with aspartame in apparently healthy children and adults, in young persons during weight reduction, and in certain special populations such as phenylketonurics (see Visek 1984). In the study of apparently healthy children and adults maintained on aspartame for 13 weeks, the only comment and report regarding behavioral factors was that "the subjective complaints which were recorded in biweekly interviews were not clinically important" (Visek 1984, p. 499). In the study of overweight young adults maintained on aspartame for 13 weeks, the participants' subjective complaints were recorded but apparently were not systematically elicited or reviewed, and such factors as sleep and mood were not examined. Furthermore, cognitive functioning was not formally evaluated in any of these studies, nor was any attempt made to ascertain whether subjects experienced any change in their memory or concentration skills. Essentially, these two studies constituted the entire body of human neurobehavioral literature on aspartame prior to its approval by the Food and Drug Administration.

Previously, a study investigating the effects of phenylalanine on learning in nonhuman primates had demonstrated deficits on some of the more difficult tests available for the Wisconsin General Testing Apparatus (WGTA) (Waisman and Harlow 1965). These deficits were not replicated in a later study using the WGTA (Suomi 1984), but in this second paradigm the primates were exposed to phenylalanine only during infancy with a lengthy withdrawal prior to testing, whereas in the original study the animals were maintained on phenylalanine during the actual learning performance evaluations.

Since approval for aspartame was granted, three reports have appeared which provide somewhat more information, while at the same time raising some cautionary signals. The first of these did not actually administer aspartame but examined the effects of elevated plasma phenylalanine on the neuropsychological performance of patients with treated phenylketonuria (PKU) (Krause et al. 1985). The literature on the cognitive performance of children who follow phenylalanine-restricted diets has thus far supported a subtle but positive effect of the diet in improving cognitive performance, and it is well known that if this syndrome is untreated in infancy, it can lead to irreversible mental retardation, often accompanied by seizures. Adolescent or early-adult presentations of PKU are rare but typically involve a clinical picture with a prominent psychosis. In brief, this experiment used a double-blind, crossover design in which subjects served as their own control on a battery of repeatable neuropsychological tests selected to evaluate both higher integrative and more fundamental cognitive functions. The most difficult task was a Computerized Forced-Choice Reaction Time procedure which assessed visual-perceptual discrimination. While simpler tests of motor speed and dexterity and of visual-spatial sequencing did not show any effect in relation to dietary manipulation of phenylalanine, slower performance on the Forced-Choice Reaction Time test was correlated with increased phenylalanine plasma levels. The authors also examined urinary dopamine excretion and found that there was a clear inverse correlation between this variable and performance on the Reaction Time test. As urinary dopamine fell, choice reaction time increased; that is, performance worsened.

In a similar attempt to examine the effects of increased plasma phenylalanine levels on the cognitive performance of normal, nonphenylketonuric humans, Lieberman et al. (1987) administered aspartame to normal, healthy, male, young adult volunteers and assessed their performance on a battery of tests shown to be sensitive to the effects of tryptophan and tyrosine, of other food constituents, of caffeine, and of sedative-hypnotic medications. They also had subjects fill out two mood scales and a measure of drowsiness. A double-blind, placebo-controlled, crossover design was used, and aspartame was administered in a single, acute, low or high dose, either alone or in combination with carbohydrates. Carbohydrates depress the levels of other amino acids that usually compete with phenylalanine and therefore potentially facilitate phenylalanine transport across the blood-brain barrier. None of these conditions resulted in any change in the subjects' cognitive performance or mood state, even though plasma phenylalanine levels were clearly elevated by the aspartame.

While the results of this study might be taken to suggest that aspartame ingestion in healthy individuals is likely to produce few acute adverse effects, several issues have still not been addressed. The study was restricted to males, and it is not known whether these findings can be generalized to women. All of the subjects were healthy and were chosen because they reported no known sensitivity to aspartame. Consequently, these results cannot be generalized to an undifferentiated sample of individuals who have no history of exposure to aspartame or who either may have a reported vulnerability to this substance or may be at risk for such a vulnerability given other neurodevelopmental deficits or medical conditions. The method of administration of the aspartame was via oral capsules which may not parallel the time course for phenylalanine effects when this substance is ingested in foods or beverages. It should also be remembered that this was an acute dosing study, a situation which is unlikely to mimic the consumption pattern for this substance, and it is unclear what the effect of chronic phenylalanine elevation may be following prolonged daily aspartame intake. Finally, while the cognitive measures used here may have been shown to be sensitive to other substances, they may be insufficiently difficult to detect subtle alterations in higher integrative functions and may not require complex perceptual judgments, decision-making or problem-solving skills, or memory at a level at which these functions may be more likely to be impaired.

The third study to investigate any relationship between behavior and aspartame assessed the response of male college students to a single dose of aspartame or phenylalanine (Ryan-Harshman et al. 1987). Cognitive functions were not tested. Using visual analog scales, the subjects rated their "stomach sensations" of emptiness, rumbling, ache, and nausea, their "head sensations" of headache, dizziness, and faintness, and their "general sensations" of feeling drowsy, weak, nervous, tense, drugged, depressed, alert, and mentally slow. There was no effect of either the aspartame or phenylalanine in producing any consistent changes in the subjects' ratings of these "sensations." This study, obviously, is subject to the same criticisms as the preceding one but is further compromised by failing to assess any cognitive functions objectively.

In summary, the experimental investigation of the potential effects of aspartame on human behavior and cognitive functioning remains inconclusive at this time and can perhaps best be characterized as inadequate. Those studies which have failed to show any effect for this substance both used acute exposure to a single dose of aspartame and preselected subjects to avoid individuals who might be more likely to show some vulnerability to aspartame consumption. The study with positive results, meanwhile, was the only study in which there was chronic elevation of blood phenylalanine. However, this study also preselected subjects for an inborn error of metabolism in relation to phenylalanine and did not actually use aspartame as the means for elavating plasma phenylalanine levels. Other studies examining chronic exposure to aspartame in healthy children and adults unfortunately did not systematically collect behavioral or cognitive data. Consequently, the potential effect of aspartame in altering human behavior and cognitive functioning remains entirely unknown at this time. What is perhaps most striking is the lack of investigation of this topic and the outstanding need which exists for a comprehensive study of this problem both in normal individuals and those potentially at risk. While it might be argued that the successful introduction of aspartame into the consumer market without apparent major repercussion may indicate that this substance is safe for human consumption, the absence of proof cannot be taken

TABLE 20.1 Neurobehavioral complaints related to aspartame consumption reported to the Center for Brain Science and Metabolism and the Massachusetts Institute of Technology (CBSM/MIT) and to the Centers for Disease Control (CDC) as reported by Bradstock et al. (1986).

CBSM/MIT	CDC
Headache	Headache
Dizziness	Dizziness
Sleep disturbance	Insomnia
Visual impairment	Visual impairment
Abdominal pain	Abdominal pain
Hyperactivity	Hyperactivity
Light-headed	Nausea
Autonomic symptoms	Fatigue
Blackouts	Numbness
Seizures	Depression
Hallucinations	Anxiety
Memory loss	Irritability
Speech deficit	Altered menses

as proof of absence. Furthermore, a recent report from the Centers for Disease Control (Bradstock et al. 1986), as well as our own experience in collecting information from individuals who have had adverse reactions to this product, suggests that aspartame may have more widespread and diverse effects on human behavior than has heretofore been contemplated. It has been suggested that the variety of symptoms reported by individuals in response to aspartame mitigates against a neurobehavioral effect of this substance because no unitary constellation of symptoms is constant to all complainants. Such a conclusion seems unjustified given the wide range of individual biological variation in response to chemical agents. For example, it could similarly be stated that there is no uniform constellation of symptoms which appears in response to low or even moderate alcohol consumption; however, one would be hard pressed to maintain that alcohol has no neurobehavioral consequences, even in minimal doses. To date, we haviors reported are summarized in Table 20.1. The types of symptoms reported to us are somewhat similar to those reported to other groups, but there is probably a greater representation of seizures in our population. In fact, our interest in the potential relationship between epilepsy, or a lowered seizure threshold, and aspartame consumption led us to design a study to investigate the effects of chronic aspartame exposure on several neurobehavioral parameters. Prior to obtaining approval for a population of complainants with seizure manifestations to go through the protocol, however, the Institutional Review Board of the Massachusetts Institute of

Technology required that we first examine a population of normal subjects. This sample yielded some surprising data which then led us to study another population of normal control subjects. We will present here the preliminary results of these two groups.
preliminary results of these two groups.

Essentially, normal volunteers were recruited from the undergraduate and graduate schools of MIT and the local Cambridge, Massachusetts area. They were screened to meet the following criteria prior to admission to the study. Subjects had to be free of any active or past history of neurological disease, have no other active physical disease, and have no history of seizures, no history of psychiatric hospitalization, and no history of developmental learning disability. Subjects could not be taking any prescription medications and had to have results within normal limits on EEG, neurological and physical examination, and neuropsychological screening and could not have any abnormal lab values on a routine battery of chemistry, hematology, and amino acid profiles. Furthermore, subjects had to have a history of moderate exposure to aspartame without any reported adverse effect. This was defined as a daily minimum of two to three cans of aspartame-sweetened soda; however, we actively sought to enroll subjects whose consumption history exceeded a liter of aspartame-sweetened soda per day. In summary, we made a conscious effort to preselect individuals who we felt would be unlikely to experience any effect from chronic aspartame exposure.

In a first phase, five subjects were admitted to an open protocol in which they were informed that they would receive aspartame up to a daily equivalent of 50 mg/kg of body weight, administered in three divided doses, in capsule form. Subjects were maintained on a controlled diet with an upper limit restriction of proteins and carbohydrates, as well as a 2-liter daily fluid restriction. Subjects were admitted to the Clinical Research Center at MIT and slept on site but were allowed to carry out their normal daily activities. They were observed for a 3-day baseline and then started on aspartame for 12 days, followed by 3 days of observation off aspartame, discharge from the Clinical Research Center, and follow-up one week later.

Subjects completed a daily symptom diary and had interval EEG telemetry, neurological and physical examinations, amino acid profiles, chemistry, hematology, and urinalysis screens, and neuropsychological testing during the baseline, aspartame exposure, and follow-up phases of the protocol. In a second phase, still in progress as of this writing, five additional normal volunteers, loosely matched for age, sex, and academic background, have been entered into the same protocol but are blind with regard to whether they are receiving aspartame or placebo. In fact, they are all receiving placebo in order to act as controls for the first five subjects. The nursing staff responsible fo the daily care of the subjects and for administering the substance is also blind for this phase. Placebo and aspartame capsules are identical in appearance and taste.

While there is much we have not yet analyzed from this protocol, we will report briefly here on the results of the cognitive, neuropsychological testing in the four, matched pairs of subjects on whom data are currently available. First, what cognitive functions were measured? Intelligence quotient estimates were obtained at baseline, but this was primarily for the purpose of matching subjects, and this type of test is recognized to be a poor measure of state-dependent effects. More important, from our perspective, given the role of serotonin, dopamine, and norepinephrine in the central nervous system, were functions such as attention and response set flexibility. More particularly, we were interested in evaluating the subjects' attention and response set by means of tasks requiring some sustained behavioral output, or alternation between various performance strategies on the same task, rather than by a simple, reactive response in a passive, signal detection paradigm. To this end, we selected tests such as Word List Generation, Form B of the Trailmaking Test, and the Stroop Color Word Naming Test which place a premium on these functions. We were also interested in memory and, finally, we wanted a task that combined attention, skilled motor output, and some immediate memory load with shifting stimuli and response demands, that is, we wanted a task that approximated the complexity at which the brain has to function in everyday life. We were fortunate in finding just such a measure in a MacIntosh software application developed in the educational realm called "Think Fast" (S. Steffin and D. Harris for Brainpower Corp., Calabasas, Calif., 1985). "Think Fast" requires the subject to alternately match, copy, or recall sequences of letters and patterns of block designs in a self-governing, self-scoring, progressively difficult series of levels. Once the program is engaged, the examiner has no influence on the pacing or selection of stimuli, which are randomly generated and time limited.

The results presented in Table 20.2 are the mean difference scores between the baseline and exposure testings on each of the measures used for four subjects under each of the two conditions, aspartame (open) and placebo (blind). The use of difference scores utilizes each subject as his own control, establishes each subject's own baseline performance as the zero point, and eliminates the effect of any baseline differences between individuals or groups. Statistical comparisons (*t* test) of the difference between the mean changes in performance during placebo versus aspartame exposure showed a trend toward significance for the Trailmaking, Stroop Interference, and all Think Fast trials, and yielded significant comparisons for the Stroop Word Reading and Motor Response Set tasks. Learning and memory for simple stimuli and more passive attentional tasks showed no significant difference between the two exposure groups.

In addition to these cognitive measures of neurobehavioral functioning, we reviewed the behavior and subjective complaints of the participants during the two exposure conditions based on their daily symptom diary and the nurses' notes. Summarizing this information and taking into consid-

TABLE 20.2 Average change in performance from baseline to exposure conditions for placebo- and aspartame-exposed subjects.[a]

Measure	Average change for subjects receiving: Placebo	Aspartame	t-Probability
Learning			
Mean increase in words recalled during learning trials	0.4	0.375	N.S. >0.25
Mean increase in words recalled during memory trials	0.19	1.19	N.S. >0.25
Attention			
Mean increase in total span	2.3	1.71	N.S. >0.25
Mean increase in words generated	4.75	1.00	N.S. >0.25
Response set			
Mean decrease in time to complete:			
Trails A	5.38	10.83	N.S. >0.25
Trails B	18.38	−1.4	$p<0.16$
Stroop Word Reading	1.62	−13.13	$p<0.002$
Stroop Interference	10.25	−8.25	$p<0.17$
Mean decrease in reciprocal motor program errors bilaterally	6.75	1.88	$p<0.04$
Think Fast			
Mean improvement in score for:			
Condition A	397.38	−140.25	$p<0.13$
Condition B	351.25	−181.25	$p<0.08$
Mean decrease in number of errors:			
Condition A	2.25	−0.75	$p<0.12$
Condition B	1.38	−0.38	$p<0.14$

[a] Pilot data: $n = 4$/group; 2 male, 2 female.

eration only those symptoms that represent a change from behaviors observed or reported during the baseline phase, it can be seen that none of the placebo subjects reported or were observed to have any significant neurobehavioral changes whereas three of the five aspartame subjects reported the appearance of at least two of the following symptoms during the exposure period. Subjects developed focal pains, autonomic symptoms, nausea, lightheadedness, sleep disruption, frontal headaches, photophobia, and visual disturbances, apparently in response to chronic, daily exposure to aspartame. Furthermore, the nursing staff spontaneously noted that two of the aspartame-exposed subjects became irritable, anxious, and complaining whereas none of the placebo-exposed subjects were ever described in these terms.

These results were somewhat surprising to us. Initially, this study was designed as a pilot project, restricted to low-risk subjects, to test the safety and efficacy of this protocol prior to examining a population of complainants who maintained that they had developed seizures in response to aspartame exposure. However, the finding of positive results in this normal group of subjects, which presumably cannot be attributed to the structure of the protocol, to dietary habits, or to taking up residence at the Clinical Research Center, has realigned our priorities. It seems that aspartame may be capable of producing significant neurobehavioral effects, some of which may not even be within the subject's awareness. This issue requires further research and must be investigated under more rigorous conditions using a double-blind, crossover design. While it would certainly be premature to conclude that chronic, high-dose aspartame ingestion interferes with cognitive functioning, it would seem wise, based on the present report, to keep an open mind to such potential brain-behavior relationships.

Acknowledgment. This work was supported by the Center for Brain Sciences and Metabolism Charitable Trust, Cambridge, MA.

References

Bradstock, M.K., Serdula, M.K., Marks, J.S., Barnard R.J., Crane, N.T., Remington, P.L., and Trowbridge, F.L. (1986). Evaluation of reactions to food additives: the aspartame experience. Am. J. Clin. Nutr. **43:**464–469.

Krause, W., Halminski, M., McDonald, L., Dembure, P., Salvo, R., Freides, D., and Elsas, L. (1985). Biochemical and neuropsychological effects of elevated plasma phenylalanine in patients with treated phenylketonuria: a model for the study of phenylalanine and brain function in man. J. Clin. Invest. **75:**40–48.

Lieberman, H.R., Caballero, B., Garfield, G.S., and Bernstein, J.G., (1987). The effects of aspartame on human mood, performance and plasma amino acid levels. Poster presentation, Symposium on Dietary Phenylalanine and Brain Function, Washington, D.C., May 8–10.

Ryan-Harshman, M., Leiter, L.A., and Anderson, H.G. (1987). Phenylalanine and aspartame fail to alter feeding behavior, mood and arousal in men. Physiol. Behav. **39:**247–253.

Stegink, L.D. (1984). Aspartame metabolism in humans: acute dosing studies. In Stegink, L.D., and Filer, L.J. (eds.), Aspartame: physiology and biochemistry. New York: Marcel Dekker, pp. 509–553.

Suomi, S.J. (1984). Effects of aspartame on the learning test performance of young stumptail macaques. *In* Stegink, L.D., and Filer, L.J. (eds.), Aspartame: physiology and biochemistry. New York: Marcel Dekker, pp. 425–445.

Visek, W.J. (1984). Chronic ingestion of aspartame in humans. *In* Stegink, L.D., and Filer, L.J. (eds.), Aspartame: physiology and biochemistry. New York: Marcel Dekker, pp. 495–508.

Waisman, H.A., and Harlow, H.F. (1965). Experimental phenylketonuria in infant monkeys. Science **147:**685–695.

CHAPTER 21

Effects of High Plasma Phenylalanine Concentration in Older Early-Treated PKU Patients:

Performance, Neurotransmitter Synthesis, and EEG Mean Power Frequency

Wilma L. Krause,* Margaret Halminski,† Mary Naglak,† Linda McDonald†, Rino Salvo,† David Freides,† Charles Epstein,† Philip Dembure,† Allen Averbook,† and Louis J. Elsas†

In two separate but related studies, we investigated the effect of increased plasma phenylalanine concentration on performance, neurotransmitter synthesis, and EEG mean power frequency (MPF) in older early-treated PKU subjects. When plasma phenylalanine was increased from a mean of 605 ± 423 μM ($\pm$ SD) to a mean of 2315 ± 1038 μM by increasing phenylalanine in the diet for a week, performance time on a computerized choice reaction test was prolonged, neurotransmitter concentration in plasma and urine was decreased, and the EEG-MPF was decreased. When plasma phenylalanine was reduced, the effects were reversible. These observations suggest that increased phenylalanine affects brain function in patients with a significant defect in phenylalanine hydroxylase function. Whether similar effects occur at lower plasma phenylalanine concentrations in individuals with milder defects or in persons heterozygous for a defective phenylalanine hydroxylase gene should be explored.

Phenylketonuria (PKU) provides a unique opportunity for studying the effect on brain function of an increase in a single amino acid. The human species appears to be unique in carrying an abnormal gene for phenylalanine hydroxylase and therefore provides the only naturally occurring model for study. A number of investigators have looked at the effect of high concentrations of brain phenylalanine on the synthesis of dopamine and serotonin (Curtius et al. 1981, Lou et al. 1985). Others have looked at

* Current address: Division of Medical Genetics, Department of Pediatrics, University of Iowa, Iowa City, IA 52242, USA
† Division of Medical Genetics, Departments of Pediatrics, Biochemistry, Psychology, and Neurology, Emory University, Atlanta, GA 30322 USA.

the deterioration in IQ in children after they were taken off phenylalanine-restricted diets (Seashore et al. 1985, Koch et al. 1982, Holtzman et al. 1986). The EEG has been investigated from the point of view of both abnormal patterns and changes in mean power frequency (Rolle-Daya et al. 1975, Donker et al. 1979). A few years ago at Emory University we decided to investigate under controlled conditions the possible effects of increased phenylalanine on the central nervous systems of early-treated PKU patients who displayed normal or near normal nervous system function; that is, their IQs were normal or near normal, and they were in school or in the work place (Krause et al. 1985, 1986). Most were still adhering to a partially restricted phenylalanine diet. A few had been off diet since about the age of six years.

Design

Initially, we set out to look for changes in performance on a group of neuropsychological tests which measured a broad range of brain functions. At the same time, we attempted to evaluate neurotransmitter synthesis by measuring urine dopamine. In a second study, we looked at changes in the mean power frequency in the EEGs of a similar group of patients and measured plasma L-DOPA at the same time. Both studies used a double-blind, crossover model. About half of the subjects received high phenylalanine the first week, low the second, and high the third. Conversely, the remaining subjects received low phenylalanine the first week, high the second, and low the third. All subjects were on a carefully calculated and monitored diet during their stay in the Clinical Research Unit at Emory University. The only manipulation of the diet for purposes of the study was the addition of tasteless L-phenylalanine to the usual "formula" during the increased phenylalanine epoch. The amount added was individualized for each subject and was based on diet information which had been accumulated over the years of monitoring and treatment. The objective was to raise the plasma concentration to nontreatment levels (approximately 1300 μM or greater). The average daily phenylalanine intake prescribed was approximately 90 mg per kg per day. The average required for achieving "control" was 12 mg per kg per day during the low periods. Each epoch lasted for seven or nine days (seven in the first study and nine in the second). The subjects, their parents and caretakers, the laboratory personnel, and the persons performing the neuropsychological tests and the EEGs were not informed of the diet changes. Plasma was obtained on the last day of the epoch in the first group and on the last three days in the second study group. The 24-hour urine collections were scheduled to overlap the blood sampling, which was done at approximately 11:00 A.M. (semi-fasting). In the second study, plasma was obtained for L-DOPA determination at the same time.

Neuropsychological tests were performed at the time of admission to the study to establish a baseline and to familiarize the patients with the equipment. The tests were then administered once (in the early afternoon) at the end of each diet epoch. In the EEG study, ten 2-second artifact-free samples were entered into the computer for calculation of mean power frequency by fast Fourier analysis. The EEGs were run in the late afternoon on the last three days of each diet epoch so that triplicate samples could be compared for day-to-day variation.

By using the crossover design, we hoped not only to strengthen the validity of the observations but also to assess the reversibility of the effects.

Results

Of 29 low plasma phenylalanine values, 23 fell between 200 and 800 μM, 3 were above (936 to 1632), and 3 were below (32 to 94). Of 22 high values, 11 fell between 1400 and 2500 μM, 7 were less than 1400 μM (797 to 1400), and 4 were greater than 2500 μM (2500 to 4405). Changes in phenylalanine concentration were compared with changes in urine dopamine, changes in plasma L-DOPA, changes in EEG mean power frequency, and changes in performance on the battery of neuropsychological tests. The results are best seen in graphic display in the four figures below. In all figures, the change in plasma phenylalanine is indicated on the x-axis and the other parameter on the y-axis.

In Figure 21.1, as the phenylalanine plasma concentration decreases, the urine dopamine concentration rises, and, conversely, as the plasma phenylalanine increases, the urine dopamine decreases. These findings are seen in all but two of the results; that is, all the results but two fall in quadrants I and III.

In Figure 21.2, as the phenylalanine concentration decreases, the plasma L-DOPA increases, and, conversely, as the plasma phenylalanine concentration increases, the plasma L-DOPA concentration falls. Again, all but two of the values fall in quadrants I and III.

Figure 21.3 depicts the results on the computerized choice reaction time. This test involved the matching of geometric figures flashed on a video screen. Either a positive or negative (match or nonmatch) response was required. Each response was recorded and timed. This was the one test in the neuropsychological test battery which consistently showed a difference regardless of the age or IQ of the patient. It is a test of discriminatory learning. As shown in Figure 21.3, as the plasma phenylalanine decreased, the response time (in milliseconds) also decreased, and as the phenylalanine increased, the response time also increased, i.e., the response was slowed. All but three results were in quadrants II and IV.

Results of the EEG mean power frequency study for channel 7 (left centro-occipital) are depicted in Figure 21.4. As the phenylalanine de-

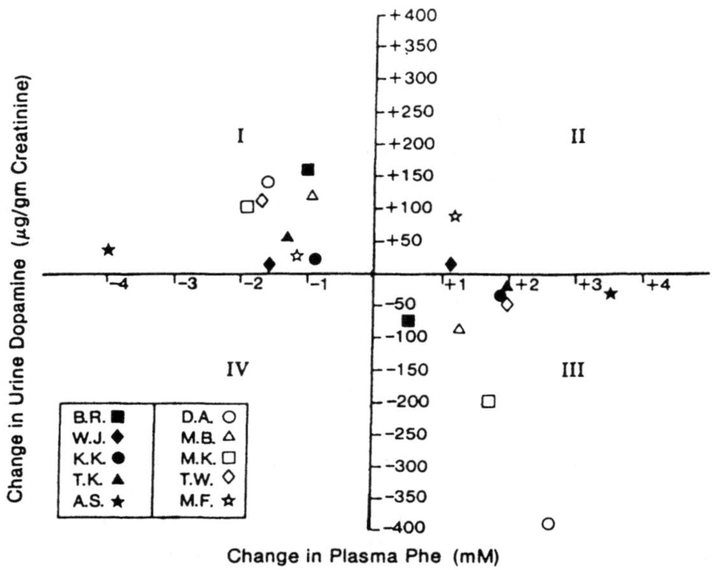

FIGURE 21.1. The relationship of changes in plasma phenylalanine to changes in urine dopamine during dietary manipulation of phenylalanine. Each patient is represented by a symbol that appears twice on the graph, indicating the difference in dopamine between the first and second week and the second and third week plotted against parallel changes in plasma phenylalanine. Reproduced from The Journal of Clinical Investigation, 1985, 75, pp. 40–48 by copyright permission of the American Society for Clinical Investigation.

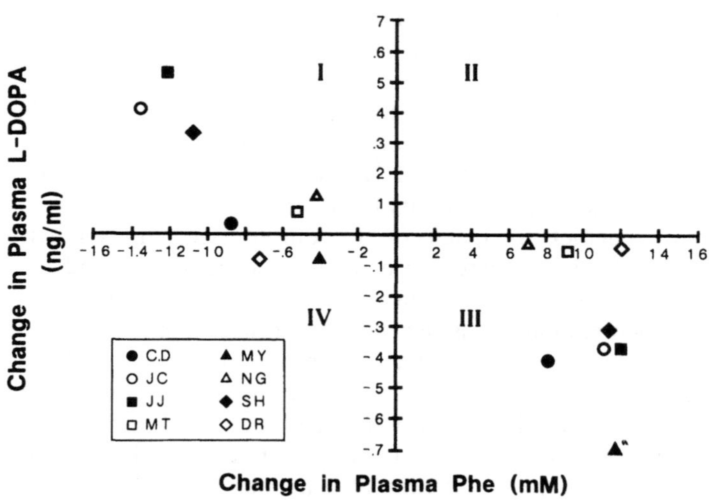

FIGURE 21.2. The relationship of changes in plasma phenylalanine to changes in plasma L-DOPA during dietary manipulation of phenylalanine. Each patient is represented by a symbol that appears twice on the graph, indicating the difference in plasma L-DOPA between the first and second week and the second and third week plotted against parallel changes in plasma phenylalanine. Reproduced with permission from Pediatric Research, 1986, 20, pp. 1112–1116.

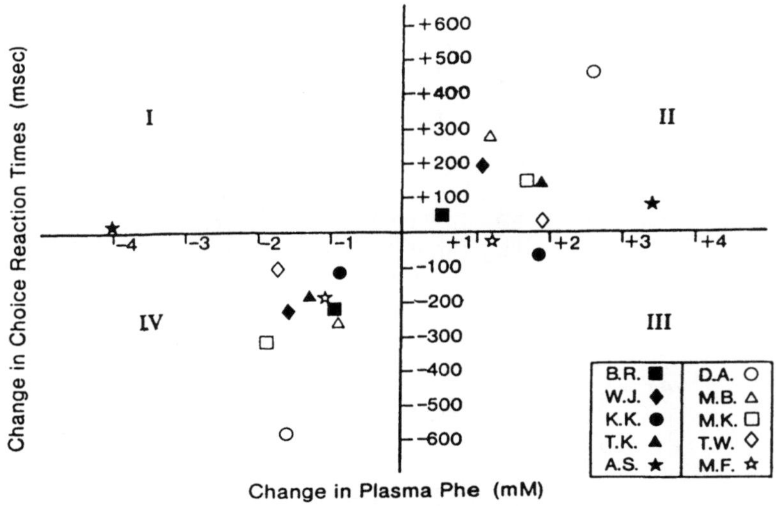

FIGURE 21.3. The relationship of changes in plasma phenylalanine to changes in choice reaction time during dietary manipulation of phenylalanine. Each patient is represented by a symbol that appears twice on the graph, indicating the difference in mean choice reaction time between the first and second week and the second and third week plotted against parallel changes in plasma phenylalanine. Reproduced from The Journal of Clinical Investigation, 1985, 75, pp. 40–48 by copyright permission of the American Society for Clinical Investigation.

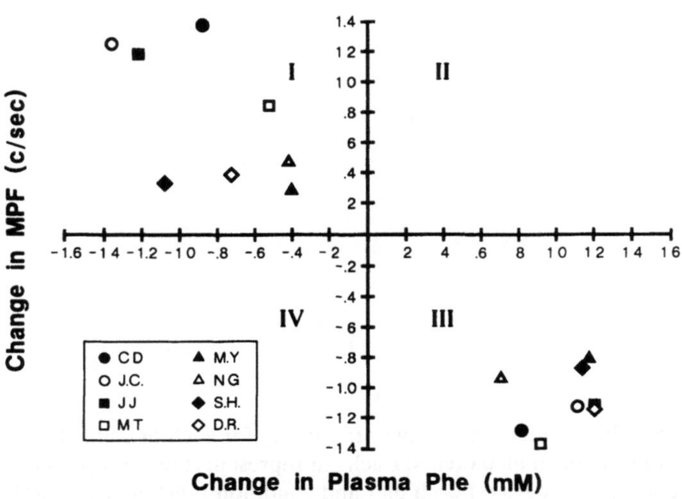

FIGURE 21.4. The relationship of changes in plasma phenylalanine to changes in mean power frequency (MPF) in channel 7 during dietary manipulation of phenylalanine. Each patient is represented by a symbol that appears twice on the graph, indicating the difference in MPF between the first and second week and the second and third week plotted against parallel changes in plasma phenylalanine. Reproduced with permission from Pediatric Research, 1986, 20, pp. 1112–1116.

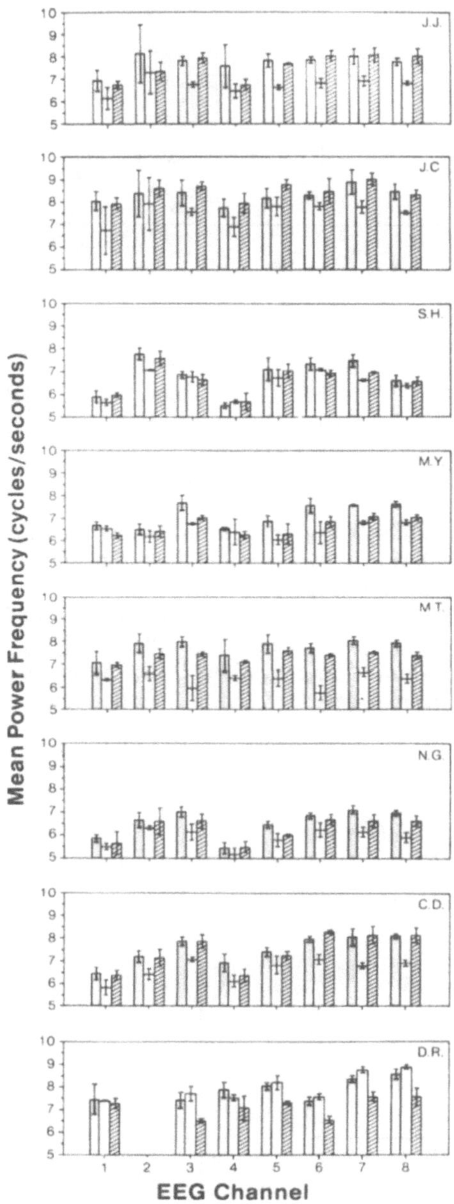

FIGURE 21.5. The average mean power frequency for each of eight channels for all three diet conditions for all patients. Each bar represents the average of three values for three consecutive days on each diet and condition bracketed by 2 SDs. Conditions 1, 2, and 3 are indicated by gray, open, and hatched bars, respectively. These conditions were indicative of the low-high-low dietary phenylalanine for the first seven patients (J.J., J.C., S.H., M.Y., M.T., N.G., and C.D.) and high-low-high protocol for the eighth patient, D.R. Reproduced with permission from Pediatric Research, 1986, 20, pp. 1112–1116.

creased, the mean power frequency (number of cycles per second) increased, and, conversely, as the plasma phenylalanine concentration rose, the mean power frequency (number of cycles per second) decreased. All of the results are in quadrants I and III. Differences in mean power frequency were seen in all eight channels and all were significant ($p < 0.01$) by two-way analysis of variance. These changes are shown in Figure 21.5.

Discussion

The results of these studies suggest that increasing the plasma phenylalanine concentration to levels that can be achieved in subjects with defective phenylalanine hydroxylase function does affect brain function and that these changes may be mediated through changes in dopamine synthesis. Clearly, the results point to the need for studying several issues. There is a need to identify a group of neuropsychological tests which is designed to assess individual facets of brain function. There is also a need for a direct measurement of neurotransmitter synthesis in the brain. The meaning of changes in mean power frequency in conscious healthy subjects must be interpreted. The question of whether there is a threshold for the effect of increasing plasma phenylalanine should be investigated. It is important to determine whether individuals heterozygous for an abnormality in the gene for phenylalanine hydroxylase might also exhibit findings similar to those seen in our patients if they achieved intermediate plasma phenylalanine concentrations, e.g., in the 250 to 500 μM range or higher on ingestion of large amounts of phenylalanine.

Acknowledgments. This work was supported in part by grant 12-110 from the March of Dimes (to L.J.E.) and grant 2-M01-RR00039-25 from the National Institutes of Health (to Emory University).

References

Curtius, H-C.H., Niederwieser, A., Viscontini, M., Leimbacher, W., Wegman, H., Biehova, B., Rey, F., Schaub, J., and Schmidt, H. (1981). Serotonin and dopamine synthesis in phenylketonuria. Exp. Med. Biol. **133**:277–291.

Donker, D.N.J., Reits, D., Van Sprang, F.J., Van Leeuwen, W.S., and Wadman, S.K. (1979). Computer analysis of the EEG as an aid in the evaluation of dietetic treatment in phenylketonuria. Electroencephalogr. Clin. Neurophysiol. **46**:205–213.

Holtzman, N.A., Kronmal, R.A., van Doorninck, W., Azen, C., and Koch, R. (1986). Effect of age at loss of dietary control on intellectual performance and behavior of children with phenylketonuria. N. Engl. J. Med. **314**:593–598.

Koch, R., Azen, C.G., Friedman, E.G., and Williamson, M.L. (1982). Preliminary report on the effects of diet discontinuation in PKU. J. Pediatr. **100**:870–875.

Krause, W., Halminski, M. McDonald, Dembure, P., Salvo, R., Freides, D., and Elsas, L. (1985). Biochemical and neuropsychological effects of elevated plasma phenylalanine in patients with treated phenylketonuria. A model for the study of phenylalanine and brain function in man. J. Clin. Invest. **75:**40–48.

Krause, W., Epstein, C., Averbook, A., Dembure, P., and Elsas, L. (1986). Phenylalanine alters the mean power frequency of electroencephalograms and plasma L-DOPA in treated patients with phenylketonuria. Pediatr. Res. **20:**1112–1116.

Lou, H.C., Guttler, F., Lykkelund, C., Bruhn, P., and Niederwieser, A. (1985). Decreased vigilance and neurotransmitter synthesis after discontinuation of dietary treatment for phenylketonuria in adolescents. Eur. J. Pediatr. **144:**17–20.

Rolle-Daya, H., Pueschel, S.M., and Lombroso, C.T. (1975). Electroencephalographic findings in children with phenylketonuria. Am. J. Dis. Child. **129:**896–900.

Seashore, M.R., Friedman, M.S., Novelly, R.A., and Bapat, V. (1985). Loss of intellectual function in children with phenylketonuria after relaxation of dietary phenylalanine restriction. Pediatrics **75:**226–232.

CHAPTER 22

Changes in Physiological Concentrations of Blood Phenylalanine Produce Changes in Sensitive Parameters of Human Brain Function

Louis J. Elsas II and James F. Trotter*

We previously demonstrated that increases of 1.0 m*M* in blood phenylalanine (Phe) decreased plasma L-DOPA, slowed brain electrical discharge, and prolonged performance on neuropsychological tests of higher integrative function. In this study, we evaluate the sensitivity of these tests of brain function during moderate elevation of plasma Phe. Six adult heterozygotes for phenylketonuria (PKU) and one homozygous normal volunteer were studied using a double-blind, double-crossover protocol of four 2-week intervals. Volunteers ingested a constant diet of 50 to 60 mg of Phe per kg per day supplemented by either 100 mg of Phe per kg per day or placebo. On the final two days of each study interval, the following parameters of brain function were evaluated: mean power frequency (MPF) of the EEG, plasma L-DOPA concentration, and cognitive function. The results indicate that plasma Phe rose in both heterozygotes and control during supplemental Phe ingestion from a mean of 99 ± 20 μM to 239 ± 166 μM. There was a wide range of interindividual variation with changes of plasma Phe between 7 μM and 403 μM, thus enabling evaluation of relative sensitivities among these parameters of brain function. Changes in EEG, L-DOPA, and cognitive function were all inversely related to plasma Phe changes between 330 and 403 μM. The MPF was the most sensitive of these parameters with changes seen with as low as 7 μM change in Phe. These changes were seen in the high-frequency α-band (8 to 12 cps) and gave a slope such that for every 100 μM change in plasma Phe, there was a 0.125-cps change in the inverse direction for brain wave function. This slope was similar to that seen in previous studies at changes in Phe concentration between 600 and 2400 μM. The plasma Phe concentrations used in these studies are attained by the chronic ingestion of 34 mg of aspartame per kg per day.

* Division of Medical Genetics, Department of Pediatrics, Emory University School of Medicine, Atlanta, GA 30322, USA.

Introduction

In previous studies, we found that large changes in plasma phenylalanine (Phe) (i.e., changes of 600 to 2400 μM) were associated with an inverse change in the following parameters of brain function: performance on neuropsychological tests of higher integrative function (Krause et al. 1985); plasma L-DOPA concentration (Krause et al. 1985, 1986); and the mean power frequency of the electroencephalogram (MPF of the EEG) (Krause et al. 1986). While these large changes in plasma Phe were clearly associated with altered brain function, the effect of changes in lower concentrations of plasma Phe remained unknown.

The purposes of this study are twofold. First, among the three parameters of brain functions discussed above, we determine which is the most sensitive. Second, we ascertain what effect, if any, changes in blood phenylalanine concentrations between 7 and 403 μM might have on these determinants.

Methods

Study Design

Six obligate heterozygotes for the classical phenylketonuria (PKU) gene and one homozygous normal were studied. Genotypes were confirmed by previously described methods (Griffin and Elsas 1975). An 8-week, double-blind, double-crossover protocol was used with each subject serving as his/her own control (Trotter et al. 1987). This protocol is outlined in Figure 22.1. A standardized diet was maintained throughout the study. Measurements of amino acids, neurotransmitters, MPF of the EEG, and cognitive functions were made in duplicate, during the final two days of each 2-week study interval.

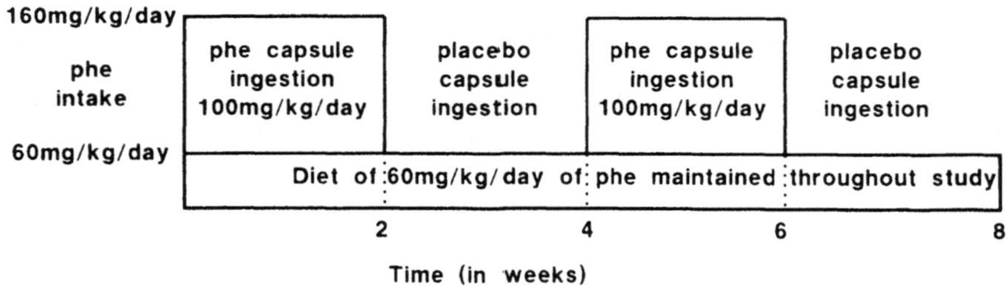

FIGURE 22.1. Double-crossover, double-blind, clinical research design. All patients were maintained on a constant Phe intake of 60 mg/kg per day. One half of the patients were "on-off-on-off," the other half "off-on-off-on," an added 100 mg of Phe per kg per day (on) or sucrose placebo (off). Capsules were given with meals and at bedtime, and neither patients nor physicians were cognizant of which protocol was in progress.

Plasma amino acids were analyzed by ion-exchange chromatography with lithium buffers using previously described methods (Krause et al. 1985, 1986, Griffin and Elsas 1975). L-DOPA was analyzed by a radioenzymatic essay (Krause et al. 1986). Blood was drawn at mid-day, in the semifasting state, following a standardized breakfast.

Electroencephalography

Eight-channel EEG studies were performed in the late morning on each of the final two days of each 2-week interval in the relaxed waking state, as previously described (Krause et al. 1986). To determine the MPF, ten 2-second samples were analyzed from each channel using an EEG fast Fourier transform (Krause et al. 1986). High-frequency (8 to 12 cps) α-bands were analyzed independently.

Neuropsychology Tests

A computerized battery of tests of higher cognitive function were performed on each subject on the final day of each 2-week period. These tests were derived from experimental cognitive psychology and reflected specific aspects of information processing (Ellis 1985). Ten tests were subjected to repetitive study in controls using the same protocol as indicated in Figure 22.1. Six tests had a high test-retest reliability and minimal artifact and were packaged into a 45-minute, computer-game format. Of these six, the Stroop task (Dyer 1973) was selected for presentation here.

Results

Effect of Phe Supplement

The relationships between mean plasma Phe concentration and mean Phe intake for all four study periods are shown in Figure 22.2. The ratio of the plasma Phe concentration to that of tyrosine, or to the summed concentrations of the other large neutral amino acids, was similarly altered in relation to the total amount of Phe ingested in the diet.

While all subjects demonstrated an increase in plasma Phe during supplemental Phe ingestion, interindividual variation is apparent in Figure 22.2. Note that D.H. had a semifasting Phe concentration of 136 μM which rose to 539 μM on Phe supplementation. She complained of emotional changes during the "on" period and withdrew from subsequent studies. G.J. is a general manager for a large microscope firm. During the "on" period, his mean plasma Phe concentration rose from 112 μM to 341 μM. He inadvertently left a microscope in an airport and complained of several episodes of "forgetfulness." Both individuals are heterozygotes for phenylketonuria but were not aware of which protocol they were receiving until

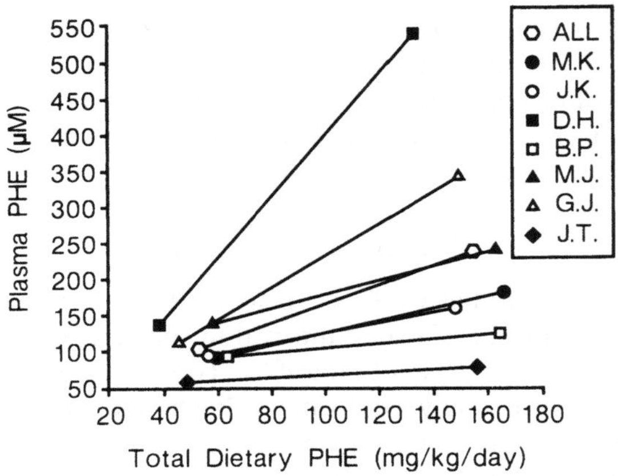

FIGURE 22.2. Effect of Phe intake on plasma Phe. Each individual is indicated by a symbol. The mean of four observations ("off" and "on" additional Phe intake in each individual are connected by a line. The average plasma Phe for all subjects "off" supplement was 99 ± 20 μM and "on" supplemental Phe was 239 ± 166 μM.

after the code was broken. Thus, the data in Figure 22.2 demonstrate differences in response to identical phenylalanine intake among different individuals. They also indicate that changes in mean plasma Phe among the seven volunteers ranged from 21 μM to 403 μM.

The absolute changes in dietary Phe, plasma Phe, and the three parameters of brain function are tabulated for each individual in Figure 22.3. A total of 17 changes were available from the seven subjects, because B.P. and D.H. completed only two of the four periodic intervals. Twenty-one changes would be expected from the protocol in Figure 22.1 if all seven subjects had undergone four periodic intervals. Chronic phenylalanine ingestion increased semifasting plasma Phe concentrations. Over a broad range of change in plasma Phe concentration, there was a change in the opposite direction of the MPF in 15 of 17 observations. For changes in serum L-DOPA, four of 17 observations were in the inverse direction, and, for cognitive function, nine of 17.

Relationship of Change in Plasma Phe to the EEG

Thus, for changes of plasma Phe between 300 and 400 μM, there were clear trends suggesting an inverse change in MPF, plasma L-DOPA, and cognitive function. The order of sensitivity for these three parameters was MPF > cognitive function > plasma L-DOPA (Figure 22.3). Of particular interest was the slowing of the MPF of the EEG when plasma Phe increased, and the converse as plasma Phe decreased. This reversible, inverse relationship held with plasma Phe changes as low as 7 μM! Since the MPF was obtained from a composite of EEG wave forms, and since statistical

Subject	Δ Dietary PHE (mg phe/kg/day)	Δ Plasma PHE (μM)	Δ MPF Channel 8	Δ DOPA (ng/ml)	Δ Stroop/Word latency score
D. H.	94.4	403.05	-0.326	-0.505	0.059
G. J.	100.4	331.11	-0.075	0.080	0.130
M. J.	-110.8	-197.92	0.085	0.360	0.039
G. J.	-111.7	-153.11	0.675	0.345	-0.066
M. K.	-98.1	-131.58	0.091	-0.675	0.110
G. J.	107.9	129.90	-0.630	-0.405	-0.127
J. K.	-106.8	-93.43	0.426	-0.320	-0.031
M. J.	103.9	78.48	-0.680	0.390	0.032
J. K.	102.6	71.83	-0.348	0.185	-0.006
J. K.	81.0	59.98	0.240	0.205	0.042
M. K.	-115.0	-50.04	-0.105	-0.355	0.073
M. K.	97.3	47.48	-0.050	0.525	-0.119
J. T.	-113.8	-27.32	0.010	-0.225	- - -
J. T.	-101.7	-13.86	0.115	-0.320	-0.025
B. P.	-102.7	-9.59	0.164	-0.645	-0.094
J. T.	106.2	7.72	-0.195	0.420	-0.071
M. J.	-99.7	-7.07	0.765	-0.225	-0.178

FIGURE 22.3. Relationships of dietary intake to changes in mean power frequency of EEG Channel 8 (MPF); plasma L-DOPA and cognitive function (Stroop Word Latency Score). A + change denotes an increase in a parameter from one 2-week interval to the next, while a − change indicates a decrease. Numbers in boxes indicate changes in the expected direction. Changes are ordered from greatest to least relative to Phe.

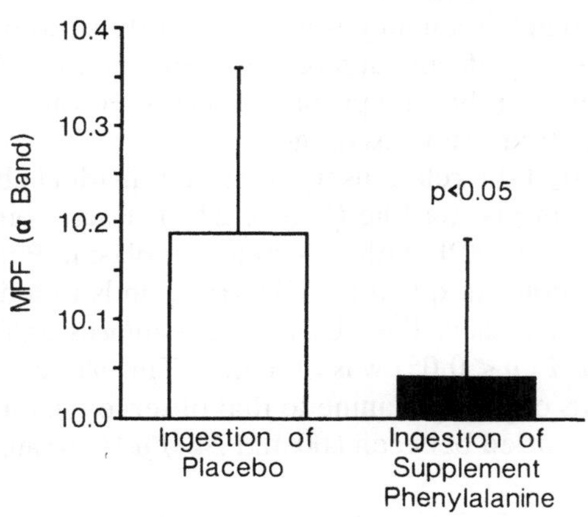

FIGURE 22.4. Mean power frequency of α-band (8–12 cps) with (on) and without (off) dietary Phe supplementation.

analysis was difficult with so few patients, it was of interest to determine which electrical wave forms were affected and to test the statistical significance of these changes (Figure 22.4).

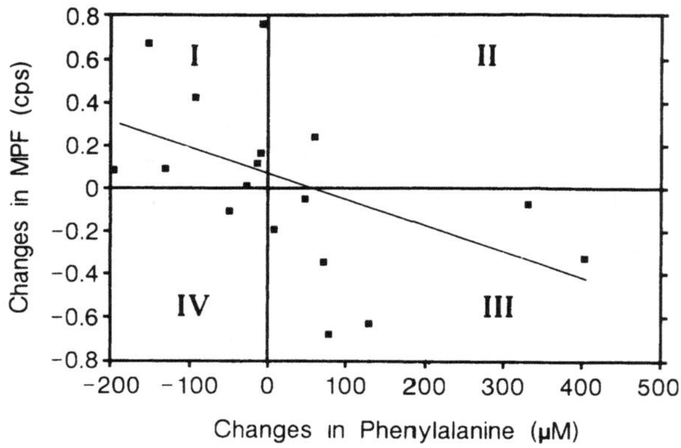

FIGURE 22.5. Relationship of each change in plasma Phe to individual changes in the mean power frequency (MPF) of electroencephalograms. Data are from Figure 22.3.

The EEG band spectrum between 8 and 12 cps (α-band) was taken as a mean for all subjects and all observations (17) on dietary Phe supplementation, and the mean was compared to that off dietary Phe supplementation. When off Phe supplementation, the mean Phe concentration was 99 ± 20 μM, and the mean band frequency was 10.19 ± 0.17 cps. When on Phe supplementation, the mean blood Phe concentration was 239 ± 166 μM, and the mean α-band frequency was significantly slowed to 10.04 ± 0.14 cps ($p < 0.05$). No significant differences were found in lower-frequency wave forms, suggesting that the primary effect of elevated plasma Phe was to slow the higher-frequency discharges.

We then evaluated the relationship of each individual change in MPF to individual change in plasma Phe (Figure 22.5). Each point in quadrant I denotes an increase in MPF with a decrease in plasma Phe concentration. Conversely, each point in quadrant III corresponds to a decrease in MPF with an increase in plasma Phe. Using least-squares analysis, a slope of -0.125 cps/100 μM ($p < 0.05$) was obtained. This slope is similar at these low concentrations of phenylalanine to that observed in our previous study when plasma Phe varied between 600 and 2400 μM (Krause et al. 1986).

Relationship to Aspartame

It was of some interest to determine to what extent chronic aspartame ingestion might affect plasma Phe in four additional heterozygotes for PKU on ad libitum diet (Figure 22.6). During control diet 1, parents of PKU patients maintained an unexpectedly low phenylalanine intake (10 to 40 mg of Phe per kg per day). This suggested that the family conformed to the low protein diet prescribed for their child. Control diet 1 was obtained

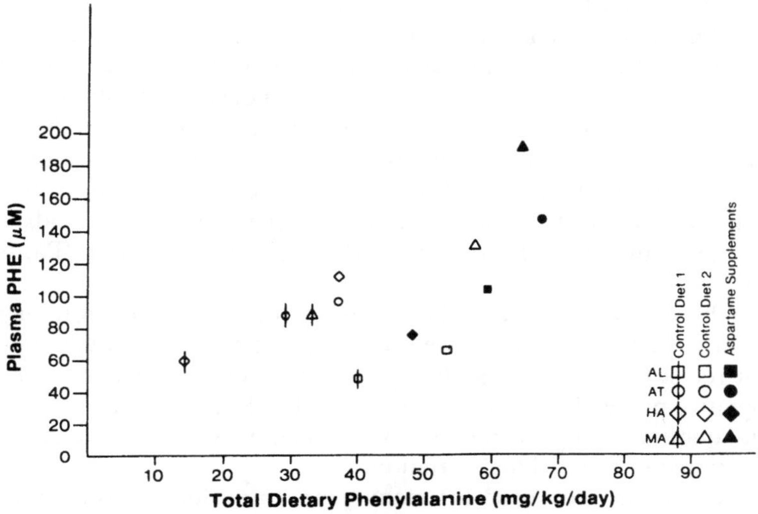

FIGURE 22.6. Relationship of plasma Phe to dietary Phe with and without aspartame supplements (34 mg/kg per day). Individual parents of classical PKU patients ingested their usual diets and encapsulated sucrose placebo or aspartame for monthly intervals. Data represent duplicate values at the end of each interval.

during late summer. Control diet 2 was during Thanksgiving and Christmas when protein and Phe intake was significantly greater (38 to 58 mg of Phe per kg per day). Aspartame was supplemented in early Fall. Thus, significant seasonal variation exists for total Phe intake and constitutes an important variable to be controlled in future studies. Of importance for this study, plasma Phe concentrations correlated directly to total dietary Phe whether ingested only as protein or supplemented by aspartame. When total Phe intake rose above 50 mg/kg per day, plasma Phe rose to concentrations between 100 and 200 μM. Our data indicate that these moderate elevations in plasma Phe produce changes in brain function which are most notable in the MPF of the EEG.

Discussion

Because humans are now ingesting free Phe in addition to dietary protein, it is of considerable importance to determine how this chronic ingestion might affect plasma Phe concentrations. Our data indicate the amount of interindividual variation in human studies of this type. Interindividual variation is probably generated in part by genetic heterogeneity of hepatic phenylalanine hydroxylase. It should be noted that in primates, as compared to rodents, this enzyme is rate-limiting in the catabolic pathway for

Phe, and thus we humans will be uniquely susceptible to the historically novel addition of phenylalanine-containing sweeteners to our diets. In fact, our data involving chronic ingestion agree with previous data on acute aspartame ingestion (Harper 1984). Both homozygous normals and heterozygotes for decreased phenylalanine hydroxylase maintain changes in plasma Phe concentrations from twice to tenfold basal concentrations. These changes in plasma Phe are unaccompanied by increases in tyrosine concentration or in the concentration of other neutral amino acids transported by common pathways across the blood-brain barrier (Pardridge 1986).

Why have studies of these Phe changes not previously been thought to affect brain function? One answer lies in the relative insensitivity and lack of test-retest reliability of psychological tests. Here we have stressed tests of cognitive processing at higher integrative levels. Intelligence or developmental quotients simply cannot be used sequentially in the same individual to determine intraindividual variation.

The EEG has been used for many years to evaluate cortical function relative to plasma Phe levels (Krause et al. 1986). We used the MPF of the EEG as a sensitive, noninvasive parameter. An inverse relationship between change in MPF and change in plasma Phe was demonstrated for changes in plasma Phe from 600 to 2400 μM, with a slope of -0.09 cps/0.1 mM (Krause et al. 1986). As shown in Figures 22.3 and 22.5, there is a slowing of the MPF for interperiod increases in plasma Phe and an increase in MPF associated with decrease in plasma Phe. This inverse relationship exists for plasma Phe concentrations as low as 7 μM. The slope determined in the present work is -0.1 cps/0.1 mM and is quite similar to the earlier value. In previous studies, this slowing in brain electrical activity corresponded to a decrease in general cognitive function and alertness (Krause et al. 1985, 1986). It is not unreasonable to assume that slowing in electrical brain wave activity during these physiological changes in plasma Phe concentrations has effects on cognition, but more observations and more sensitive tests are needed to observe statistically significant responses.

Central to the debate over the safety of the artificial sweetener aspartame is whether smaller concentration changes in plasma Phe have effects on brain function. The recent approval of aspartame by several biopolitical organizations was based on an assumption that only very large changes in plasma Phe (> 900 μM) are associated with a decrease in brain function. In the present studies, we observed an increase in plasma Phe levels in PKU heterozygotes who chronically ingested 34 mg of aspartame per kg per day for one month (Figure 22.6). When total intake of Phe rose above 50 mg/kg per day, plasma Phe rose to 100 to 200 μM. In the latter range of Phe intake, a logarithmic rather than linear relationship may hold for further dietary increases (Figure 22.6). Thus, "approved" aspartame intake superimposed on even relatively restricted dietary protein can produce increases in blood phenylalanine which slow electrical activity of the human brain.

In light of these data and the numerous reports of adverse reactions to aspartame, the effects of aspartame on brain function need further study.

Acknowledgments. Many investigators assisted in these studies. We are grateful to Norman Ellis for the development of the tests of cognitive function; Bahjat Faraj and Vernon Camp for assaying L-DOPA; Charles M. Epstein for coordinating EEG analysis; Edward Strumlauf for statistical and computer assistance; Philip Dembure and Jean Drumheller for amino acid analysis; Marcia Dunning for typing this manuscript; and Steven Yanicelli, Mary Naglak, and Phyllis Acosta, for their advice on nutrition.

References

Dyer, F.N. (1973). The Stroop phenomenon and its use in the study of perceptual, cognitive and response processes. Memory and Cognition **1:**106–120.

Ellis, N. (1985). Tests of cognitive impairment (TCI). University of Alabama, Tuscaloosa, Alabama.

Griffin, R.F., and Elsas, L.J. (1975). Classic phenylketonuria diagnosis through heterozygote detection. J. Pediatr. **86:**572–577.

Harper, A.E. (1984). Phenylalanine metabolism. *In* Stegink, L.D., and Filer, L.J. (eds.), Aspartame: physiology and biochemistry. New York: Marcel Dekker, pp. 77–111.

Krause, W., Halminski, M., McDonald, L., Dembure, P., Salvo, R., Freides, D., and Elsas, L. (1985). Biochemical and neuropsychological effects of elevated plasma phenylalanine in patients with treated phenylketonuria. J. Clin. Invest. **75:**40–48.

Krause, W., Epstein, C., Averbook, A., Dembure, P., and Elsas, L. (1986). Phenylalanine alters the mean power frequency of electroencephalograms and plasma L-DOPA in treated patients with phenylketonuria. Pediatr. Res. **20:**1112–1116.

Pardridge, W. (1986). Potential effects of the dipeptide sweetener aspartame on the brain. *In* Wurtman, R.J., and Wurtman, J.J. (eds.), Nutrition and the brain. New York: Raven Press, pp. 199–241.

Trotter, J., Epstein, C., Dembure, P., and Elsas, L. (1987). The effect of moderately elevated plasma phenylalanine on brain function. Clin. Res. **35:**61A.

Chapter 23

The Effects of Aspartame on Human Mood, Performance, and Plasma Amino Acid Levels

Harris R. Lieberman, Benjamin Caballero, Gail G. Emde, and Jerrold G. Bernstein*

Consumption of the artificial sweetener aspartame raises plasma phenylalanine levels, thereby increasing brain phenylalanine and, conceivably, affecting the syntheses of monoaminergic brain neurotransmitters known to underlie various types of behavior. We have thus assessed the effects of single aspartame doses on certain types of behavior, particularly those relating to mood and performance. In a double-blind, placebo-controlled, crossover study, a single dose of aspartame (60 mg/kg), or the same dose in combination with 37 g of carbohydrate, was administered to 20 male volunteers. [Carbohydrates enhance the entry of circulating phenylalanine into the brain by lowering plasma levels of competing large neutral amino acids (LNAA)]. A lower dose of aspartame (20 mg/kg) was also tested. Aspartame alone, or in combination with carbohydrate, did not alter any aspect of behavior that we assessed, nor did it produce detectable side effects. The ratios of plasma phenylalanine and tyrosine concentrations to those of the other LNAA were significantly increased by administration of aspartame. Since anecdotal reports of aspartame-associated neurological or behavioral side effects almost always describe effects as occurring after multiple aspartame exposures, it would be important to repeat our study using a protocol involving repeated aspartame administration.

Introduction

It has been suggested that ingestion of large quantities of aspartame could alter certain aspects of brain neurotransmission and, consequently, human behavior (Wurtman 1983, Maher and Wurtman 1983, Yokogoshi et al. 1984). One mechanism by which aspartame could alter brain function is the effect its ingestion may have on brain phenylalanine and other large neutral amino acids (LNAA) (Wurtman 1983). The LNAA—phenylalanine,

* Department of Brain and Cognitive Sciences and the Clinical Research Center, Massachusetts Institute of Technology, Cambridge, MA 02139, USA.

tyrosine, tryptophan, leucine, isoleucine, and valine—are usually found together, along with other amino acids, in protein-containing foods. Two of these, tryptophan and tyrosine, have essential central nervous system (CNS) functions—as precursors for the synthesis of a variety of monoamine neurotransmitters (Wurtman et al. 1981).

When aspartame is consumed in high doses, it can increase the plasma levels of phenylalanine beyond the normal physiological range (Stegink et al. 1980, Wurtman 1983). Changes in the peripheral concentration of any LNAA can directly affect brain LNAA levels because all of these amino acids are actively transported from the plasma into the brain by a single carrier system. Aspartame-induced changes in brain phenylalanine are likely to be greatest when it is consumed alone, without the other LNAA that are usually ingested with phenylalanine when actual foods are eaten, or when it is given in combination with carbohydrate-rich foods.

This study was conducted to determine whether acute doses of aspartame alone and with a carbohydrate could modify certain human behaviors, specifically those previously shown to be associated with monoaminergic neurotransmission.

Methods

This study was conducted using a double-blind, placebo-controlled, crossover design. Twenty-three healthy male subjects, aged 18 to 38, were enrolled after each gave written informed consent. Twenty subjects completed the study: two had to be excluded as a result of procedural errors and one subject voluntarily withdrew. None of the subjects reported, prior to enrollment in the study, having experienced any adverse reactions to aspartame. Each subject who completed the study participated in six sessions. The first was a practice session to acquaint him with the behavioral tests. The next five were treatment conditions during which he received, in a counterbalanced order determined by a Latin-square design: aspartame (60 mg/kg) and carbohydrate placebo; aspartame (60 mg/kg) and 36 g of carbohydrate; aspartame (20 mg/kg) and carbohydrate placebo; aspartame placebo and carbohydrate; aspartame placebo and carbohydrate placebo. All doses of aspartame and its placebo were administered in the same quantity of identical capsules.

Prior to each test session, subjects fasted from midnight and then ingested a 120-kcal granola bar at 7:30 A.M. No additional food was consumed until ingestion of the experimental agents at 11:00 A.M. Performance and mood were then assessed until mid-afternoon. In addition, two 10-ml blood samples, to measure plasma amino acid levels, were drawn, one at the start of each session and one three hours later. After completing a side effects questionnaire, the subjects were discharged at 2:30 P.M.

Behavioral Tests Administered

The eight behavioral tests used in this study were:

1. Short-duration auditory reaction time: This test was presented automatically by a microcomputer (Lieberman et al. 1983). After 25 practice trials, data from 100 further trials were recorded.
2. Four-choice reaction time: In this test, subjects are repeatedly presented, for approximately 10 minutes, with one of four similar visual stimuli on a CRT screen (Lieberman et al. 1986b).
3. Wilkinson auditory vigilance test: In this modified version of the original test, subjects were required to detect the occurrence of short tones embedded in slightly longer tones and random noise (Lieberman et al. 1986b). Since this test requires one hour to complete, it is difficult to maintain the necessary concentration.
4. Digit Symbol Substitution Test (DSST): The DSST is a timed (90 seconds) task taken from the Wechsler IQ test. The subject is presented with a series of digits and a code which identifies each digit with a particular symbol.
5. Tapping: For this motor task, the subject must alternately tap, for two minutes, two wedge-shaped targets with a metal stylus.
6. Profile of mood states (POMS): The POMS is a self-report mood questionnaire yielding factor-analytically-derived scales for Tension-Anxiety, Depression-Dejection, Anger-Hostility, Vigor-Activity, Fatigue-Inertia, and Confusion-Bewilderment (McNair et al. 1971).
7. Visual analogue mood scales (VAMS): The VAMS consists of 32 scales that have been used to assess changes in mood after various treatments (Spring et al. 1983).
8. Stanford sleepiness scale (SSS): The SSS is a seven-point scale that was developed to quantify the progressive stages of sleepiness (Hoddes et al. 1972).

Amino Acid Assays

Plasma amino acids were assayed by high-performance liquid chromatography with fluorometric detection.

Results

Latin-square analyses of variance (ANOVA) were performed on the scores of each of the mood and performance tests. No effects of aspartame alone or aspartame in combination with carbohydrate on any of these tests were detected. Additionally, the side effects questionnaire administered found no significant effects of aspartame.

To determine whether the treatments altered plasma levels of phenylalanine and the other LNAA, four of the subjects were randomly selected and their samples analyzed. The ratio of phenylalanine, tyrosine, and tryptophan concentrations to those of the other LNAA was each derived since these ratios are the best peripheral indicators of LNAA transport across the blood-brain barrier, and thus of brain LNAA levels. Individual repeated-measure ANOVAs were then performed on each of the three ratios computed.

Aspartame significantly increased plasma phenylalanine ratio when administered in a dose of 60 mg/kg alone ($p < 0.01$), when this dose was combined with carbohydrate ($p < 0.01$), and in a dose of 20 mg/kg ($p < 0.05$) compared to either placebo alone or placebo plus carbohydrate. Carbohydrate potentiated the effects of aspartame on phenylalanine ratio, i.e., the combination of 60 mg of aspartame and 37 g of carbohydrate induced a greater change than this dose of aspartame alone ($p < 0.05$). Tyrosine ratio also significantly increased after aspartame administration but only when the highest dose (alone and with carbohydrate) was administered ($p < 0.05$). Tryptophan ratio was not altered by any of the treatments administered. However, plasma samples were taken well after peak phenylalanine levels and alterations in tryptophan ratio resulting from carbohydrate administration had occurred (Lieberman et al. 1986a).

Discussion

No behavioral effects of aspartame were detected by any of the performance tests or mood questionnaires administered nor were any adverse effects detected by the side effects questionnaire. Since a high dose of aspartame was administered in this study (both with and without a moderate amount of carbohydrate) and no behavioral changes or side effects were observed, it appears likely that more moderate levels of aspartame ingestion will have few acute adverse effects on most individuals. (Of course, the present study excluded individuals who claimed to be sensitive to this substance.)

It should be noted, however, that although we tested several behaviors and used a high dose of aspartame, it is impossible to rule out the possibility that aspartame alters some other aspect of brain function. Behavioral effects of aspartame might be observed, for example, if different doses were administered, or if other behavioral tests were employed, or if the sweetener were consumed repeatedly for days or weeks. The tests we used are sensitive to the effects of a variety of drugs and food constituents on human behavior, particularly those that appear to exert their effects by altering plasma amino acid levels (Lieberman et al. 1986a). However, aspartame could affect some parameter not typically evaluated in studies of food constituents or drugs, or a particularly sensitive subgroup of the population. Most of the anecdotal reports describing aspartame-associated neurological or behavioral adverse reactions involved subjects who had consumed the sweetener on numerous consecutive days, and not only on one occasion, as was the case in our study. It is therefore important that behavioral studies be conducted using a multiple-treatment protocol.

Acknowledgment. This work was supported by the Center for Brain Science and Metabolism Charitable Trust, Cambridge, MA.

References

Hoddes, E., Dement, W., and Zarcone, V. (1972). The history and use of the Stanford Sleepiness Scale. Psychophysiology **9:**150.

Lieberman, H.R., Corkin, S., Spring, B.J., Growdon, J.H., and Wurtman, R.J. (1983). Mood, performance, and pain sensitivity: changes induced by food constituents. J. Psychiatr. Res. **17:**135.

Lieberman, H.R., Caballero, B., and Finer, N. (1986a). The composition of lunch determines afternoon plasma tryptophan ratios in humans. J. Neural Transm. **65:**211–217.

Lieberman, H.R., Spring, B.J., and Garfield, G.S. (1986b). The behavioral effects of food constituents: strategies used in studies of amino acids, protein, carbohydrate and caffeine. Nutr. Rev. **44**(suppl.):61–70.

Maher, T.J., and Wurtman, R.J. (1983). High doses of aspartame reduce blood pressure in spontaneously hypertensive rats. N. Engl. J. Med. **309:**1125.

McNair, P.M., Lorr, M., and Droppleman, L.F. (1971). Profile of Mood States Manual, Educational and Industrial Testing Service, San Diego, California.

Spring, B.J., Maller, O., Wurtman, J., Digman, L., and Cozolino, L., (1983). Effects of protein and carbohydrate meals on mood and performance: interactions with sex and age. J. Psychiatr. Res. **17:**155.

Stegink, L.D., Filer, L.J., Baker, G.L., and McDonnell, J.E. (1980). Effect of an abuse dose of aspartame upon plasma and erythrocyte levels of amino acids in phenylketonuric heterozygous and normal adults. J. Nutr. **110:**2216–2224.

Wurtman, R.J. (1983). Neurochemical changes following high-dose aspartame with dietary carbohydrate. N. Engl. J. Med. **309:**429–30.

Wurtman, R.J., Hefti, F., and Melamed, E. (1981). Precursor control of neurotransmitter synthesis. Pharmacol. Rev. **32:**315–335.

Yokogoshi, H., Roberts, C.H., Caballero, B., and Wurtman, R.J. (1984). Effects of aspartame and glucose administration on brain and plasma levels of large neutral amino acids and brain 5-hydroxyindoles. Am. J. Clin. Nutr. **40:**1–7.

Chapter 24

Aspartame and Behavior in Children

Mark L. Wolraich*

The purpose of the present chapter is to review what actual evidence exists on the relationship between aspartame and behavior in children. Nine challenge studies employing about 10 mg/kg of aspartame found no difference in children's behavior or cognitive performance challenged with aspartame or sugar (sucrose or fructose). One study found no effect even with a larger dose of 30 mg/kg. These studies do not support a concern that aspartame has adverse effects on the behavior of children, but they were limited because they only examined short-term effects. The effects of chronic ingestion of large doses of aspartame remains unknown.

The theoretical concerns about and the results of research into the effects of aspartame on animals are presented in other chapters of this book. The purpose of the present chapter is to review what actual evidence exists on the relationship between aspartame and behavior in children. It is this information, dealing with the effects of aspartame on humans, that is essential for making any decisions about any potentially harmful effects.

In examining the existing evidence on the effects of aspartame on the behavior of children, it is helpful to first review the process by which these effects are studied. Concerns are frequently raised from anecdotal reports of parents or clinicians who suspect that effects they are seeing are due to the ingestion of aspartame. While personal observation is an important first step in raising concerns, such reports do not constitute proof of adverse effects. Particularly in studies of human behavior, the power of suggestion is so great that almost any effect can be seen if the patients, family, or clinicians believe it exists. For example, even in trials with stimulant medication for hyperactivity, a proven form of treatment (Wolraich 1977), a 20% placebo response rate can be anticipated (Ullmann and Sleator 1986). Therefore, it is essential that any reports of behavioral effects that are to be considered as evidence be limited to double-blind controlled studies.

* Division of Developmental Disabilities, University of Iowa, Iowa City IA 52242, USA.

Double-blind controlled studies employ objective research designs which attempt to control systematic biases as much as possible. There are several elements essential to the design. To begin with, the studies need to employ a control group. This is particularly important in studying children, because they tend to mature over time in their normal course of development. A control group consists of a sample of subjects who are observed with the same dependent measures, but who do not receive the intervention under question. A matched group of subjects can be employed as controls or, if the intervention is not anticipated to have lasting effects, the subjects can be employed as their own controls in a crossover design.

The subjects and the researchers need to be unaware of which subject receives the intervention. This is accomplished by employing a placebo intervention and a double-blind design, in which neither the subjects nor the researchers know when the subjects receive the actual or the placebo interventions. In addition, it is important to utilize assessment procedures that are generally accepted, and as objective as possible. The studies reviewed in this chapter have all met these criteria.

Studies attempting to assess an intervention, such as aspartame use, can be accomplished by one of two methods. One method would be to vary the diet of the subjects over a period of time, employing diets with high, low, or no aspartame content, with assessments undertaken during and at the end of each diet period. The diet that has little or no aspartame would serve as the control diet, and the diets would have to appear sufficiently similar so that the subjects and their families would not be able to identify those with aspartame-sweetened foods. This type of study would have the advantage of being ecologically valid, for it approximates the real world. Unfortunately, no such studies have been completed to date. They are expensive studies, since to ensure subject compliance requires providing the food to the families and monitoring the family intake.

The other method employed to assess aspartame is to challenge subjects with aspartame or placebo and then measure the effects for a period of time after the challenge. These studies are less expensive to undertake, but they can only measure short-term effects. Nine challenge studies have been completed, employing aspartame or sugar as the challenge agents. Some caution has to be taken in interpreting the results of these studies, because they were designed to assess the behavioral effects of sugar where the aspartame was actually employed as the placebo. In addition to these, one study (Kruesi et al. 1986) was undertaken specifically to study aspartame effects.

There are two particular findings that suggest the results of these studies of the behavioral effects of sugar are also applicable to aspartame effects. First, two studies (Behar et al. 1984, Goss 1984) employed saccharin as the placebo instead of aspartame. The results of these studies were similar to those employing aspartame. This suggests that it is not likely that aspartame caused behavioral changes that also masked changes due to sugar. It is possible that saccharin also results in equivalent behavioral changes, but it would be highly unlikely to find equivalent changes for all three substances. Second, the two studies undertaken by Wolraich et al. (1985), in addition to measuring behaviors on the two challenge days when sucrose and aspartame were administered, also measured behaviors on a baseline day when there were no challenges. No significant differences were observed. The two results therefore suggest that a lack of observed behavioral differences in children when they were challenged with sugar, and when they were challenged with aspartame, are due to an absence of behavioral effects for both of these substances, rather than an equivalent change.

The two studies by Wolraich et al. (1985) examined the behavior of 16 hyperactive boys in each of the studies. They were between the ages of 7 and 12 in the first study, and 8 and 12 in the second. The first day was used to establish baseline levels, while on the second and third days, the subjects were challenged with a sucrose dose of 1.75 g/kg, or aspartame in the range of 6.37 to 8.75 mg/kg depending upon the flavor of beverage chosen (mean dose, 197.3 mg). The order of presentation was counterbalanced and concealed. The assessment battery started one-half hour after the boys received the drink and included cognitive and laboratory tests and behavioral observations which took two and one-half hours to complete. In the first study, the drink was administered one and one-half hours after the subjects completed lunch. In the second study, the drink was administered in the morning after an overnight fast. Thirty-seven dependent variables were employed, including playroom observations, examiner ratings, learning and memory tasks, and measures of impulsivity. No significant differences were found between the aspartame and sucrose challenges for any of the 37 variables, and as was stated previously, no differences were found between the two challenge days and the baseline day.

A similar study of 16 hyperactive boys (Milich and Pelham 1986) was performed in a classroom setting. The criteria for the subjects, the doses of sucrose and aspartame, and the overall design were similar to the previous two studies (Wolraich et al. 1985). The dependent variables included measures of classroom behavior, academic productivity and accuracy, noncompliance with adult requests, and positive and negative peer interactions. No significant differences were found between the challenges of sucrose and aspartame on any of the measures. The importance of this study was that it was performed in the children's natural setting and employed important functional measures.

Two studies were undertaken with normal preschool children (3 to 5 years of age) (Ferguson et al. 1986, Goldman et al. 1984). The study by Ferguson et al. (1986) examined 18 normal preschool children who were challenged with apple juice containing either 30 g of sucrose or 167 mg of aspartame, administered in a random-order, double-blind design. The measures collected included behavioral observations during the laboratory testing, actometer readings, parent and teacher ratings, and developmental assessments (including drawings, peg boards, and coordination tasks). The results revealed no significant effects for any of the measures although three boys were noted to produce poorer developmental drawing on the sucrose challenge days.

The second study (Goldman et al. 1984) employed 8 preschool subjects who were observed in a free-play situation and tested with a continuous performance test after challenges with sucrose (2 g/kg) or aspartame. The dose of aspartame was not specified, but since the drinks were matched for sweetness, it would be safe to expect that the aspartame doses were at least in a range similar to that of the previous studies described. On the continuous performance test, the children were found to increase their errors over time—in other words, their performance worsened—on the days when challenged with sucrose, while they decreased their errors over time—or improved their performance—on the days when challenged with aspartame. Neither of these studies provide evidence for behavior effects of aspartame in normal preschool children.

Two studies (Mahan et al. 1984, Ferguson et al. 1986) examined children who were thought by their parents to be adversely affected by sugar. These were elementary-school children. The findings of these studies were similar to the studies cited previously. Two other studies examined a group of children with a mixture of psychiatric disorders (Conners 1983). These studies obtained conflicting results. In one study, motor activity was significantly higher on sugar as compared to aspartame challenges; in the other, the findings were reversed. The two studies employed a number of other dependent variables besides motor activity, none of which demonstrated significant differences. In addition, while the motor changes were significantly different, they were not striking clinical changes, given the absence of other effects. It is important for changes to be clinically significant as well as statistically significant. Further, one of the studies which employed saccharin (Behar et al. 1984) also found a decrease in motor activity after several hours. This finding would suggest that the changes were more likely due to effects of the sugar rather than the effects of aspartame.

All of the studies described employed doses of aspartame in the range of about 10 mg/kg. This is considerably less than the quantities suggested as possible by Pardridge (1986). Only one study (Kruesi et al. 1986) tested children with a higher dose (30 mg/kg). This study also considered aspartame as a potential causative agent from the outset, rather than utilizing it as the placebo. It tested 30 preschool boys. Nineteen were boys whose parents or teachers felt their behavior worsened after they ingested sugar, or who felt there was a history of improved behavior when sugar in their diet was restricted. The remaining 11 boys were familiar playmates of the other subjects, and were not regarded as responders. All of the 11 non-responders and 8 of the 19 responders had no psychiatric diagnoses, while the other 11 had a mixture of psychiatric diagnoses, with 36% of them diagnosed as having an attention deficit disorder with hyperactivity.

The children participated in eight challenge days, separated by a day to allow for washout. The challenges consisted of sucrose and glucose at 1.75 g/kg, aspartame at 30 mg/kg, and saccharin in an unspecified dose. The challenges were presented in a random-order, double-blind design after a standardized high-carbohydrate breakfast.

One of the challenges occurred in a playroom setting with a responder and peer nonresponder present. The observation included a free-play situation for the first 90 minutes, followed by a cleanup time, a 15-minute storytime, and placement in a standard object conflict situation. Aggression and emotional reactivity were viewed on videotape and rated by independent raters who were unaware of the challenge conditions. Motor activity was measured by actometers, and the playroom teacher also completed a rating scale. The second challenge was undertaken at home and was administered by the parents, who also completed a behavior rating scale.

No significant differences between any of the challenges were found by the independent observers, teachers, or parents in either of the settings. The only measure to show any significant difference was motor activity. There was significantly less motor activity following aspartame ingestion compared to either glucose or sucrose, but not when compared to saccharin. The findings of the challenge with a higher dose of aspartame were similar to the previous studies, and again did not support a relationship between aspartame ingestion and behavioral changes in children.

Conclusion

The results of all the empirical studies to date do not support a causal relationship between aspartame and behavior in children. These studies are limited in that all were short-term challenges. Long-term chronic ingestion could have effects that would not be detected by these studies. As stated earlier, all but one of the studies were not originally designed to examine aspartame effects; only one study compared the behavior of the children after ingestion of another artificial sweetener, and only two studies reported baseline measures. It is also possible that the studies have not yet identified the parameters affected by aspartame, or identified children who are vulnerable to its effects. However, even when given these limitations, at this point one would have to conclude that present empirical evidence from clinical studies does not provide support for the concern that aspartame has an adverse effect on the behavior of children.

References

Behar, D., Rapoport, J., Adams, A., Berg, C., and Cornbath, M. (1984). Sugar challenge testing with children considered behaviorally "sugar reactive." J. Nutr. Behav. **1**:277–288.

Conners, C.K. (1983). The effect of sucrose, fructose, and aspartame on behavior, cognitive performance, and brain function in hospitalized children. Final report of a project supported by the Sugar Association, Washington, D.C., October.

Ferguson, H.B., Stoddart, C., and Simeon, J.G. (1986). Double blinded challenge studies of behavioral and cognitive effects of sucrose-aspartame ingestion in normal children. Nutr. Rev. **44**:144–150.

Goldman, J.A., Lerman, R.H., Contois, J.H., and Udall, J.N. (1984). The behavior of preschool children following ingestion of sucrose. Paper presented at the annual meeting of the American Psychological Association, Toronto, Canada.

Gross, M. (1984). Effect of sucrose on hyperkinetic children. Pediatrics **74**:876–878.

Kruesi, M., Rapoport, J.L., Cummings, E.M., Berg, C., Ismond, D., Zahn-Waxler, C., Flament, M., and Yarrow, M. (1986). Aspartame and children's behavior. *In* International Aspartame Workshop Proceedings, Nov. 17–21, 1986, Aspartame Technical Committee, International Life Sciences Institute-Nutrition Foundation, Washington, D.C.

Mahan, L.K., Chase, M., Furukawa, C.T., Shapiro, G.G., Pierson, W.E., and Bierman, W. (1984). Sugar "allergy" and children's behavior. Presented at the Annual Meeting of the American Academy of Pediatrics, Washington, D.C.

Milich, R., and Pelham, W.E. (1986). The effects of sugar ingestion on the classroom and playgroup behavior of attention deficit disordered boys. J. Consult. Clin. Psychol. **54**:714–718.

Pardridge, W.M. (1986). The safety of aspartame. J. Am. Med. Assoc. **256**:2678.

Ullmann, P.K., and Sleator, E.K. (1986). Responders, nonresponders, and placebo responders among children with attention deficit disorder. Clin. Pediatr. **25**:594–599.

Wolraich, M.L. (1977). Stimulant drug therapy in hyperactive children: research and clinical implications. Pediatrics **60**:512–518.

Wolraich, M.L., Milich, R., Stumbo, P., and Schultz, F. (1985). Effects of sucrose ingestion on the behavior of hyperactive boys. J. Pediatr. **106**:675–681.

CHAPTER 25

Responses to Carbohydrate Consumption Among Insulin-Dependent Diabetics

Bonnie Spring,* Michael J. Bourgeois,† Margarette Harden,‡ Robert Garvin,* and Gary Chong*

Healthy adults have been found to report heightened fatigue and sleepiness two hours after eating an unbalanced, high-carbohydrate meal. A plausible underlying mechanism is that the insulin secretion triggered by such a meal permits enhanced brain influx of tryptophan and serotonin synthesis by lessening plasma levels of the competing branched-chain amino acids. We tested the hypothesis that, after consuming an unbalanced carbohydrate meal, insulin-dependent diabetics would show evidence of insulin resistance and would fail to report fatigue. Self-reported fatigue, serum insulin and plasma glucose were assessed before and after healthy subjects ($n = 7$) and insulin-dependent diabetics ($n = 10$) ate a 799-calorie lunch supplying 105 g carbohydrate and 0.7 g protein. Unlike normals, diabetics failed to exhibit fatigue after consuming the carbohydrate lunch. Diabetics also showed evidence of insulin insensitivity, failing to suppress glucose to clinically acceptable levels even in the presence of elevated insulin levels. Findings suggest that insulin insensitivity may blunt behavioral responses to carbohydrate consumption.

Rationale

Many consumers ingest beverages sweetened by aspartame together with carbohydrate-rich, protein-poor snacks. By triggering insulin secretion, unbalanced carbohydrate-rich snacks cause the branched-chain amino acids to leave the bloodstream and be taken up by muscle. With these competing plasma amino acids lessened, brain influx of phenylalanine and tryptophan are enhanced, the latter with partially known behavioral consequences, and the former with unknown but possible behavioral effects.

Insulin is believed to "drive" the elevation in the plasma tryptophan ratio that triggers behavioral change after carbohydrate consumption. Presumably, a rise in the plasma phenylalanine ratio is also facilitated

* Psychology Department, Texas Tech University, Lubbock, TX 79409, USA.
† Pediatrics Department, Texas Tech University, Lubbock TX 79409, USA.
‡ Department of Food & Nutrition, Texas Tech University, Lubbock, TX 79409, USA.

by carbohydrate-induced insulin secretion, when aspartame is ingested together with carbohydrate snacks. We therefore studied the response to carbohydrate consumption in insulin-dependent diabetics who are unable to produce endogenous insulin. Our aim was to determine whether this population responds similarly to normal controls after consuming a high-carbohydrate, low-protein meal.

Healthy adults have been found to report heightened fatigue and sleepiness two hours after eating an unbalanced, high-carbohydrate meal (Hartmann et al. 1977, Spring et al. 1986). Females show drowsiness more consistently than males (Spring et al. 1983). In contrast, obese adults who exhibit carbohydrate craving report little change in fatigue (Lieberman et al. 1986). In addition, many obese individuals show below-normal plasma tryptophan ratios (Heraief et al., 1983). Their ratios are characterized by elevated levels of the branched-chain amino acids, which result, presumably, from insulin resistence. Like obese adults with non-insulin-dependent diabetes, juvenile-onset insulin-dependent diabetics may manifest insulin resistance, although the pathophysiology appears to differ. Whereas non-insulin-dependent diabetes with obesity is characterized by receptor insensitivity due to a decreased number of insulin receptors, in insulin-dependent diabetes, insulin-bound antibodies prevent insulin from attaching to receptors. We tested the hypothesis that, after consuming an unbalanced carbohydrate meal, insulin-dependent diabetics would show evidence of insulin resistance and would fail to report increased fatigue.

Subjects

Subjects were 17 female students aged 16 to 29: 10 juvenile-onset, insulin-dependent diabetics and 7 normal controls. Experimental subjects had received treatment for diabetes for at least two years. None endogenously produced insulin, as evidenced by an absence of C peptide. Prior to entry into the protocol, subjects were examined by a physician to establish that they were in good physical health and not at risk from study procedures.

Procedure

Subjects fasted from 8:00 P.M. and came to the laboratory the following morning. At 7:00 A.M., diabetic subjects took their usual doses of both regular insulin and intermediate-acting NPH or lente insulin. Subjects ate breakfast at 7:30 A.M. (2 pieces whole wheat toast, 1 tbs jam, 4 oz orange juice, 1 cup skim milk), and then remained in the laboratory engaged in sedentary activities throughout the morning. At 10:30 A.M., an intracatch was inserted into the subject's forearm vein to permit periodic 10-ml blood samples to be taken 30 minutes prior to lunch, and 45, 90, 135, and 165 minutes after lunch. Mood was assessed by a self-report questionnaire, the

TABLE 25.1. POMS fatigue, serum insulin, and plasma glucose pre- and post-lunch in juvenile diabetics and normals.

		Mean (SD) values for pre- and post-lunch samples				
Measure	Group	Pre	Post 1	Post 2	Post 3	Post 4
Fatigue	Normal	3.0 (3.1)	5.0 (3.8)	7.4 (6.0)	10.0 (8.6)	5.3 (5.1)
	Diabetic	4.4 (4.2)	3.2 (2.3)	4.1 (4.0)	4.7 (3.3)	3.8 (2.3)
Insulin (micro IU/ml)	Normal	3.7 (1.3)	32.0 (23.0)	37.6 (24.4)	28.2 (17.9)	23.6 (12.5)
	Diabetic	124.1 (96.4)	116.0 (94.6)	109.4 (83.9)	116.9 (94.4)	124.6 (119.2)
Glucose (mg/dl)	Normal	80.0 (2.3)	123.0 (20.4)	121.1 (18.9)	108.9 (18.4)	102.9 (17.6)
	Diabetic	145.1 (72.9)	239.2 (50.7)	270.0 (45.4)	282.2 (55.2)	282.0 (61.8)

Profile of Mood States (POMS), 45 minutes prior to lunch, and 30, 75, 120, and 150 minutes after lunch. Lunch, served between 12:15 and 12:45 P.M., was six lunch bars supplying 105 g carbohydrate, 0.7 g protein, 42.7 g fat, and 799 calories.

Each blood sample was divided into two aliquots: one with EDTA anticoagulant and one without anticoagulant. The anticoagulated aliquot was centrifuged and the plasma collected for glucose determination performed within 10 minutes of the separation. The remaining aliquot was allowed to clot at room temperature, centrifuged, and the serum removed and frozen at −20°C until insulin determination. Plasma glucose was determined by the glucose oxidase method with a Beckman Glucose Analyzer 2. Serum total insulin levels were determined by solid-phase ^{125}I radioimmunoassay using a commercially available kit (Coat-A-Count, Diagnostic Products Corp, TKIN2).

Data were analyzed by split-plot analysis of variance with the diagnostic group as the between factor and time as the repeated measures factor.

Results

A planned comparison supported the hypothesis that two hours after consuming carbohydrate, normal controls report greater fatigue than diabetic subjects [$t(15) = 1.79$, $p < 0.05$, one-tailed test). Considering all time samples, the Group X Time interaction for fatigue only approached significance [$F(4,60) = 2.10$, $p < 0.10$]. Although both groups displayed substantial variability, inspection of Table 25.1 indicates that diabetic subjects reported low and stable fatigue levels throughout the protocol, in contrast to normals whose mean fatigue score more than tripled two hours after lunch.

Serum total insulin levels were significantly higher for diabetics than controls throughout the protocol [$F(1,15) = 6.79$, $p < 0.02$], and the magnitude of this difference did not vary with time. Differences between the two groups on plasma glucose did vary with time [$F(4,60) = 16.67$, $p = 0.001$]. Prior to lunch, the groups did not differ in plasma glucose. After lunch at all time samples, diabetics showed significantly elevated glucose levels ($p < 0.05$, Neuman-Keuls test).

Discussion

The diabetics' significantly elevated insulin levels, coupled with their failure to suppress glucose to clinically acceptable levels, are consistent with the likelihood of insulin insensitivity in this population. The absence of carbohydrate-induced fatigue in insulin-dependent diabetics, even in the presence of high serum insulin levels, suggests that insulin insensitivity may blunt behavioral responses to carbohydrate consumption. If carbohydrates normally potentiate the behavioral effects of phenylalanine, findings suggest the hypothesis that insulin-resistant individuals may prove relatively resilient to such effects.

References

Hartmann, E., Spinweber, C., and Fernstrom, J. (1977). Diet, amino acids and sleep. Sleep Res. **6**:61.

Heraief, E., Burckhardt, P., Mauron, P., Wurtman, J., and Wurtman, R.J. (1983). The treatment of obesity by carbohydrate deprivation suppresses plasma tryptophan and its ratio to other large neutral amino acids. J. Neural Transm. **57**:187–195.

Lieberman, H.R., Wurtman, J.J., and Chew, B. (1986). Changes in mood after carbohydrate consumption among obese individuals. Am. J. Clin. Nutr. **44**:772–778.

Spring, B., Chiodo, J., Harden, M., Bourgeois, M., Lutherer, L., Harner, D., Crowell, S., and Swope, G. (1986). Effects of noon meals varying in nutrient composition on plasma amino acids, glucose, insulin and behavior. Psychopharm. Bull. **22**:1026–1029.

Spring, B., Maller, O., Wurtman, J., Digman, L., and Cozolino, L. (1983). Effects of protein and carbohydrate meals on mood and performance: interactions with sex and age. J. Psychiatr. Res. **17**:155–167.

Part V Behavioral and Neurochemical Effects of Aspartame and Phenylalanine in Patients with Phenylketonuria and Hyperphenylalaninemia

CHAPTER 26

Epidemiology and Natural History of Phenylketonuria and Other Hyperphenylalaninemias

Flemming Güttler*

Millions of newborn infants all over the world are screened today for phenylketonuria (PKU), and many thousands of early-treated children and young persons are bright and completely normal. Research workers from many disciplines have cooperated to elucidate the biochemical and genetic background of PKU and other hyperphenylalaninemias and to prevent these inherited metabolic errors before clinical symptoms appear. This is a brief summary of the history of hyperphenylalaninemia covering a period of five decades from the time of Følling's first publication in 1934.

Asbjørn Følling's Discovery of a Recessive Disorder in the Metabolism of Phenylalanine Associated with Mental Retardation

In Spring 1934, a mother consulted the Norwegian physician and biochemist, Dr. Asbjørn Følling, because her daughter and son were both mentally retarded. The matter of a peculiar and clinging odor of both her children was distressing. By clinical examination of the children, Asbjørn Følling found no valuable signs except that the children were definitely mentally retarded. Ordinary urine analyses were normal, but after adding ferric chloride to the urine to test for diacetic acid, a green color appeared instead of the expected red-brown color. Intrigued by this finding, Asbjørn Følling set out to isolate and identify the unknown substance. During the extraction procedures, the green color with ferric chloride indicated in which fraction the substance was present. During the following six weeks, Asbjørn Følling succeeded in tentatively identifying the substance as phenylpyruvate. The purified substance and synthesized phenylpyruvate both had the same properties and melting point. Having identified the abnormal substance, Asbjørn Følling hypothesized an association between

* The John F. Kennedy Institute, DK-2600 Glostrup, Copenhagen, Denmark.

the mental defect and the phenylpyruvate excretion. A survey of 430 children in institutions for the mentally retarded led to the discovery of eight more cases including another two pairs of siblings. In the space of five months, Asbjørn Følling published his famous report (Følling 1934) on a new metabolic disease. He correctly suggested that the new disease was an inherited disorder of phenylalanine metabolism, and he called it "Imbecillitas phenylpyrouvica."

Identification of the Metabolic Error in Phenylketonuria

In 1947, Jervis showed that the administration of phenylalanine to normal humans led to a prompt rise in blood tyrosine, whereas no increase in blood tyrosine could be detected in patients with phenylketonuria (PKU), indicating both the normal pathway of phenylalanine metabolism and the metabolic error in PKU. Jervis (1953) later showed that postmortem liver samples from control individuals could catalyze the conversion of phenylalanine to tyrosine in vitro, whereas liver samples from two patients with PKU could not.

Initiation of Dietary Treatment and Blood Screening

Exactly 20 years after Følling's first paper, Bickel et al. (1954) published the result of dietary treatment. It was a bold experiment which radically changed the disease to a preventable form of mental retardation of interest both to the pediatrician as well as to the public authorities. Treatment with a low-phenylalanine diet did not only lead to a fall in the level of phenylalanine in blood and urine, but also to an improvement of neurological symptoms, such as seizures, whereas the improvement in IQ was less dramatic. This first report was quickly followed by others, and the final conclusion was that therapy would probably be most effective if diet was initiated as early as possible.

So, it soon became obvious that the best results were achieved by preventive therapy, which meant screening of all newborn for the metabolic error. The earliest methods used followed Følling's lead, i.e., measurement of phenylpyruvate in the urine with ferric chloride. While this measurement is still useful in the diagnosis of the disease in children and adults, too many false negative results were obtained in the neonates. This stimulated a drive toward developing a diagnostic method for measuring phenylalanine in the blood. Guthrie and Susi (1963) met this demand by the development of a bacterial "inhibition assay." Their method is a sensitive, specific, inexpensive, and rapid method for the determination of blood phenylalanine in a large number of samples.

Dietary Treatment of PKU

Numerous reports have appeared describing the regime for dietary treatment of PKU, with the results judged by the intellectual and physical development of the children. In order to assess the consequences of dietary therapy more conclusively, nationwide collaborative studies were initiated in the United States, the United Kingdom, and Germany. In 1971, the first report appeared describing the observations from these large centers (Bickel et al. 1971). It became evident that children with PKU who were treated within the first month, if possible within the first two weeks, made normal developmental progress. The median IQ of 57 early-treated PKU children examined at age 7 to 15 years was 115, with a range of 80 to 145. Dietary treatment in these children was initiated within the first month of life (median age, 11 days) (Güttler 1984).

Wherever the treatment of phenylketonuria is undertaken, construction of the low-phenylalanine diet follows the same basic principle, namely, the provision of enough protein (and other essential food) to ensure maximal growth and development, while at the same time reducing the phenylalanine intake sufficiently to ensure that the patient's serum phenylalanine is kept at approximately 5 times the normal level, i.e., 5 mg/dl (300 μmol/l), employing a phenylalanine-free hydrolysate or an amino acid mixture.

Discovery of Different Phenotypes of PKU

The shift in the screening procedure of neonates for hyperphenylalaninemia led to the discovery of different PKU phenotypes. The recognition that not all cases of neonatal hyperphenylalaninemia are caused by the disorder described by Asbjørn Følling led to a state of semantic turmoil evidenced by the plethora of terms that are used to describe the various types of phenylalanine hydroxylase deficiency in childhood. Generally, children with neonatal blood phenylalanine levels above 20 mg/dl (1200 μmol/l) and a phenylalanine tolerance at 5 years of age of 10 to 20 mg/kg per day are classified as "*classical PKU.*" Children with this type of PKU show phenylalanine hydroxylase activities below 1% of normal (Bartholomé et al. 1975, Kaufman 1976). Children who at the age of 5 years tolerate 20 to 50 mg of phenylalanine per kg per day are often designated "*mild PKU.*" This phenotype usually shows phenylalanine hydroxylase activities of 1% to 3% of normal (Bartholomé et al. 1975).

The question of whether dietary management should be initiated in all hyperphenylalaninemic children arose when it was recognized that children with serum phenylalanine concentrations persistently equal to or below 10 mg/dl (600 μmol/l) on a normal diet appeared to show a normal intellectual and behavioral development without treatment (Levy et al. 1971). The phenylalanine hydroxylase activity in liver biopsies of these children is 3% to 6% of normal. This residual enzyme activity may explain why these children are able to eliminate a completely dissolved phenylalanine load of 0.1 g per kg body weight within 24 hours vs. 3 to 5 days for PKU children (Güttler 1971, Güttler and Wamberg 1972). The phenotype is named "benign hyperphenylalaninemia or HPA."

Regardless of how many different phenotypes of phenylalanine hydroxylase deficiency we may be able to distinguish, the ability to differentiate newborn children who require dietary therapy from those who do not is essential. Therefore, frequent reevaluations of the phenylalanine tolerance of the infant are urgent.

Correlation Between Polymorphic DNA Haplotypes of the Phenylalanine Hydroxylase Gene and Clinical Phenotypes of PKU

DNA analysis of the phenylalanine hydroxylase gene has revealed different alleles and offered the possibility for haplotype analysis (Lidsky et al. 1985b, Ledley et al. 1986, Daiger et al. 1986). Any given individual in the population will be either homozygous or heterozygous with respect to the length of the DNA fragments obtained after digestion with a restriction endonuclease (restriction fragment length polymorphism, RFLP). The combination of the different-length fragments of DNA obtained after digestion of an individual's genomic DNA with each of seven restriction enzymes and a cDNA copy of the phenylalanine hydroxylase gene as the probe defines the RFLP haplotypes of the phenylalanine hydroxylase alleles of this individual. By comparing the RFLP haplotypes of the mutant phenylalanine hydroxylase alleles in a PKU child with those in the respective parents, each of whom has a normal and a mutant allele, it is possible to determine for each parent the haplotype associated with the normal allele and the haplotype associated with the mutant phenylalanine hydroxylase allele (Güttler and Woo 1985, Ledley et al. 1986). The RFLP haplotypes of 74 normal phenylalanine hydroxylase alleles and 74 mutant alleles have been determined in 37 Danish families. In total, 12 RFLP haplotypes associated with normal and mutant phenylalanine hydroxylase alleles were identified. Of the 74 mutant alleles analyzed, 67 (91%) were associated with only four haplotypes. Affected children who were homozygous for two of these haplotypes had classical PKU. On the other hand, children who had inherited mutant alleles associated with one of the two other haplotypes had a milder clinical course (mild PKU or benign HPA) (Güttler et al. 1987). These observations indicate that there may be a minimum of four different mutant alleles at the phenylalanine hydroxylase locus, which could lead to patients with 10 different phenotypes.

TABLE 26.1. Screening incidence in Denmark of the various phenotypes of persistent hyperphenylalaninemia per 100,000.

Phenotype	Incidence per 100,000 (95% confidence limits)
Classical PKU	7 (4–10)
Mild PKU	3 (1–5)
Benign HPA	4 (2–6)

From Güttler (1980).

Results of Neonatal Screening

The first collective results of mass screening for PKU in eight western European countries were published in 1973 (Bickel et al. 1973). The average screening incidence of PKU was about 1 per 8000. During the following years, collective results of mass screening for inborn errors of metabolism were regularly published. The incidence rate varies considerably in different countries from 1 per 3000 in Ireland to 1 per 28,000 in Belgium. The incidence of PKU in six western European countries was 1 per 9300 and thus similar to the average incidence in the United States. The mean screening incidence of benign hyperphenylalaninemia in the western European countries was on the order of 7 per 100,000. However, the screening incidence of this disorder seems to vary greatly, with 1 per 8000 in northwest Germany and 1 per 81,000 in the Manchester region. As discussed in the reports, these variations may partly be due to genetic differences between the countries, a hypothesis supported by the observations of Szeinberg et al., Levy et al., and Thalhammer (cf. Güttler 1980) of a significantly different geographic and/or ethnic distribution of PKU families as compared to families with benign hyperphenylalaninemia. The screening incidence of the various phenotypes of persistent hyperphenylalaninemia in Denmark is shown in Table 26.1.

Detection of Heterozygotes

The first attempts at detecting individuals heterozygous for PKU were performed by Hsia et al. in 1956. Their method, determination of plasma phenylalanine concentrations after an oral load of phenylalanine, with its obvious practical and theoretical importance, has been modified numerous times. So far, the efforts to detect carriers for PKU have been based on two methods: determinations of plasma phenylalanine and tyrosine concentrations after a load of deuterium-labeled or unlabeled phenylalanine administered either orally or intravenously; and determination of the concentrations of phenylalanine and tyrosine in a single specimen obtained from fasting or semifasting individuals (cf. Güttler 1980). These methods of detection, however, suffer from an overlap of the values for the heterozygotes and the normal homozygotes which does not permit the heterozygotes to be classified with certainty.

The advances in recombinant DNA technology mentioned above, including restriction endonuclease analysis of leukocyte DNA and a gene-specific probe, have recently shown that carrier detection for PKU can be offered by establishing linkage of PKU alleles with RELP haplotypes (Woo et al. 1983, Daiger et al. 1986, Ledley et al. 1986, Güttler and Woo 1986a).

The Complex Phenylalanine Hydroxylase System

Twenty years after Følling's discovery, essentially nothing was known about the enzyme system responsible for the conversion of phenylalanine to tyrosine, except for the important assay developed by Udenfriend and Cooper in 1952. In 1956, Mitoma found that two protein fractions were required, and Kaufman (1957, 1959) demonstrated that one of the proteins was dihydropteridine reductase, which catalyzes the reduction of dihydropteridine to its active reduced form, tetrahydrobiopterin (cf. Figure 26.1). The entire reaction system consists of four essential components: the two enzymes, phenylalanine hydroxylase and dihydropteridine reductase, and the two coenzymes, tetrahydrobiopterin and reduced pyridine nucleotide (NADH) (cf. Kaufman 1976). Recent observations indicate that tetrahydrobiopterin is a rate-limiting factor in the reaction (Lykkelund et al. 1985).

Identification of Hyperphenylalaninemia Due to Dihydropteridine Reductase Deficiency or Defective Biopterin Synthesis

The first case of hyperphenylalaninemia due to dihydropteridine reductase deficiency was described by Kaufman and collaborators in 1975. In 1976, Leeming et al. reported a different defect in a hyperphenylalaninemic infant, who in spite of a well-controlled phenylalanine-restricted diet, developed athetoid movements of the arms, myoclonia, and an abnormal EEG. This patient had reduced concentrations of biopterines in the urine and blood. Additional reports by Kaufman et al. (1978) and by Curtius et al. (1979) confirmed that this was a new form of PKU due to a defect in the synthesis of tetrahydrobiopterin.

Dihydropteridine reductase can easily be measured in cultured skin fibroblasts of the neonate. Defects in the synthesis of tetrahydrobiopterin are detected by measurements of the urinary excretion of pterines, especially biopterin and neopterin, using the HPLC method described by Niederwieser et al. (1982). Today it is estimated that 1% to 3% of neonatal

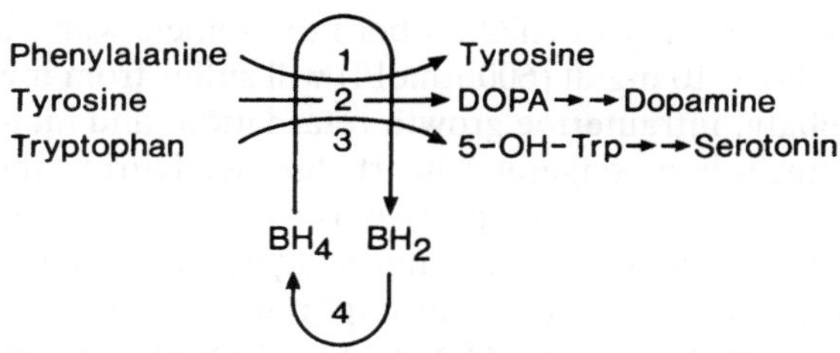

FIGURE 26.1. The tetrahydrobiopterin (BH_4)-requiring aromatic amino acid hydroxylases responsible for the formation of tyrosine, dopamine, and serotonin: 1, phenylalanine hydroxylase; 2, tyrosine hydroxylase; 3, tryptophan hydroxylase; 4, dihydropteridine reductase.

hyperphenylalaninemia is caused by one of these defects. Early detection and early start of therapy are important in order to avoid irreversible brain damage in children affected by these disorders. Therefore, it is recommended that the determination of urinary pterines should be performed in all hyperphenylalaninemic neonates, even those with borderline elevated blood phenylalanine levels.

Tetrahydrobiopterin is an essential coenzyme not only for the hydroxylation of phenylalanine to tyrosine, but also for the hydroxylation of tyrosine to L-DOPA and hence dopamine, and for the hydroxylation of tryptophan to 5-hydroxytryptophan and hence serotonin (Figure 26.1). Defects in the synthesis of these neurotransmitters as well as of adrenaline and noradrenaline are likely to be responsible for the neurological symptoms mentioned above. Both Kaufman et al. (1982) and Niederwieser et al. (1984) have reported on patients in whom all symptoms disappeared after oral administration of 10 to 20 mg of tetrahydrobiopterin per kg body weight. However, other patients do not respond as well to these high doses of tetrahydrobiopterin. So, most patients are still treated with neurotransmitter replacement consisting of L-DOPA, 5-hydroxytryptophan, and a decarboxylase inhibitor like carbidopa, combined with small doses of tetrahydrobiopterin to keep serum phenylalanine at normal levels. Finally, it should be mentioned that although the biochemical pathway of tetrabiopterin in man is still under discussion, it is clear that more than two enzymes are involved in the synthesis from GTP. Today, defects in two enzymes have been discovered to cause a defective biopterin synthesis (Niederwieser et al. 1984). In addition, milder variants with partially defective biopterin synthesis have been described (Güttler et al. 1984).

Maternal PKU

In 1973, Perry et al. presented a survey summarizing the data published through 1972 concerning 104 children born to 35 women with hyperphenylalaninemia. The survey concludes that despite ascertainment bias, it is likely that a great majority of children born to mothers with plasma phenylalanine levels above 10 mg/dl (600 μmol/l) will suffer from mental retardation, microcephaly, intrauterine growth retardation, and major congenital abnormalities including congenital heart disease. During the past years, Levy et al. have published retrospective as well as prospective studies on the effect of treated and untreated maternal hyperphenylalaninemia on the fetus (Levy and Waisbren 1983). An important question about maternal PKU is whether dietary treatment begun after conception is helpful. Lenke and Levy (1980) summarized the outcome of 121 pregnancies in 155 women with PKU or hyperphenylalaninemia treated after conception and three pregnancies treated before conception. They conclude that it is not clear whether dietary treatment begun after conception is helpful, and that treatment may have to be initiated before.

A problem concerned with maternal PKU is the fact that many PKU treatment programs are small parts of a large genetic or pediatric unit, so that patients who are not regularly at clinics are lost. Registers for PKU should be developed for the followup of girls who have discontinued the diet sometime during childhood, so that they can learn about maternal PKU. A prospective study performed by Levy and Waisbren (1983) revealed a frequency of unrecognized maternal PKU or hyperphenylalaninemia of approximately 1 per 33,000. Such a prevalence may justify antenatal screening for maternal PKU or that at least all women with a family story of PKU, women with low intelligence of uncertain origin, and women who have previously given birth to a microcephalic infant are tested for PKU at their first antenatal visit.

Dietary Termination

While there is no doubt about the excellent effect of early treatment of PKU children with a low-phenylalanine diet, nobody knows, however, at what age diet discontinuation might be safe. Cabalska et al. (1977) showed a fall in IQ and abnormalities in EEG and behavior among 32 PKU children evaluated over four years, their diets having been stopped before the age of 5. Isabel Smith in London and Horst Bickel in Heidelberg and their colleagues carried out a study to compare the experiences of these two European treatment centers (Smith et al. 1978). This study revealed a drop in IQ over two years from stopping the diet at about the age of 8, which was statistically significant for the 47 London children, but did not reach statistical significance for the 22 Heidelberg children, who went on a relaxed low-phenylalanine diet. In 1980, Waisbren et al. reviewed 19 published studies on diet termination and psychological outcome. About half reported significant loss in IQ scores after termination. The U.S. Collaborative Study randomly selected children to stop or continue their diet at the age of 6. This study shows that the 28 children who stopped their diet had a fall in IQ whereas the 27 who continued on the diet had no changes in IQ. These results reached statistical significance (Koch et al. 1982).

These data indicate that subtle changes in cerebral function may occur in children with PKU when the phenylalanine diet has been discontinued. The reports conclude that it thus would appear wise to delay the age of diet discontinuation, perhaps to adolescence or later, until more conclusive data are available. There are a number of biochemical candidates for the clinical features of "late-onset phenylalanine intoxication" including a decreased synthesis of dopamine and serotonin.

The Effect of Diet Discontinuation in PKU on Brain Neurotransmitters

Thirty years ago, it was observed that the synthesis of serotonin, dopamine, and norepinephrine was impaired in untreated PKU as judged either by a decreased concentration in the blood or decreased excretion in the urine of these neurotransmitters, or their metabolites, 5-hydroxyindoleacetic acid (5-HIAA) and homovanillic acid (HVA) (Armstrong and Robinson 1954, Pare et al. 1957, Weil-Malherbe 1955, Nadler and Hsia 1961). When early treatment of PKU with a phenylalanine-restricted diet was routinely introduced, an inverse relationship was found between phenylalanine levels and the urinary excretion of dopamine and serotonin (Curtius et al. 1981, Krause et al. 1985). An inverse relationship between blood phenylalanine levels and cerebrospinal fluid (CSF) concentrations of HVA and 5-HIAA has also been reported (McKean 1972, Butler et al. 1981, Lou et al. 1985).

Recently, the effect of the discontinuation of diet in PKU on the synthesis of dopamine, norepinephrine, and serotonin has been examined, and the possible relationship between low levels of these neurotransmitters and impaired performance on neuropsychological tests has been evaluated (Brunner et al. 1983, Krause et al. 1985). In some PKU patients, the performance on neuropsychological tests of higher integrative function is impaired after discontinuation of diet, especially when blood phenylalanine values exceed 1200 μmol/l (Krause et al. 1985), and the patients often complain of lack of concentration and emotional instability.

Quite recently, Krause et al. (1986) found a statistically significant decrease in the mean power frequency of the electroencephalogram and in plasma L-DOPA when plasma phenylalanine increased. When the PKU patients return to a phenylalanine-restricted diet, or are given tyrosine and tryptophan as supplements to a free diet (Lou et al. 1987), the impairment of neuropsychological and behavioral functions appears to be reversible (Krause et al. 1985, 1986, Lou et al. 1985, Güttler and Lou 1986). The improvement is accompanied by a significant increase in plasma L-DOPA (Krause et al. 1986), an increase in dopamine and serotonin excretion

(Krause et al. 1985), and an increase in CSF concentrations of HVA and 5-HIAA (Lou et al. 1985). These findings indicate reversible effects of elevated plasma phenylalanine on electrical function of the brain which may be mediated in part through inhibition of catecholamine synthesis (Krause et al. 1986).

The Human Phenylalanine Hydroxylase Gene

Even though phenylalanine hydroxylase is only synthesized in liver cells, the human phenylalanine hydroxylase gene is present in all nucleated cells on chromosome 12q22→q24.1 (Lidsky et al. 1985c). The gene product can be isolated from liver tissue (Woo et al. 1974), and antibodies against the enzyme protein will precipitate not only the enzyme but also the polysomes containing mRNA for phenylalanine hydroxylase (Robson et al. 1982). The nucleotide sequence of mRNA is an exact complementary copy of the nucleotide sequence of the coding regions (the exons) of the phenylalanine hydroxylase gene. Using mRNA as the template and reverse transcriptase, a complete full-length complementary copy (cDNA) of these important sequences of the gene has been synthesized and cloned (Kwok et al. 1985). Analysis of this cDNA has shown that the exons of the gene consist of 1353 nucleic acids corresponding to an enzyme monomer of 451 amino acids (Kwok et al. 1985). Furthermore, the cDNA copy of the gene codes for the synthesis of authentic human phenylalanine hydroxylase when inserted into an expression vector and transfected into cultured skin fibroblasts which normally do not synthesize this enzyme (Ledley et al. 1985).

Having isolated the cDNA of the gene, it was possible to isolate and characterize the normal human phenylalanine hydroxylase gene. The gene appeared to be about 90,000 nucleotide base pairs (90 kb) in length and contains 13 exons interrupted by rather large noncoding intervening sequences (introns) (DiLella et al. 1986a).

Prenatal Diagnosis of PKU

Restriction fragment length polymorphism (RFLP) analysis can be used to track mutant genes in PKU families (e.g., by comparing the parental and the proband DNAs), and the data derived from such analyses can be used for prenatal diagnosis (Woo et al. 1983, Lidsky et al. 1985a, Güttler and Woo 1986). RFLP analysis of DNA obtained from leukocytes from 33 Danish families with PKU demonstrated that in families with two affected children, the segregation of the mutant allele and disease state was concordant. Allelic segregation between the proband and unaffected siblings was discordant in all families (Woo et al. 1983, Woo et al. 1984, Daiger et al. 1986).

Mutations Responsible for Classical PKU

The observation that the haplotypes associated with the mutant alleles can be correlated to the patients' phenotypes (Güttler et al. 1987) supports the hypothesis that there may be a tight linkage between phenylalanine hydroxylase haplotypes and specific PKU mutations. Molecular cloning and sequence analyses have defined the specific mutation responsible for classical PKU associated with two of the common haplotypes, which comprise 60% of the PKU alleles in Denmark (DiLella et al. 1986b, and data to be published). Using mutant-specific oligonucleotide probes, preliminary studies have demonstrated that these mutations are also present in England, Ireland, Scotland, Switzerland, and Italy. The data suggest the spread of PKU chromosomes in the Caucasian race by founder effect (DiLella et al. 1986b).

If studies confirm the hypothesis regarding linkage of specific mutations and RFLP haplotypes of the phenylalanine hydroxylase gene, it should be possible to detect 90% of the mutant chromosomes in the Caucasian race using the appropriate oligonucleotide probes. Furthermore, appropriate oligonucleotide probes provide an important potential resource for designing the proper therapeutic regime by early prediction of the phenotype of the hyperphenylalaninemic newborn child, and for carrier detection in families with PKU and in couples in which one partner has no family history of PKU. The incidence of PKU is on average 1 in 10,000 newborn children in most European countries as well as in the United States, Canada, and Australia. So, there is a 2% chance that an individual with no family history of PKU is a carrier of the trait. The new diagnostic probes provide a unique possibility for genetic counseling of couples at risk.

Concluding Remarks

One can foresee that importance will be attached to the problems concerned with dietary termination, the outcome of planned pregnancies in maternal PKU, further investigation at the molecular level of the enzymatic defect, and clarification of the nature and pathogenesis of brain damage in PKU. Recent results indicate that it should be possible to demonstrate the molecular basis for the various phenotypes of phenylalanine hydroxylase deficiency (Ledley et al. 1986, Güttler et al. 1987) and that gene therapy for PKU is within reach (Ledley et al. 1985). The widespread interest in PKU involving research workers from many disciplines has been useful so that instead of focusing on a specific disease, we are now investigating the biochemical basis responsible for the various disorders causing hyperphenylalaninemia. As beneficial returns, we have had the opportunity to look at the biochemical mechanisms behind some important neurological dysfunctions induced by phenylalanine due to our present knowledge of the interaction of the components of the pterin-requiring mixed-function oxygenases, i.e., the hydroxylation systems of the aromatic amino acids necessary for the synthesis of both catecholamine neurotransmitters and serotonin.

Acknowledgments. The Danish Medical Research Council, the Danish Health Insurance Foundation, the NOVO Foundation, Frantz Hoffmann's Memorial Fund, the Egmont Fund, and P. Carl Petersen's Fund have supported the author's contribution to this review.

References

Armstrong, M.D., and Robinson, K.S. (1984). On the excretion of indole derivatives in phenylketonuria. Arch. Biochem. **52:**287–288.

Bartholomé, K., Lutz, P., and Bickel, H. (1975). Determination of phenylalanine hydroxylase activity in patients with phenylketonuria and hyperphenylalaninemia. Pediatr. Res. **9:**899–903.

Bickel, H., Gerrard, J., and Hickmans, E.M. (1954). The influence of phenylalanine intake on the chemistry and behavior of a phenylketonuria child. Acta Paediatr. Scand. **43:**64–77.

Bickel, H., Hudson, F.P., and Woolf, L.I. (1971). Phenylketonuria and some other inborn errors of metabolism. Stuttgart: Georg Thiese Verlag.

Bickel, H., Beckers, R.G., Wamberg, E., Schmid-Rüter, E., Feingold, J., Cahalane, S.F., Bottine, E., Jonxis, J.H.P., Colombo, J.P., and Carson, N. (1973). Collective results of mass screening for inborn metabolic errors in eight European countries. Acta Paediatr. Scand. **62:**413–16.

Brunner, R.L., Jordan, M.K., and Berry, H.K. (1983). Early treated phenylketonuria: neuropsychologic consequences. J. Pediatr. **102:**831–835.

Butler, L.J., O'Flynn, M.E., Seifert, W.E., and Howell, R.R. (1981). Neurotransmitter defects and treatment of disorders of hyperphenylalaninemias. J. Pediatr. **78:**729–733.

Cabalska, B., Duszynska, N., Borzymowska, J, Zorska, K., Koslacz-Folga, A., and Bozkowa, K. (1977). Termination of dietary treatment in phenylketonuria. Eur. J. Pediatr. **126:**126–253.

Curtius, H.-Ch., Vollmin, J.A., and Baerlocher, K. (1972). The use of deuterated phenylalanine for the elucidation of the phenylalanine-tyrosine metabolism. Clin. Chim. Acta **37:**277–285.

Curtius H.-Ch., Niederwieser, A., Viscontini, M., Otten, A., Schaub, J., Scheibenreiter, S., Schmidt, H. (1979). Atypical phenylketonuria due to tetrahydrobiopterin deficiency, diagnoses and treatment with tetrahydrobiopterin, dihydrobiopterin and sepiapterin. Clin. Chem. Acta. **93:**251–262.

Curtius, H.-Ch., Niederwieser, A., Viscontini, M., Leimbacher, W., Wegman, H., Blehova, B., Rey, F., Schaut, J., and Schmidt, H. (1981). Serotonin and dopamine synthesis in phenylketonuria. Adv. Exp. Med. Biol. **133:**277–291.

Daiger, S., Lidsky, A.S., Chakraborty, R., Koch R., Güttler, F., and Woo, S.L.C. (1986). Effective use of polymorphic DNA haplotypes at the phenylalanine hydroxylase locus in prenatal diagnosis of phenylketonuria. Lancet **i:**229–232.

DiLella, A.G., Kwok, S.C.M., Ledley, F.D., Marvit, J., and Woo, S.L.C.

(1986a). Molecular structure and polymorphic map of the human phenylalanine hydroxylase gene. Biochemistry **25:**743–749.

DiLella, A.G., Marvit, J., Lidsky, A.S., Güttler, F., and Woo, S.L.C. (1986b). Tight linkage between a splicing mutation and a specific DNA haplotype in phenylketonuria. Nature **322:**799–803.

Følling, A. (1934). Uber Ausscheidung von Phenylbrenztraubensäure in den Harn als Stoffwechselanomalie in Verbindung mit Imbezillität. Z. Physiol. Chem. **227:**169–176.

Guthrie, R., and Susi, A. (1963). A simple phenylalanine method for detecting phenylketonuria in large populations of newborn infants. Pediatrics **32:**338–43.

Güttler, F. (1971). Persistent hyperphenylalaninemia. Scand. J. Clin. Lab. Invest, Suppl. **118:**48.

Güttler, F. (1980). Hyperphenylalaninemia: diagnosis and classification of the various types of phenylalanine hydroxylase deficiency in childhood. Acta Pædiatr. Scand. Suppl. **280:**1–80.

Güttler, F. (1984). Phenylketonuria: 50-Years since Følling's discovery and still expanding our clinical and biochemical knowledge. Acta Pædiatr. Scand. **73:**705–716.

Güttler, F., and Lou, H. (1986). Dietary problems of phenylketonuria: effect on CNS transmitters and their possible role in behaviour and neuropsychological function. J. Inher. Metab. Dis. **9**(Suppl. 2):169–177.

Güttler, F., and Wamberg, E. (1972). Persistent hyperphenylalaninemia. Acta Pædiatr. Scand. **62:**333–337.

Güttler, F., and Woo, S.L.C. (1985). Molecular genetics of PKU: prenatal diagnosis and carrier detection by gene analysis. *In* Bicke, H., and Wachtel, H. (eds.), Recent progress in the understanding, recognition and management of inherited diseases of amino acid metabolism. Stuttgart: Georg Thieme Verlag, pp. 18–36.

Güttler, F., and Woo, S.L.C. (1986). Molecular genetics of PKU. J. Inher. Metab. Dis. **9**(Suppl. 1):58–68.

Güttler, F., Lou, H., Lykkelund, C., and Niederwieser, A. (1984). Combined tetrahydrobiopterin-phenylalanine loading test in the detection of partially defective biopterin synthesis. Eur. J. Pediatr. **141:**136–139.

Güttler, F., Ledley, F.D., Lidsky, A.S., DiLella, A.G., Sullivan, S.E., and Woo, S.L.C. (1987). Correlation between polymorphic DNA haplotypes at the phenylalanine hydroxylase locus and clinical phenotypes at the phenylketonuria. J. Pediatr. **110:**68–71.

Hsia, D.Y.Y., Driscoll, K.W., Troll, W., and Knox, W.E. (1956). Detection by phenylalanine tolerance tests of heterozygous carriers of phenylketonuria. Nature **178:**1239–40.

Jervis, G.A. (1953). Phenylpyruvic oligophrenia deficiency of phenylalanine-oxidizing system. Proc. Soc. Exp. Biol. Med. **82:**514–515.

Kaufman, S. (1957). The enzymatic conversion of phenylalanine to tyrosine. J. Biol. Chem. **226:**511–524.

Kaufman, S. (1959). Studies on the mechanism of the enzymatic conversion of phenylalanine to tyrosine. J. Biol. Chem. **234:**2677–2682.

Kaufman, S. (1976). Phenylketonuria: biochemical mechanisms. *In* Agranoff, B.W., and Aprison, M.H. (eds.), Advances in neurochemistry, Vol 2. 1–132. New York: Plenum Press, pp. 1–132.

Kaufman, S., Holtzman, N.A., Milstien, S., Butler, I.J., and Krumholtz, A.

(1975). Phenylketonuria due to a deficiency of dihydropteridine reductase. N. Engl. J. Med. **293:**673–79.

Kaufman, S., Berlow, S., Summer, G.K., Milstien, S., Schulman, J.D., Orloff, S., Spielberg, S., and Pueschel, S. (1978). Hyperphenylalaninemia due to a deficiency of biopterin. A variant form of phenylketonuria. N. Engl. J. Med. **299:**673–79.

Kaufman, S., Kapatos, G., McInnes R.R., Schulman, J.D., and Rizzo, W.B. (1982). Use of tetrahydropterins in the treatment of hyperphenylalaninemia due to defective synthesis of tetrahydrobiopterin: evidence that peripherally administered tetrahydropterins enter the brain. Pediatrics **70:**376–380.

Koch, R., Azen, C.G., Friedman, E.G., and Williamson, M.L. (1982). Preliminary report on the effects of diet discontinuation in PKU. Pediatrics **100:**870–75.

Krause, W., Halminski, M., McDonald, L., Dembure, P., Salvo, R., Freides, D., and Elsas, L. (1985). Biochemical and neuropsychological effects of elevated plasma phenylalanine in patients with treated phenylketonuria. J. Clin. Invest. **75:**40–48.

Krause, W., Epstein, C., Averbook, A., Dembure, P., and Elsas, L. (1986). Phenylalanine alters the mean power frequency of electroencephalograms and plasma L-DOPA in treated patients with phenylketonuria. Pediatr. Res. **20:**1112–1116.

Kwok, S.C.M., Ledley, F.D., DiLella, A.G., Robson, K.J.H., and Woo, S.L.C. (1985). Nucleotide sequence of a full-length complementary DNA clone and amino acid sequence of human phenylalanine hydroxylase. Biochemistry **24:**556–61.

Ledley, F.D., Grenett, H.E., DiLella, A.G., Kwok, S.C.M., and Woo, S.L.C. (1985): Gene transfer and expression of human phenylalanine hydroxylase. Science **228:**77–79.

Ledley, F.D., Levy, H.L., and Woo, S.L.C. (1986). Molecular analysis of the inheritance of phenylketonuria and mild hyperphenylalaninemia in families with both disorders. N. Engl. J. Med. **314:**1276–1280.

Leeming, R.J., Blair, J.A., and Rey, F. (1976). Biopterin derivates in atypical phenylketonuria. Lancet **i:**99.

Lenke, R.R., and Levy, H.L. (1980). Maternal phenylketonuria and hyperphenylalaninemias. An international survey of the outcome of untreated and treated pregnancies. N. Engl. J. Med. **309:**1269–1274.

Levy, H.L., and Waisbren, S.E. (1983). Effects of untreated phenylketonuria and hyperphenylalaninemia on the fetus. N. Engl. J. Med. **309:**1269–1274.

Levy, H.L., Shih, V.E., Karolkewicz, V., French, W.A., Carr, J.R., Cass, V., Kennedy, J.L., and MacCready, R.A. (1971). Persistent mild hyperphenylalaninemia in the untreated state. A prospective study. N. Engl. J. Med. **285:**424–429.

Lidsky, A.S., Güttler, F., and Woo, S.L.C. (1985a). Prenatal diagnosis of classic phenylketonuria by DNA analysis. Lancet **i:**549–551.

Lidsky, A.S., Ledley, F.D., DiLella, A.G., Kwok, S.C.M., Daiger, S.P., Robson, K.J.H., and Woo, S.L.C. (1985b). Extensive restriction site polymorphism at the human phenylalanine hydroxylase locus and application in prenatal diagnosis of phenylketonuria. Am. J. Hum. Genet. **37:**619–634.

Lidsky, A.S., Law, M.L., Morse, H.G., Kao, F.T., and Woo, S.L.C. (1985c). Regional mapping of the human phenylalanine hydroxylase gene and the PKU locus on chromosome 12. Proc. Natl. Acad. Sci. USA **82:**6221–6225.

Lou, H.C., Güttler, F., Lykkelund, C., Bruhn, P., and Niederwieser, A. (1985). Decreased vigilance and neurotransmitter synthesis after discontinuation of dietary treatment for phenylketonuria in adolescents. Eur. J. Pediatr. **144:**17–20.

Lou, H.C., Lykkelund, C., Gerdes, A.-M., Udesen, H., and Bruhn, P. (1987). Increased vigilance and dopamine synthesis by large doses of tyrosine or phenylalanine restriction in phenylketonuria. Acta Pædiatr. Scand. **76:**560–565.

Lykkelund, C., Lou, H.C., Rasmussen, V., and Güttler, F. (1985). Biopterin, neopterin and tyrosine responses to combined oral phenylalanine and tetrahydrobiopterin loading tests in two normal children and in a girl with partial biopterin deficiency. J. Inher. Metab. Dis. **8**(Supp. 2):95–96.

McKean, C.M. (1972). The effects of high phenylalanine concentrations on serotonin and catecholamine metabolism in the human brain. Brain Res. **47:**469–476.

Mitoma, C. (1956). Studies on partially purified phenylalanine hydroxylase. Arch. Biochem. Biophys. **60:**476–484.

Nadler, H.L., and Hsia, D.Y.Y. (1961). Epinephrine metabolism in phenylketonuria. Proc. Soc. Exp. Biol. Med. **107:**721–722.

Niederwieser, A., Curtius, H.-Ch., Wang, M., and Leupold, D. (1982). Atypical phenylketonuria with defective biopterin metabolism. Monotherapy with tetrahydrobiopterin or sepiapterin: screening and study of biosynthesis in man. Eur. J. Pediatr. **138:**110–112.

Niederwieser, A., Blau, N., Wang, M., Joller, P., Atarés, M., and Cardesa-Garcia, J. (1984). GTP cyclohydrolase I deficiency, a new enzyme defect causing hyperphenylalaninemia with neopterin, biopterin, dopamine, and serotonin deficiencies and muscular hypotonia. Eur. J. Pediatr. **141:**208–214.

Pare, C.M., Sandler, M., and Stacey, R.S. (1957): 5-Hydroxytryptamine deficiency in phenylketonuria. Lancet 551–553.

Perry, T.L., Hansen, S., Tischler, B., Richards, F.M., and Sokol, M. (1973). Unrecognized adult phenylketonuria. Implications for obstetrics and psychiatry. N. Engl. J. Med. **289:**395–398.

Robson, K.J.H., Chandra, T., MacGillivray, R.T.A., and Woo, S.L.C. (1982). Polysome immunoprecipitation of phenylalanine hydroxylase mRNA from rat liver and cloning of its cDNA. Proc. Natl. Acad. Sci. USA, **79:**4701–4705.

Smith, I., Lobascher, M.E., Stevenson, J.E., Wolff, O.H., Schmidt, H. Grubel-Kaiser, S., and Bickel, H. (1978). Effect of stopping low-phenylalanine diet on intellectual progress of children with phenylketonuria. Br. Med. J. **II:**723–726.

Udenfriend, S., and Cooper, J.R. (1952). The enzymatic conversion of phenylalanine to tyrosine. J. Biol. Chem. **194:**503–511.

Waisbren, S.E., Schnell, R.R., and Levy, H.L. (1980). Diet termination in children with PKU. A review of psychological assessments used to determine outcome. J. Inher. Metab. Dis. **3:**149–153.

Weil-Malherbe, H. (1955). Blood adrenaline and intelligence. J. Ment. Sci. **101:**733–745.

Woo, S.L.C., Gilliam, S.S., and Woolf, L.I. (1974). The isolation and properties of phenylalanine hydroxylase from human liver. Biochem. J. **139:**741–749.

Woo, S.L.C., Lidsky, A.S., Güttler, F., Chandra, T., and Robson, K.J.H. (1983). Cloned human phenylalanine hydroxylase gene allows prenatal diagnosis and carrier detection of classical phenylketonuria. Nature **306:**151–155.

Woo, S.L.C., Lidsky, A.S., Güttler, F., Chandra, T., and Robson, K.J.H. (1984). Prenatal diagnosis of classical phenylketonuria by gene mapping. J. Am. Med. Assoc. **251:**1998–2002.

CHAPTER 27

Reconsidering the Genetics of Phenylketonuria: Evidence from Molecular Genetics

Fred D. Ledley and Savio L.C. Woo*

Recent studies of the molecular genetics of phenylketonuria (PKU) have confirmed and complicated prior notions about the genetics of PKU. These studies have demonstrated that the PKU genotype is highly heterogeneous. Two mutations account for 60% of European "PKU" genes. These mutations are associated with relatively rare phenylalanine hydroxylase haplotypes, suggesting that the high incidence of PKU may reflect a "founder" effect or positive selection for these two mutant alleles.

While the metabolic lesion in phenylketonuria (PKU) was identified as a genetic deficiency of phenylalanine hydroxylase (PAH) as early as 1948, a detailed understanding of the defects in PKU and the mild forms of hyperphenylalaninemia (MHP) remains elusive. It has been difficult to critically define the deficits underlying PKU and MHP because these disorders are heterogeneous, and because PAH is expressed only in the liver and there has rarely been sufficient pathological material for thorough analysis. In this report, we reassess classical genetic and biochemical data (reviewed by Scriver and Clow 1980, Güttler 1980, Kaufman 1983) in light of recent molecular genetic analysis which both confirms and complicates prior notions about the genetics of PKU.

Classical Genetics

Early genetic studies demonstrated that PKU was inherited in an autosomal recessive fashion and established segregation ratios for both PKU and MHP characteristic of autosomal recessive inheritance. Newborn screening for PKU has provided a wealth of information about the incidence of PKU (reviewed by Levy 1973, Thalhammer 1975). Several curious observations are inherent in these data. First, the incidence of PKU varies widely (from 1:4500 in Ireland to <1:100,000 in Japan) and is independent of the incidence of MHP (from 1:6000 in Arab populations in Israel

* Department of Cell Biology and Institute of Molecular Genetics, Baylor College of Medicine, Houston, TX 77030, USA.

to <1:100,000 in England and New Zealand). Second, the frequency of the mutant allele (approximately 1:100 in the United States and western Europe) is unexpectedly high for a disorder which, while not fatal, is usually severe enough to prevent procreation. The "PKU" mutation does not appear to be in "Hardy-Weinberg equilibrium," where the loss of alleles through infertility is matched by heterozygote procreation and spontaneous mutation. The unusually high frequency of "PKU" genes is thought to reflect either a "founder effect" or positive selective pressure which increases the reproduction of carriers of the mutant gene.

Third, there appears to be a gradient in the frequency of PKU across Europe with highest prevalence in Celtic populations of Ireland, Scotland, and western Europe, and lower prevalence to the east. Elegant studies have demonstrated that PKU is more common in western Norway and western Denmark than eastern Norway, Sweden, or eastern Denmark. This is consistent with a "founder" effect (reviewed by Güttler 1980) but does not account for the incidence of PKU in places such as China (1:13,000) and Czechoslovakia (1:8000).

Finally, there is a complicated genetic relationship between PKU and MHP. Excluding the rare disorders of biopterin metabolism, all of the hyperphenylalaninemias involve mutations in the PAH gene (Kaufman 1983). Classical genetic studies suggested that there are at least three classes of "PKU genes" or "MHP genes" carrying different mutations (reviewed by Güttler 1980, Scriver and Clow 1980).

Biochemical Genetics

Characterization of the enzymatic defect in PKU and MHP on a biochemical level remains incomplete and controversial. There is general consensus that PKU is associated with absent or extremely low levels of residual PAH activity while MHP is associated with higher levels of residual PAH activity. In MHP, immunoreactive PAH protein and enzymatic activity is uniformly present though enzyme activity is reduced to between 3% and 5% of normal and exhibits abnormal kinetics (reviewed by Kaufman 1983). Studies with PKU yield inconsistent result. Some individuals have no immunoreactive protein or enzymatic activity; others have immunoreactive protein and extremely low levels of enzyme activity (less than 1%). This variability reflects both individual differences in methodology (reviewed by Kaufman 1983).

Molecular Genetics

The PAH locus, located on human chromosome 12 (12q22$\rightarrow$q24.1) (Lidsky et al. 1985a), comprises 13 exons spanning 90,000 bases (DiLella et al. 1986a). The chromosomal gene is transcribed into a mRNA of 2400 bases which contains a 1353-base open reading frame coding for a protein of 451 amino acids (MW = 51,672) (Kwok et al. 1985). Gene transfer of the full-length cDNA into cultured cells leads to the production of immunoreactive PAH protein and enzymatic activity, indicating that expression of a single genetic locus is necessary and sufficient to constitute PAH apoenzyme activity (Ledley et al. 1985, 1987).

Cloned PAH genes have been used as probes to study the genetics of PKU. Initial studies were based on normal variation (or polymorphism) in the PAH gene in different individuals. Variation in sequences which are target sites for specific restriction endonucleases creates a "polymorphic site" which will yield restriction fragments of different length (RFLPs) in different individuals when genomic DNA is cut with that restriction endonuclease. RFLPs represent markers within the PAH gene which are powerful tools for studying the genetics of PKU (Woo et al. 1984).

Eight polymorphic sites were identified at the PAH locus at frequencies of 31% to 69% (Lidsky et al. 1985b). The presence or absence of the eight sites comprises a "haplotype" which is a highly specific marker for different PAH alleles. Identification of the haplotypes of the four PAH alleles in a family makes it possible to trace the inheritance of each allele among different family members. Such family studies have demonstrated concordant segregation of mutant genes (genes in PKU probands) with individual PAH alleles, confirming that PKU is linked to the PAH gene (Daiger et al. 1983). Identifying the haplotypes of a proband, parents, and siblings also enables prenatal diagnosis for PKU. A sibling who inherits the same PAH alleles as the proband will be affected with PKU (Woo et al. 1983).

Classification of PAH haplotypes is a powerful tool for population studies. Analysis of haplotypes in PKU families in Denmark identified the haplotyes of normal and mutant alleles in 33 families with PKU (Table 27.1) (Güttler et al. 1987, DiLella et al. 1986b). The haplotypes designated 1 to 4 were found in 75% of normal alleles and 90% of mutant alleles. Interestingly, 58% of the mutant PAH alleles (PKU) were haplotype 2 or 3 while only 8% of normal alleles had these haplotypes (Table 27.1).

Studies of the β-globin locus in thalassemia have demonstrated a correlation between haplotypes at the β-globin allele and thalassemic mutation on that allele. In 80% of individuals, a single mutation is associated with a particular haplotype, and different mutations are associated with different haplotypes (Kazazian et al. 1984).

In order to determine whether a similar relationship existed between haplotypes and mutations at the PAH locus, PAH chromosomal genes were cloned from PKU patients homozygous for haplotypes 2 and 3. The exons and intron-exon boundaries of these genes were sequenced and compared with normal PAH sequences. A single mutation was identified on haplotype 3 chromosomes (DiLella et al. 1986b): a *g* to *a* transition in the consensus *gt* sequence at the intron-exon boundary adjacent to exon 12. A single mutation was identified on haplotype 2 chromosomes: a *c* to *t* transi-

TABLE 27.1. Association of mutations with different haplotypes in Denmark and northern Europe.

Haplotype	Enzyme								Denmark					Europe
	BglII	PvuIIa	PvuIIb	EcoRI	MspI	XmnI	HindIII	EcoRV	Normal alleles n	Normal alleles %	Mutant alleles n	Mutant alleles %	Mutant frequency[a]	Mutant frequency[b]
1	−	+	−	−	+	−	−	−	23	34.6	12	18.2	2.5×10^{-3}	—[c]
2	−	+	−	−	+	−	+	+	3	4.5	13	19.7	2×10^{-2}	1.9×10^{-2}
3	−	+	−	+	−	+	−	−	2	0.3	25	37.9	6×10^{-2}	4×10^{-2}
4	−	+	−	+	−	+	+	+	21	31.8	9	13.6	2×10^{-3}	—[c]
Other									17	25.9	7	13.6	2×10^{-3}	3.5×10^{-3}[c]
All									66	100	66	100	4.6×10^{-3}	6×10^{-3}

[a] Mutant frequency was computed for 66 normal alleles and 66 mutant alleles from 33 PKU families in the Danish population, assuming an incidence of PKU of 1:12,000 (Levy 1973).

[b] Frequencies for Europe were computed from data for Germany, Scotland, Denmark, and Switzerland using the incidence of PKU given by Levy (1973) averaged for the population of each country [New York Times Atlas of the World (1978)].

[c] Values for haplotypes 1 and 4 are incorporated in the calculation of "other."

tion which results in the substitution of an arginine for tryptophan at amino acid 408 (DiLella et al. 1987).

In order to characterize the protein product of these mutant genes, cDNA clones containing the mutant sequences were introduced into cultured cells by DNA-mediated gene transfer (Marvit et al. 1987). Cells transformed with the mutant clones produced mRNA containing the mutant sequences but did not produce detectable PAH protein or PAH enzymatic activity. These results suggest that the protein product of these mutant genes is unstable and produces a CRM^- phenotype with no enzymatic activity. Oligonucleotide probes were designed for the normal and mutant sequences in order to investigate the distribution of these mutations. In European populations, there was uniform linkage of these mutations with their respective haplotypes. Thus, these two mutations account for more than 50% of the "PKU genes" in European populations. Three other mutations have been identified in isolated individuals with PKU in Scotland, China, and Germany. Studies are currently under way to identify mutations associated with haplotypes other than 2 and 3.

The relationship between PKU and MHP has been investigated by haplotype analysis of two especially informative families in which some siblings have "classical" PKU and others have MHP (Ledley et al. 1986). This study demonstrates that allelic PAH mutations cause hyperphenylalaninemic phenotypes of differing severity, and that the same mutant allele can be associated with "classical" PKU in some individuals, and MHP in others (Figure 27.1a).

The Genetics of PKU

These preliminary findings provide new perspectives on the genetics of PKU and MHP. Molecular studies have confirmed, and perhaps extended, the appreciation that PKU and MHP are heterogeneous disorders arising from multiple mutant alleles. Two mutations account for 60% of "classical PKU" alleles in Europe. Three other mutations have been identified in rare individuals. Forty percent of the "PKU" mutations (in Europe) have not yet been characterized.

The heterogeneity of the PKU genotype complicates the interpretation of biochemical and clinical data. Previous characterizations of so-called "homozygous" individuals with PKU or MHP tacitly assumed that such an individual was "homozygous" and thus had two identical "PKU" alleles. Molecular analysis demonstrates that most PKU patients are compound heterozygotes. In the Danish PKU population, 75% of PKU patients are heterozygous for PAH haplotypes. Thus, the clinical phenotype and biochemical properties of hepatic PAH reflect the composite activity and interactions of the gene products from two mutant alleles and may not provide consistent information about any particular mutation.

Molecular studies demonstrate that the same alleles in different diploid

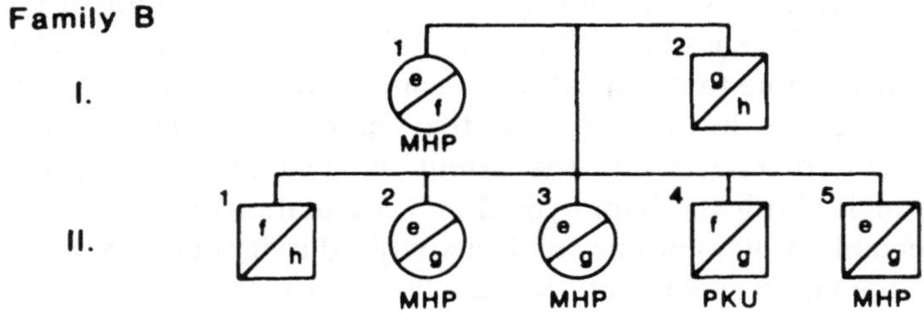

(a)

Haplotype	Haplotype 1	2	3	4
1	Mild			
2	Classic Mild Mild Mild	Classic Classic		
3	Mild Mild Mild Mild	Classic Classic	Classic Classic Classic† Classic†	Classic Classic Mild Mild Mild Mild
4	Mild Mild			

(b)

FIGURE 27.1. Genotypic and phenotypic heterogeneity in PKU. (a) Haplotype analysis of a family in which one family member (II.4) has classic PKU, and one parent (I.1) and several siblings (II.2, II.3, II.5) have MHP. This analysis demonstrates the interaction of alleles in causing PKU and MHP, respectively. The four parental haplotypes are designated e, f, g, and h. Reprinted by permission of The New England Journal of Medicine, 314, pp. 1276–1280, 1986. (b) Correlation of genotype and phenotype for individuals with haplotypes 1 to 4 in the Danish population. Table compares the haplotypes with the severity of the phenotype. Classification is based on dietary phenylalanine tolerance: classic, tolerance of 1 to 20 mg/kg per day; mild, tolerance of 21 to 50 mg/kg per day; †, siblings. Reproduced from Journal Pediatrics, 1987, 110, pp. 68–71, Güttler et al., by copyright permission of the C.V. Mosby Company.

combinations can cause either "mild" or "classic" forms of PKU (Figure 27.1b) (Güttler et al. 1987), or even MHP (Ledley et al. 1986). Thus, it is inadequate to classify mutant PAH genes as "classical PKU genes," "atypical PKU genes," or "MHP genes" (Ledley et al. 1986). The designation of genotype based on specific mutations and the activity of mutant genes in cultured cells or homozygous patients should supplant the designation of "PKU genes" and "MHP genes" based on clinical phenotype. It is entirely possible that clinical classifications will correlate with particular molecular lesions in the same way that the different clinical forms of α-thalassemia correlated with the extent of the molecular genetic lesion. Interestingly, both the haplotype 2 and 3 mutations express mRNA but no detectable protein or enzymatic activity (Marvit et al. 1987).

The population genetics of PKU can be reconsidered based on individual mutant alleles rather than phenotypes. The most striking finding is the large number of haplotype 2 and 3 alleles in the study population which contained mutations. The proportion of all haplotype 2 and 3 alleles which contain mutations can be extrapolated from the haplotypes of normal and mutant alleles in PKU families and the general incidence of PKU. The frequency of mutations on haplotype 2 and 3 alleles is 2×10^{-2} and 6×10^{-2}, respectively, in Denmark (1.9×10^{-2} and 4×10^{-2} among northern European populations studied). In contrast, the frequency of mutations on all other alleles is only 2×10^{-3} in Denmark (3.5×10^{-3} in northern Europe). These latter alleles alone would cause an incidence of PKU of 1×10^{-5}, an unremarkable incidence which is not obviously incompatible with "Hardy-Weinberg equilibrium" pressures. Thus, consideration of genetic pressures contributing to the unusually high prevalence of PKU can be restricted to analysis of haplotype 2 and 3 alleles.

Existing data do not indicate the nature of this genetic pressure. The association of a large proportion of mutant alleles with minor haplotypes 2 and 3 is consistent with a founder effect in which the spread of mutant alleles passively follows the spread of haplotype 2 and 3 alleles in the European population. In order to account for the spread of two different mutations by a founder effect, it is necessary to postulate that two independent mutations occurred in a founder population, or that these two mutations arose in two independent founder populations. No crossover events have been detected between haplotype 2 and 3 mutations or with any other alleles in several European populations. This result suggests that the admixture of these alleles is relatively recent since crossover events can be expected to accumulate over time. Furthermore, if the haplotype 2 and 3 mutations arose in a single founder population, both must represent recent mutational events that occurred shortly before their introduction in the general European population. Alternatively, the absence of crossovers may indicate that these mutations arose in separate populations and have had limited opportunity for recombination.

The "founder" hypothesis may be tested by seeking "founder" populations where haplotypes 2 and 3 are the predominant normal alleles. Yet if the ratio of normal:mutant alleles observed for haplotypes 2 and 3 in Europe were preserved in a founder population, the incidence of PKU in this population would be 1:4 to 1:600. Thus, even if haplotype 2 and 3 mutations arose in founder populations, it is likely that some form of selective pressure increased the gene frequency in the founder population, or facilitated the spread of the mutant allele in European populations.

Haplotype data suggest that such positive selection specifically involves haplotypes 2 and 3 rather than any component of the PKU (or heterozygote) phenotype since alleles 1 and 4, which cause a similar phenotype, do not exhibit unusually high gene frequency. One way that such selection might occur is if mutations on haplotype 2 and 3 alleles are genetically linked to adjacent genes, unrelated to PAH, which confer a reproductive advantage. Positive selection for such an adjacent gene could carry along the mutant PAH gene by so-called "hitch-hiking" or coevolution. The insulin-like growth factor-1 gene, which is adjacent to PAH on the long arm of chromosome 12, might be a candidate locus for such positive selection. It will be interesting to investigate whether variations at linked loci may confer a selective advantage on haplotype 2 and 3 mutations.

Conclusion

Molecular data concerning the genetics of PKU should reinvigorate, not supplant, efforts to delineate the clinical and biochemical characteristics of this disorder. In particular, it is now possible to reinterpret previous clinical and biochemical data describing the clinical outcome of PKU, the efficacy of diet discontinuation, the biochemical characteristics of the mutant enzyme, and mechanisms for positive selection based on the specific genotype of individuals with PKU. Just as sickle cell disease, thalassemia, and hemoglobin C disease represent distinct phenotypes within anemia associated with different point mutations at the globin locus, a variety of distinct phenotypes associated with specific mutations should be delineated for mutations at the PAH locus.

Acknowledgments. F. Güttler, A.G. DiLella, J. Marvit, A. Lidsky, and S. Sullivan made major contributions to this work. This work was supported by grant HD-17711 from the National Institutes of Health. F.D.L. is an Assistant Investigator and S.L.C.W. an Investigator of the Howard Hughes Medical Institute.

References

Daiger, S., Lidsky, A.S., Chakraborty, R., Koch, R., Güttler, F., and Woo, S.L.C. (1983). Effective use of polymorphic DNA haplotypes at the phenylalanine hydroxylase locus in prenatal diagnosis of phenylketonuria. Lancet **i**:229–231.

DiLella, A.G., Kwok, S.C.M., Ledley, F.D., Marvit, J., and Woo, S.L.C. (1986a). Molecular structure and polymorphic map of the human phenylalanine hydroxylase gene. Biochemistry **25**:743–749.

DiLella, A.G., Marvit, J., Lidsky, A.S., Güttler, F., and Woo, S.L.C. (1986b). Tight linkage between a splicing mutation and a specific DNA haplotype in phenylketonuria. Nature **322:**799–803.

DiLella, A.G., Marvit, J., Brayton, K., and Woo, S.L.C. (1987). An amino acid substitution in phenylketonuria is in linkage disequilibrium with DNA haplotype 2. Nature **327:**333–336.

Güttler, F. (1980). Hyperphenylalaninemia: diagnosis and classification of the various types of phenylalanine hydroxylase deficiency in childhood. Acta. Paediatr. Scand. **280:**1–80.

Güttler, F., Ledley, F.D., Lidsky, A.S., DiLella, A.G., Sullivan, S.E., and Woo, S.L.C. (1987). Correlation between polymorphic DNA haplotypes at phenylalanine hydroxylase locus and clinical phenotypes of phenylketonuria. J. Pediatr. **110:**68–71.

Kaufman, S. (1983). Phenylketonuria and its variants. Adv. Hum. Genet. **13:**217–297.

Kazazian, H.H., Jr., Orkin, S.H., Markham, A.F., Chapman, C.R., Youssoufian, H., and Waber, P.G. (1984). Quantification of the close association between DNA haplotypes and specific β-thalassaemia mutations in Mediterraneans. Nature **310:**152–154.

Kwok, S.C.M., Ledley F.D., DiLella, A.G., Robson, K.J.H., and Woo, S.L.C. (1985). Nucleotide sequence of a full-length complementary DNA clone and amino acid sequence of human phenylalanine hydroxylase. Biochemistry **24:**556–561.

Ledley, F.D., Grenett, H.E., DiLella, A.G., Kwok, S.C.M., and Woo, S.L.C. (1985). Gene transfer and expression of human phenylalanine hydroxylase. Science **228:**77–79.

Ledley, F.D., Levy, H.L., and Woo, S.L.C. (1986). Molecular analysis of the inheritance of phenylketonuria and mild hyperphenylalaninemia in families with both disorders. N. Engl. J. Med. **314:**1276–1280.

Ledley, F.D., Hahn, T., and Woo, S.L.C. (1987). Selection for phenylalanine hydroxylase activity in cells transformed with recombinant retroviruses. Som. Cell Mol. Genet. **13**(2)**:**145–154.

Levy, H.L. (1973). Genetic screening for inborn errors of metabolism. U.S. Dept. of Health, Education, and Welfare, Washington, D.C.

Lidsky, A.S., Law, M.L., Morse, H.G., Kao, F-T., Rabin, M., Ruddle, R.H., and Woo, S.L.C. (1985a). Regional mapping of the phenylalanine hydroxylase gene and the phenylketonuria locus in the human genome. Proc. Natl. Acad. Sci. USA **82:**6221–6225.

Lidsky, A.S., Ledley, F.D., DiLella, A.G., Kwok, S.C.M., Daiger, S.P., Robson, K.J.H., and Woo, S.L.C. (1985b). Extensive restriction site polymorphism at the human phenylalanine hydroxylase locus and application in prenatal diagnosis of phenylketonuria. Am. J. Hum. Genet. **37:**619–634.

Marvit, J., DiLella, A.G., Brayton, K., Ledley, F.D., Robson, K.J.H., and Woo, S.L.C. (1987). GT to AT transition at a splice donor site causes shipping of the preceding exon in phenylketonuria. Nucleic Acids Research **15:**5613–5628.

Scriver, C.R., and Clow, C.L. (1980). Phenylketonuria and other phenylalanine hydroxylation mutants in man. Annu. Rev. Genet. **14:**179–202.

Thalhammer, O. (1975). Frequency of inborn errors of metabolism. Humangenetik **30:**273–286.

Woo, S.L.C., Lidsky, A.S., Güttler, F., Chandra, T., and Robson, K.J.H. (1983). Cloned human phenylalanine hydroxylase gene allows prenatal diagnosis and carrier detection of classical phenylketonuria. Nature **306:**151–155.

Woo, S.L.C., Lidsky, A.S., Güttler, F., Thrumalachary, C., and Robson, K.J.H. (1984). Prenatal diagnosis of classical phenylketonuria by gene mapping. J. Am. Med. Assoc. **251:**1998–2002.

CHAPTER 28

Effect of Phenylalanine on Brain Maturation: Implications for the Treatment of Patients with Phenylketonuria

Frits A. Hommes and Kiyosato Matsuo*

High levels of phenylalanine inhibit the central nervous system specific ATP-sulfurylase, leading to a decreased synthesis of sulfatides. The hypothesis is developed that this in turn leads to an increased turnover of central nervous system myelin which is not compensated by an increased rate of synthesis. As a consequence, synaptic contacts regress, resulting in fewer stable synaptic contacts, which is a contribution to brain dysfunction observed in not treated or poorly treated PKU patients.

There is perhaps no inborn error of metabolism which has been studied more intensely than phenylketonuria (PKU) and of which so little is understood as far as the consequences of the genetic defect on brain function are concerned.

Any theory to explain brain dysfunction in PKU must take into account the current theories on synaptogenesis, recognizing the proper formation of synaptic contacts as the basis of the dynamic integrative capacities of the central nervous system. There is presently no generally accepted theory as to how these processes take place (Hopkins and Brown 1985). However, there is ample evidence for and agreement on the fact that initially excess synaptic connections are made which are subsequently either stabilized to permanent connections or eliminated. It has been postulated that genetic factors control the proper interaction between the main categories of neurons, including the overproduction of synapses, while the survival of certain synapses is controlled by epigenetic factors (Changeux et al. 1973, Changeux and Danchin 1976). An excessive level of phenylalanine during the period of synaptogenesis (the vulnerable period; Dobbing 1972) might therefore interfere with the stabilization process, resulting in regression of the labile contacts. As a consequence, the adult brain will be equipped with fewer synaptic contacts, which contributes to the mental retardation. A possible scenario leading to such a decreased number of synaptic contacts in the adult is depicted in Figure 28.1.

* Department of Cell and Molecular Biology, Medical College of Georgia Augusta, GA 30912, USA.

Mutated phenylalanine hydroxylase gene

↓

Increased blood phenylalanine levels

↓

Increased central nervous system phenylalanine levels

↓

Inhibition of oligodendroglial cell specific ATP-sulfurylase

↓

Decreased availability of PAPS

↓

Decreased protection of myelin proteins against proteolytic degradation

↓

Increased turnover of myelin

↓

Disintegration of myelin

↓

Regression of labile synaptic contacts

↓

Fewer remaining stabile synaptic contacts = contribution to mental retardation

FIGURE 28.1. Proposed sequence of events leading to brain dysfunction in phenylketonuria.

The key event in this scenario is the inhibition by phenylalanine of ATP-sulfurylase, the first enzyme of sulfate activation. Chase and O'Brien (1970) have identified the formation of 3′-phosphoadenosine-5′-phosphosulfate (PAPS) as a site of action of phenylalanine. It was subsequently shown (Hommes 1985) that it is indeed the PAPS-forming system that is inhibited by phenylalanine, and more specifically the ATP-sulfurylase rather than the adenosine-5′-phosphosulfate(APS)-kinase, the second enzyme of sulfate activation (Matsuo et al. 1987a). Those enzymes of liver are not affected by phenylalanine as would be expected since PKU patients show central nervous system pathology, not liver pathology. It implies that the liver and brain enzymes are different. That was suspected from genetic experiments on the brachymorphic mouse (Sugahara and Schwartz 1982a,b), and this genetic evidence has furthermore been complemented by kinetic and chemical analyses of the enzyme systems of brain and liver (Table 28.1).

The kinetics of ATP-sulfurylase are complex and so is the inhibition by phenylalanine. Sulfate is an allosteric activator of the brain enzyme with a Hill coefficient of 1.0. (Figure 28.2). The brain ATP-sulfurylase follows Michaelis-Menten kinetics at low ATP (<0.1 m*M*) but is activated by ATP at high ATP.

Phenylalanine inhibits both phases (Figure 28.3), probably competitively at the low-ATP level. It is not likely that this is of physiological significance, since this takes place at unphysiologically low ATP concentrations. However, the activation by ATP takes place at physiological ATP levels.

Half-maximum activation is observed at 2 m*M* ATP. Phenylalanine interferes with this activation. A full understanding of this inhibition has to

TABLE 28.1. Properties of ATP-sulfurylase and APS-kinase of liver and brain.[a]

	ATP-sulfurylase		APS-kinase	
	Brain	Liver	Brain	Liver
Molecular weight	300,000	68,000	300,000	60,000
Activation by ATP	+	−	−	+
$K_m/K_{0.5}$ for ATP	17 μM/2 mM	0.38 mM/NA[b]	0.2 mM/NA	NA/2.7 mM
Activation by SO_4^{2-}	+	−		
$K_m/K_{0.5}$ for SO_4^{2-}	NA/3.9 mM	2.5 mM/NA		
Activation/inhibition by APS			+/+	+/+

[a] Studies on the brain enzymes were carried out on the enzyme from fetal calf brain, those on the liver enzymes from rat liver (after Burnell and Roy 1978, Hommes et al. 1987, Matsuo et al. 1987a).
[b] NA: not applicable.

TABLE 28.2. Inhibitors of brain ATP-sulfurylase and some of their kinetic constants.[a]

Inhibitor	Apparent K_i at low ATP (mM)	Apparent K_i at high ATP (mM)
Phenylalanine	1.95	0.92
Leucine	4.26	3.79
Arginine	3.53	1.17
NH_4Cl	2.80	2.39
Lactate	2.29	5.40

[a] Experimental conditions were those described in the legend of Figure 28.3.

wait until the mechanism of activation by ATP is understood. This inhibition is not limited to phenylalanine but is also observed—with the same kind of kinetics as shown in Figure 28.1 for phenylalanine—with arginine, leucine, ammonia, and lactate. The kinetics suggests a competitive type of inhibition. Assuming this to be the case, K_i values for the two phases of inhibition can be calculated (Table 28.2). These are well within the ranges of concentrations observed in the inborn errors of metabolism associated with these compounds. These inborn errors of metabolism do show "demyelination" (Prensky et al. 1968, Schumann et al. 1978, Harding et al. 1984).

The phenylalanine-sensitive ATP-sulfurylase is most active in those areas of the brain most heavily affected by the PKU condition (Matsuo and Hommes 1987), it is active in oligodendroglial cells, and its activity closely parallels the rate of myelination (Matsuo et al. 1987b). These are necessary requirements for the involvement of this enzyme in the etiology of brain dysfunction in PKU.

The increased turnover of myelin, specifically due to the hyperphenylalaninemic condition, has been demonstrated to occur (Berger et al.

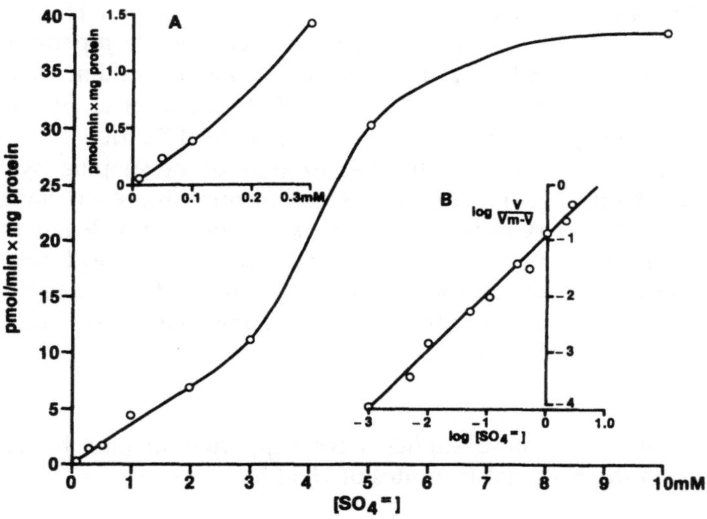

FIGURE 28.2. Lineweaver-Burk plot for ATP-sulfurylase of fetal calf brain with sulfate as the variable substrate. ATP-sulfurylase was assayed as described by Hommes and Moss (1986). As enzyme source was used the fraction not retarded by Sephadex G-150 of a 100,000 × *g* supernatant of a fetal calf brain homogenate. The ATP concentration was 13.3 m*M*. Inset A shows an enlargement at lower sulfate concentrations, inset B the Hill plot.

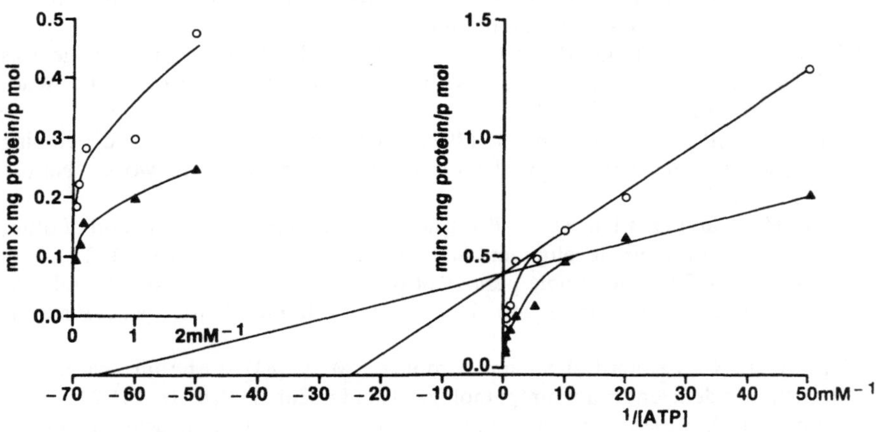

FIGURE 28.3. Lineweaver-Burk plot for ATP-sulfurylase of fetal calf brain with ATP as the variable substrate in the absence (▲) and in the presence (○) of 3.25 m*M* phenylalanine. The inset shows an enlarged portion of the graph at higher ATP concentrations. Experimental conditions were the same as those described in the legend of Figure 28.2. The sulfate concentration was 1.8 m*M*. The apparent K_m for ATP at low ATP concentrations can be calculated to be 15 μM, while the apparent K_m for ATP at high ATP concentration is 2.2 m*M*.

1980, Hommes et al. 1982a,b, Taylor 1982, Taylor and Hommes 1983, Hommes 1985). Since this increased turnover is not compensated by an increased rate of synthesis, the net result is a loss of myelin, which is precisely what has been observed, both in PKU (Shah et al. 1972a) and in experimental hyperphenylalaninemia (Shah et al. 1972b).

An important factor in the stabilization of developing synapses is thought to be the positive feedback from neurotransmitters (Changeux and Danchin 1976, Burry et al. 1984). An axon functionally degenerating because of the breakdown of its insulating myelin sheath will only allow a minor positive feedback by the neurotransmitter for synapse stabilization—if any at all. This is a further contributing factor in the regression of the synapses, still in the labile state.

Acknowledgment. These studies were supported in part by grant HD 17822 from the National Institutes of Health.

References

Berger, R., Springer, J., and Hommes, F.A. (1980). Brain protein and myelin metabolism in young hyperphenylalaninemic rats. Mol. Cell Biol. **26:**31–36.

Burnell, J.N., and Roy, A.B. (1978). Purification and properties of the ATP-sulfurylase of rat liver. Biochim. Biophys. Acta **527:**239–248.

Burry, R.W., Knirs, D.A., and Scribner, L.R. (1984). Mechanisms of synapse formation and maturation. *In* Jones, G. (ed.), Current topics in research in synapses, Vol. I. New York: Alan Liss, pp. 1–51.

Changeux, J.P., Couriege, P., and Danchin, A. (1973). A theory of the epigenesis of neuronal networks by selective stabilization of synapses. Proc. Natl. Acad. Sci. USA **70:**2977–2978.

Changeux, J.P., and Danchin, A. (1976). Selective stabilization of developing synapses as a mechanism for the specification of neuronal networks. Nature **267:**705–712.

Chase, H.P., and O'Brien, D. (1970). Effects of excess phenylalanine and of other amino acids on brain development in infant rats. Pediatr. Res. **4:**96–102.

Dobbing, J. (1972). Vulnerable periods of brain development. *In* von Muralt, A. (ed.), Lipids, malnutrition and the developing brain. Amsterdam: Elsevier, pp. 9–29.

Harding, B.N., Leonard, J.S., and Erdohasi, M. (1984). Ornithine carbamoyl transferase deficiency: a neuropathological study. Eur. J. Pediatr. **141:**215–220.

Hommes, F.A. (1985) Myelin turnover at later stages of brain development in experimental hyperphenylalaninemia. *In* Bickel, H., and Wachtel, U. (eds.), Inherited diseases of amino acid metabolism. Stuttgart: Georg Thieme Verlag, pp. 67–85.

Hommes, F.A., Eller, G.A., and Taylor, H.E. (1982a). The effect of phenylalanine on myelin metabolism in adolescent rats. *In* Cockburn, F., and Gitzelmann, R. (eds.), Inborn errors of metabolism in humans. Lancaster: MTP Press, pp. 193–199.

Hommes, F.A., Eller, G.A., and Taylor, H.E. (1982b). Turnover of the fast component of myelin and myelin proteins in experimental hyperphenylalaninemia. Relevance to termination of dietary treatment. J. Inher. Metab. Dis. **5**:21–27.

Hommes, F.A., and Moss, L. (1986). The assay of ATP-sulfurylase. Anal. Biochem. **154**:100–104.

Hommes, F.A., Moss, L., and Touchston, J. (1987). Purification and some properties of liver APS-kinase. Biochim. Biophys. Acta **924**:270–275.

Hopkins, W.G., and Brown, M.C. (1985). Development of nerve cells and their connections. Cambridge: Cambridge University Press.

Matsuo, K., and Hommes, F.A. (1987). Regional distribution of the phenylalalnine sensitive ATP-sulfurylase in brain. J. Inher. Metab. Dis. **10**:62–65.

Matsuo, K., Moss, L., and Hommes, F.A. (1987a). Properties of the 3′-phosphoadenosine 5′-phosphosulfate (PAPS) synthesizing systems of brain and liver. Neurochem. Res. **11**:1–10.

Matsuo, K., Moss, L., and Hommes, F.A. (1987b). The development of ATP-sulfurylase and APS kinase in rat cerebrum and liver. Develop. Neurosci. **9**:128–132.

Prensky, A.L., Carr, S., and Moser, H.W. (1968). Development of myelin in inherited disorders of amino acid metabolism. Arch. Neurol. **19**:552–559.

Schumann, R.M., Leech, R.W., and Scott, C.R. (1978). The neuropathology of the non-ketotic and ketotic hyperglycinemias: three cases. Neurology **28**:139–144.

Shah, S.N., Peterson, N.A., and McKean, C. (1972a). Lipid composition of human cerebral white matter and myelin in phenylketonuria. J. Neurochem. **19**:2369–2376.

Shah, S.N., Peterson, N.A., and McKean, C. (1972b). Impaired myelin formation in experimental hyperphenylalaninemia. J. Neurochem. *19*:479–485.

Sugahara, K., and Schwartz, N.B. (1982a). Defect in 3′-phosphoadenosine 5′-phosphosulfate synthesis in brachymorphic mice. I. Characterization of the defect. Arch. Biochem. Biophys. **214**:589–601.

Sugahara, K., and Schwartz, N.B. (1982b). Defect in 3′-phosphoadenosine 5′-phosphosulfate synthesis in brachymorphic mice. II. Tissue distribution of the defect. Arch. Biochem Biophys. **214**:602–609.

Taylor, E.H. (1982). Effect of experimental hyperphenylalaninemia on myelin metabolism and neurotransmitter synthesis at later stages of brain development. Thesis, Medical College of Georgia.

Taylor, E.H., and Hommes, F.A. (1983). Effect of experimental hyperphenylalaninemia on myelin metabolism at later stages of brain development. Int. J. Neurosci. **20**:217–228.

CHAPTER 29

Patterns of Phenylalanine Metabolites, Vitamin B_6 Status, and Learning Disabilities in Phenylketonuria Children: Modeling for Diet Criteria

Annie Prince*

Early detection and careful dietary management has eliminated the mental retardation and neurological damage characteristic of untreated or poorly managed phenylketonuria (PKU). Despite such achievements, approximately 50 percent of PKU children undergo significant declines in intelligence quotient test scores at school age, and one-third experience learning disabilities. Measurements of blood phenylalanine remain the only laboratory test to routinely monitor dietary compliance and metabolic control of the disorder. Individual variations in quantities and types of metabolites of phenylalanine produced in PKU children have been documented that are not explained by blood phenylalanine levels. Production of these metabolites requires vitamin B_6. Measurements of vitamin B_6 status are abnormal in some PKU children, with a wide individual variation. This project is designed to explore relationships between metabolites of phenylalanine, vitamin B_6 status, and intelligence and achievement test scores of PKU children. Sixteen subjects between 6 and 14 years of age will be studied over an 8-month period. The goal is to better understand if there is a metabolic basis for clinical individuality which can be used to predict and monitor degrees of dietary restrictions necessary for optimal control.

Significance of the Project

Phenylketonuria (PKU) is an inborn error of amino acid metabolism in which there is deficient activity of the enzyme phenylalanine hydroxylase, necessary for degradation of phenylalanine (Phe) through its primary pathway. Phe, an essential amino acid and part of all natural proteins, must be provided in the diet for optimal growth and development. In PKU, dietary Phe in excess of needs accumulates in body fluids and tissues and is forced into secondary metabolic pathways for its elimination from the body. The two major pathways utilized in PKU are transamination to phenylpyruvate

* Senior Intructor of Medicine, Division of Endocrinology, Diabetes and Clinical Nutrition, School of Medicine, Oregon Health Sciences University, Portland, OR 97201, USA.

and decarboxylation to phenylethylamine. Both the transaminase (aminotransferase) and decarboxylase (aromatic decarboxylase) enzymes utilize pyridoxal-5′-phosphate (PLP), the active form of vitamin B_6, as a necessary coenzyme. Biochemical abnormalities described in children with PKU include elevations in blood Phe, increases in urinary Phe metabolites, and the presence in urine of a compound formed between vitamin B_6 and a Phe metabolite. Each of these abnormalities has been suggested to be a causative neurotoxic factor in the irreversible brain damage which characterizes untreated PKU.

Doubt no longer exists surrounding the essentiality of strict dietary Phe restriction during infancy through preschool years to prevent mental retardation. The necessary degree and duration of treatment during school years remains controversial, with no established standard diet policy. When surveyed as a whole, the published results indicate that after diet relaxation, the performance of some children does not change, whereas the performance of other children decreases substantially. Specific deficits in conceptual and visual spatial skills with relatively preserved long-term memory and language skills suggest a pattern to neuropsychological dysfunction.

Because of an inability to differentiate between those children who should remain on strict diet and those who may safely undergo dietary relaxation, it has been recommended that all PKU children be maintained on the Phe restricted diet at least through the high school years (Fishler et al. 1985). Other disagree, citing evidence that a strict dietary regimen may elicit psychosocial abnormalities in the children as well as in their families (Matthews and Barabas 1986).

Statement of Need

There is a need to provide criteria for individualized dietary advice to promote a positive impact on neuropsychological function, school achievement, and behavioral outcomes in PKU children beyond 6 years of age. Two literature reports describe individual variations in quantities and types of urinary Phe metabolites that are not explainable by blood Phe levels (Michals and Matalon 1985, Michals et al. 1986). One of the metabolites, phenylethylamine, has a pharmacologic action similar to amphetamine. This compound, when combined with vitamin B_6, results in production of a separate metabolite, pyridoxylidene-β-phenylethylamine, which has been shown to produce noxious effects on the central nervous system in a PKU animal model.

Utilizing this model supports an association between PKU and vitamin B_6 function (Loo 1967). In the single human study which has addressed the vitamin B_6 status of children with PKU, their levels of PLP were statistically significantly higher than those of control subjects; however, the range was wide (Anderson 1986). These results support the possibility of abnormalities in vitamin B_6 status with large individual variation.

Objective

The purpose of this project is to develop a model to determine optimal dietary control of PKU in school age children. The model will utilize and integrate biochemical and clinical individual variability.

Question: What biochemical factors predict optimal dietary control?

1. Is there a relationship between urinary Phe metabolite patterns and intelligence and achievement test scores?
2. Does vitamin B_6 status correlate with biochemical or clinical patterns of Phe metabolites and learning disabilities?
3. Are blood Phe levels related to urinary Phe metabolite patterns and/or vitamin B_6 status?

The study will have a descriptive phase with plans to move to a hypothesis-testing phase as theoretical and methodological knowledge is advanced.

Study Design, Data Collection, and Analyses

The Learning Disorders Clinic Interdisciplinary Evaluation Team has agreed to provide follow-up for all PKU children from ages 6 to 14 years. Testing will follow an ongoing protocol designed specifically to evaluate children with learning problems (Table 29.1). A fasting blood sample, two 24-hour urine collections, and 4-day diet diaries will be obtained on the clinic day for the biochemical and nutritional measurements (Table 29.2). Correlation coefficients will be determined by cluster analysis of the data. This statistical test results directly in identification of several mutually exclusive subgroups within which individuals are relatively similar and between which individuals are relatively different (Overall and Klett, 1972).

TABLE 29.1 Clinical data collection.

Discipline	Instrument used for measurements
Pediatrics	Pediatric Examination of Educational Readiness At Middle Childhood
Psychology	Wechsler Intelligence Scale for Children Revised
Special Education	Woodcock-Johnson Psycho-Educational Battery
Occupational Therapy	Beery Test of Visual-Motor Integration; Tremor Quantification; STOTT Test of Handwriting
Speech Pathology	Peabody Picture Vocabulary Test, Form M; Matching Familiar Figures, Form A; Token Test for Children

TABLE 29.2. Biochemical and nutritional data collection.

Analytical method	Parameter to be measured
Blood	
Fluorimetric assay	phenylalanine (McCaman and Robins 1962)
High performance liquid chromatography	phenylethylamine (Brossat et al. 1983); phenylacetic acid (Gusovsky et al. 1985)
Enzymatic assay	pyridoxal 5-'phosphate (Chabner and Livingston 1970)
Gas-liquid chromatography/mass spectrophotometry	phenylpyruvic acid (Langenbeck et al. 1978; Loo et al. 1973); phenylacetylglutamine (Langenbeck et al. 1978); phenyllactic acid (Langenbeck et al. 1978); ortho-hydroxyphenyl-pyruvic acid (Langenbeck et al. 1978; Loo et al. 1973; Nielson 1963).
Urine	
Autoanalyzer	creatinine (Pino et al. 1965)
High-performance liquid chromatography	4-pyridoxic acid (Gregory and Kirk 1979); phenylethylamine (Brossat et al. 1983)
Gas-liquid chromatography/mass spectrophotometry	phenylpyruvic acid (Langenbeck et al. 1977 and 1978; Loo et al. 1973); phenylacetic acid (Langenbeck et al. 1977 and 1978; Wadman et al. 1971); phenyllactic acid (Langenbeck et al. 1977 and 1978); ortho-hydroxyphenyl-pyruvic acid (Langenbeck et al. 1977 and 1978; Wadman et al. 1971; Nielson 1963); pyridoxylidene-β-phenylethylamine (Loo 1967).
Diet	
Computerized analysis	protein, vitamin B_6, carbohydrate, calories (Schaum et al. 1973; AOAC 1980).

Acknowledgment. These studies are being supported in part by the National Association of Research Nurses and Dietitians, The Crippled Children's Division of the Oregon Health Sciences University, and the Pediatric Practice Group of the American Dietetic Association.

References

Anderson, K. (1986). Vitamin B_6 status of children with phenylketonuria. Nutr. Rep. Intl. **34**(3):387–392.

Association of Official Analytical Chemists (1980). Vitamin B_6 (pyridoxine, pyridoxal, pyridoxamine) in food extracts. Official Methods of Analysis, 13th ed. Wash. D.C., 768–769.

Brossat, B., Straczek, J., Belleville, F., and Nabet, P. (1983). Determination of free and total polyamines in human serum and urine by ion-pairing high performance liquid chromatography using a radial compression model. J. Chromatogr. **277**:87–89.

Chabner, B., and Livingston, D. (1970). A simple enzymatic assay for plasma pyridoxal phosphate. Anal. Biochem. **34**:413–423.

Gregory, J.F., and Kirk, J.R. (1979). Determination of urinary 4-pyridoxic acid using high performance liquid chromatography. Am. J. Clin. Nutr. **32:**879–883.

Gusovsky, F., Fawcett, J., and Javaid, J.I. (1985). A high-pressure liquid chromatographic method for plasma phenylacetic acid, a putative marker for depressive disorders. Anal. Biochem. **145:**101–105.

Fishler, K., Friedman, E.G., Azen, C.G., and Koch, R. (1985). Psychoeducational outcome in treated PKU children. In: Inherited diseases of amino acid metabolism. Bickel, H., and Wachtel, U. (eds.). New York: Georg Thieme Verlag, 163–169.

Langenbeck, U., Hoinowski, A., Mantel, K., and Möhring, H.-U. (1977). Quantitative gas chromatography and single-ion detection of aliphatic α-Keto acids from urine as their o-trimethylsilylquinoxalinol derivatives. J. Chromatogr. **143:**39–50.

Langenbeck, U., Mench-Hoinowski, A., Dieckmann, K.P., Möhring, H.-U., and Peterson, M. (1978). O-trimethylsilylquinoxalinol derivatives of aromatic α-Keto acids. J. Chromatogr. **145:**185–193.

Loo, Y.H. (1967). Characterization of a new phenylalanine metabolite in phenylketonuria. J. Neurochem. **4:**813–821.

Loo, Y.H., Scotto, L., and Horning, M.G. (1973). Gas chromatographic determination of aromatic acid metabolites of phenylalanine in brain. Clin. Chim-Acta. **48:**279–285.

Matthews, W.S., and Barabas, G. (1986). Social quotients of children with phenylketonuria before and after discontinuation of dietary therapy. Am. J. Ment. Def. **91**(1)**:**92–94.

McCaman, M.W., and Robins, E. (1962). Fluorimetric method for the determination of phenylalanine in serum. J. Lab. Clin. Med. **59:**885–890.

Michals, K., Lopus, M., Gashkoff, P., and Matalon, R. (1986). Phenylalanine metabolites in treated phenylketonuric children. J. Inher. Metab. Dis. **9**(suppl. 2)**:**212–214.

Michals, K., and Matalon, R. (1985). Phenylalanine metabolites, attention span and hyperactivity. Amer. J. Clin. Nutr. **92:**361–365.

Nielson, K.H. (1963). Paper chromatographic determination of aromatic α-Keto acids. J. Chromatogr. **10:**463–472.

Overall, J.E., and Klett, C.J. (1972). Empirical methods for developing classification typologies. In: Applied multivariate analysis. Overall, J.E. and Klett, C.J. (eds.). New York: McGraw Hill, 180–239.

Pino, S., Bennotti, J., and Gordyna, H. (1965). An automated method for urine creatinine which does not require a dialyzer module. Clin. Chem. **11:**664–665.

Schaum, K.D., Mason, M., and Sharp, J.L. (1973). Patient-oriented dietetic information system. J. Amer. Diet. Assoc. **63:**39–41.

Wadman, S.K., Van Der Heiden, C., Ketting, D., and Van Sprang, F.J. (1971). Abnormal tyrosine and phenylalanine metabolism in patients with tyrosyluria and phenylketonuria; gas-liquid chromatographic analysis of urinary metabolites. Clin. Chim. Acta. **34:**177–287.

CHAPTER 30

The Neurotoxic Metabolite of Phenylalanine in Phenylketonuria

Y.H. Loo, K.E. Wisniewski, K.R. Hyde, T.R. Fulton, Y.Y. Lin, and H.M. Wisniewski*

L-Arginine and calcium pantothenate significantly lowered plasma phenylalanine levels, dramatically reduced phenylacetate production, and counteracted the myelin deficit in the young suckling rat in which sustained hyperphenylalaninemia was induced. In experimental maternal hyperphenylalaninemia, offspring of mothers fed a semisynthetic diet displayed microcephaly, a deviant ganglioside pattern, and a significant deficit of sialoglycoproteins in the cerebral hemispheres. These noxious effects were prevented by replacing the semisynthetic diet with a protein diet that provided bound sialic acid, aminohexoses, *N*-acetylated aminohexoses in skim milk powder, and supplements of L-arginine, calcium pantothenate, and D-glucosamine.

Furthermore, phenylketonuric human subjects responded positively to our protein-controlled diet with the supplements of L-arginine, calcium pantothenate, and vitamins. All displayed much improved behavior and mental function.

Our research efforts have been directed toward determining whether the mental deficit in untreated phenylketonuria (PKU) is induced by phenylalanine (Phe) per se or by one or more of the metabolic products formed during sustained hyperphenylalaninemia (HP). Results of our experiments, performed with two reliable models of the disease, demonstrate that abnormally high levels of phenylacetate (PA) in the immature brain interfere with its normal development and maturation. We were able to simulate certain characteristics of PKU in our rat model, which was injected with PA only and had normal plasma and brain levels of Phe and tyrosine (Tyr). Undernutrition was ruled out as a contributory factor.

We first identified and quantitated the aromatic acids in the rat brain during sustained HP, induced by a potent hydroxylase inhibitor, P-chlorophenylalanine (*p*-ClPhe), with L-Phe. The results are shown in Table 30.1.

*New York State Office of Mental Retardation and Developmental Disabilities, Institute for Basic Research in Developmental Disabilities, 1050 Forest Hill Road, Staten Island, NY 10314, USA.

TABLE 30.1. Concentration of aromatic acids in the rat brain during sustained hyperphenylalaninemia.

	Concentration of aromatic acids (nmol/g)			
Phe concn (μmol/g)	Phenylacetic	Mandelic	Phenyllactic	Phenylpyruvic
1.11 ± 0.07	4.0 ± 0.4	0.5 ± 0.1	4.7 ± 0.9	2.5 ± 0.4
1.53 ± 0.09	17.0 ± 1.3	2.1 ± 0.8	5.7 ± 0.2	6.4 ± 0.4
1.89 ± 0.16	96.2 ± 22	7.5 ± 2.6	9.1 ± 1.3	18.1 ± 1.9

The free passage of PA from the periphery into the immature brain coupled with its slow rate of excretion account for the cumulation of this metabolic product in the brain. PA is not excreted in the free form as are the other aromatic acids, but must be converted to phenylacetylglycine (PAGY) in the rat and phenylacetylglutamine (PAGU) in the human for excretion. The brain forms the intermediate PA-CoA but lacks the enzyme for its further transformation to PAGY (glutamine in the human). This detoxification process develops postnatally. It is barely detectable in the newborn rat and does not reach adult level of activity until 40 days of age. The presence of PA-CoA in the brain is of special significance.

We next tested, individually, each of the four metabolic products found in brain for their effect on brain weight and myelin. Only PA produced microcephaly and a deficit in myelin, simulating human PKU. In the animal injected with PA, Wen et al. (1980) observed fewer myelinated axons and fewer spirals around an axon in the optic nerve.

We assessed the effect of PA on neuronal development biochemically by (1) measuring the velocity of synaptosomal uptake of neurotransmitters, (2) examining the pattern of cerebral gangliosides in the very young rat, and (3) measuring the cerebral content of gangliosides and sialoglycoproteins (SGP), which function in cellular reactions and interactions that are essential to the proper development of the brain. We found a significant reduction of cholinergic and GABAergic terminals, a disruption of ganglioside pattern, and a significant decrease in cerebral gangliosides and SGP. A decrease in dendritic growth of Purkinje cells and a deficit of synapses per neuron have been reported (Robain et al. 1983). That our PA model simulates human PKU is supported by the report of Bauman and Kemper (1982), who described a paucity of arborization of cortical neurons in human PKU. Cerebral cortical neurons of a 19-year-old untreated PKU patient had fewer dendritic branches than those of a normal 2-year-old. PA also produced behavioral deficits (Fulton et al. 1980). Utilization of acetyl-CoA for certain metabolic reactions, such as fatty acid, cholesterol, and sialic acid synthesis, is inhibited in the brain of the PA-treated rat.

Logically, to prevent brain dysfunction in PKU, it is necessary to minimize PA production, stimulate its detoxification, and ameliorate the cerebral deficiencies of myelin, gangliosides, and SGP. We observed that

TABLE 30.2.

Supplement	Plasma amino acids (μmol/ml)[a]			Phenylacetate in urine (μmol/mg creatinine)[a]		
	Phe	*p*-ClPhe	Tyr	Unconjugated	Conjugated	Total
none	4.14 ± 0.80 (18)	0.68 ± 0.093 (18)	0.45 ± 0.031 (18)	1.19 ± 0.25 (6)	5.72 ± 0.09 (6)	6.91 ± 1.10 (6)
Arginine and Ca-pantotherate	2.98 ± 0.47 (15)[b]	0.72 ± 0.13 (15)	0.38 ± 0.023 (15)	0.51 ± 0.16 (6)[b]	2.14 ± 0.80 (6)[b]	2.65 ± 0.73 (6)[b]

[a] Values are the mean ± SD of samples taken from the number of rats shown in parentheses.
[b] Significantly lower than value obtained without supplement.

TABLE 30.3. "Protein" diet provides protection against biochemical abnormalities in rat pups previously exposed to maternal hyperphenylalaninemia (MHP).[a]

Group[b]	Treatment	Diet	Plasma Phe (mol/ml)	Weight (mg) of cerebral hemisphere[c]	Sialo glycoprotein (μg/g tissue)	Gangliosides[d] (% distribution[e])		
						GD_{1a}	GD_3	GM_1
A	Control (5)	Synthetic		213 ± 19 (22)	248 ± 9 (8)			
B	MHP (8)	Synthetic	2.67 ± 0.70 (8)	156 ± 18 (39)[f]	209 ± 17 (13)[f]	34.70 ± 2.24 (13)[f]	4.38 ± 0.78 (13)[g]	7.97 ± 1.03 (13)[f]
C	Control (4)	Protein		222 ± 24 (20)	247 ± 10 (8)			
D	MHP (7)	Protein	1.55 ± 0.45 (7)	215 ± 24 (49)	252 ± 13 (20)	37.69 ± 2.58 (9)	3.76 ± 0.41 (9)	9.70 ± 0.78 (9)
E	Control (8)	Purina Chow	0.10–0.25	209 ± 20 (35)	248 ± 9 (30)	37.08 ± 2.36 (13)	3.90 ± 0.49 (13)	10.41 ± 0.94

[a] Values are the mean ± SD for the number of samples shown in parentheses.

[b] Evaluation of differences between B and A, C, D, and E by Student's t test gave p values of statistical significance. No significant difference was found between D and all three control groups (A, C, E).

[c] 2-day-old rat.

[d] 7-day-old rat.

[e] No significant alterations were found in the distribution of gangliosides Q_{1b}, T_{1b}, D_{1b}, and D_2.

[f] $p < 0.001$.

[g] $p < 0.05$.

TABLE 30.4. Counteracting the myelin deficit in HP rats with L-arginine + Ca-pantothenate.

Treatment	Supplement	Cerebrosides and sulfatides in cerebral hemisphere (mg/g wet wt)[a]
Saline control	None	3.40 ± 0.55 (8)
p-ClPhe, Phe (HP)	None	2.86 ± 0.25[b] (8)
p-ClPhe, Phe (HP)	L-Arg + Ca-pantothenate	3.52 ± 0.25 (7)

[a] Values are the mean ± SD of samples taken from the number of rats shown in parentheses.
[b] Significantly lower than control and HP + supplements; $p < 0.01$ by analysis of variance.

dietary supplements of L-arginine and pantothenate significantly reduced PA production, stimulated its excretion (Table 30.2), and restored myelin content to normal in HP suckling rats (Table 30.4). Offspring of HP mothers (Loo et al. 1985) displayed normal cerebral ganglioside pattern and content of SGP when the pregnant mothers were fed the combination of skim milk powder, L-arginine, and pantothenate (Table 30.3).

Employing the same strategy we had used in animals, we devised a dietary therapy for human PKU individuals. Supplements of L-arginine and Ca-pantothenate and a complete vitamin preparation were included. The subjects who formerly were hyperactive, irritable, and uncooperative now became more calm, cheerful, and cooperative (unpublished).

References

Bauman, M.L., and Kemper, T.H.L. (1982). Morphologic and histoanatomic observations of the brain in untreated human phenylketonuria. Acta Neuropathol. **58**:55–63.

Fulton, T.R., Triano, T., Rabe, A., and Loo, Y.H. (1980). Phenylacetate and the enduring behavioral deficit in experimental phenylketonuria. Life Sci. **27**:1271–1281.

Loo, Y.H., Hyde, K.R., Lin, F.H., and Wisniewski, H.M. (1985). Cerebral biochemical abnormalities in experimental maternal phenylketonuria: gangliosides and sialoglycoproteins. Life Sci. **37**:2099–2109.

Robain, O., Wisniewski, H.M., Loo, Y.H., and Wen, G.Y. (1983). Experimental phenylketonuria: effect of phenylacetate intoxication on number of synapses in the cerebellar cortex of the rat. Acta Neuropathol. **61**:313–315.

Wen, G.Y., Wisniewski, H.M., Shek, J.W., Loo, Y.H., and Fulton, T.R. (1980). Neuropathology of phenylacetate poisoning in rats: an experimental model of phenylketonuria. Ann. Neurol. **7**:557–566.

CHAPTER 31

Increased Vigilance and Dopamine Synthesis Effected by Large Doses of Tyrosine in Phenylketonuria

Hans C. Lou*

In a group of nine patients with classical phenylketonuria (PKU) aged 15 to 24 years, we examined the effect of a phenylalanine-restricted diet on vigilance, as judged by the continuous visual reaction times, and neurotransmitter synthesis, as judged by cerebrospinal fluid (CSF) homovanillic acid (HVA) and 5-hydroxyindoleacetic acid (5-HIAA) levels. HVA and 5-HIAA levels decreased significantly with increase in plasma phenylalanine concentration on free diet ($p < 0.01$ and $p < 0.0005$, respectively). Vigilance was normal in six patients in whom dietary treatment had been initiated at or before 1 month of age, even at plasma phenylalanine concentrations up to 1877 μmol/1. Vigilance improved on the phenylalanine-restricted diet in six of seven patients with abnormally long reaction times on the free diet. Addition of tyrosine (160 mg/kg per 24 h) to the free diet resulted in an increased HVA-5-HIAA ratio in CSF in the six patients examined. In 14 patients on the free diet supplemented with tyrosine, an improvement in vigilance (reaction times at the 90th percentile) was seen in all 12 patients with values longer than the normal mean (264 ms) ($p < 0.001$). Tyrosine treatment may be a therapeutic alternative when phenylalanine restriction is impractical.

Introduction

Early institution of a phenylalanine-restricted diet is effective in the prevention of severe mental and neurological deficits in patients with phenylketonuria (Güttler 1980). The aim is usually to achieve plasma phenylalanine levels below 425 μmol/1 during early childhood. Thereafter, a gradual liberalization of the diet is accepted.

A recent case report from our institution indicates that cessation of treatment even in adolescence may hamper mental function, at least in some patients. This has been demonstrated by decreased performance on tests of vigilance, such as continuous recording of reaction times (RT). These patients also developed impaired central nervous system (CNS) synthesis

*Department of Neuropediatrics, The John F. Kennedy Institute, DK-2600 Glostrup, Denmark.

of dopamine and serotonin, as judged by decreased concentration of homovanillic acid (HVA) and 5-hydroxyindoleacetic acid (5-HIAA) in cerebrospinal fluid (Lou et al. 1985). In a preliminary report, we suggested that tyrosine and tryptophan supplementation may be useful in the treatment of PKU (Lou 1985). The aims of the present study have been to assess vigilance in PKU patients and to make a preliminary evaluation of the possible role of high doses of tyrosine in the treatment of such patients.

Patients and Methods

Vigilance

As a functional measure of cerebral efficiency, vigilance was tested by recording continuous reaction times. A small red stimulus light was switched on at random time intervals (15 per min) for 6.5 min. The patient was required to extinguish the light as quickly as possible by pushing a button, and a total of 100 reaction times were recorded. The reaction times were measured in 1/100 s, and 10th, 50th, and 90th percentiles were calculated. Reaction times and their variability (90th – 10th percentile difference) are regarded as a valuable indicator of both structural and biochemical derangement of the brain (Elsass et al. 1985). To measure the motor component, finger tapping tests were carried out and the results were calculated in the same way.

Neurotransmitter Metabolites and Amino Acids

Cerebrospinal fluid (CSF) levels of HVA and 5-HIAA have been determined as they reflect the rate of synthesis of dopamine and serotonin in the central nervous system. Other factors, i.e., transport mechanisms, also influence such levels. Lumbar puncture was done in a standardized way: always at 10 A.M., two hours after breakfast, while the patient was still in bed. The first 0.5 ml was discarded, and the next 1.5 ml used for analysis of HVA, 5-HIAA, phenylalanine, tyrosine, and tryptophan. A 2-ml sample of venous blood was drawn simultaneously for amino acid analysis. The analyses were carried out by high-pressure liquid chromatography using electrochemical detection for the determination of HVA and 5-HIAA and fluorometric detection for the quantitative measurement of the amino acids.

Examinations Performed

The following sets of examinations were performed:

A. In a sample of nine patients, reaction times and levels of neurotransmitter metabolites and amino acids were determined after at least three weeks on the usual "relaxed" phenylalanine-restricted diet (low protein) supplemented by an amino acid mixture without phenylalanine (Phenyldon, Aminogran) or a low-phenylalanine hydrolysate (Albumaid). Plasma levels to about 1200 μmol were accepted. These examinations were also carried out after three weeks of unrestricted diet. The order of the two regimens was random.

B. In six patients who had abnormal reaction times on the free diet, reaction time and finger tapping tests were carried out in a double-blind crossover study with the free diet and 160 mg of tyrosine per kg (divided into three daily doses) or placebo tablets of similar appearance (calcium lactate) for three days.

C. In an additional sample of eight patients, reaction times were measured on the free diet and the free diet supplemented with tyrosine (160 mg/kg divided into three daily doses). The two regimens were administered for at least three days each and in a random order not known by the psychologist. In six of the patients in this group, lumbar puncture and determination of CSF transmitter metabolites and amino acids were carried out on the two regimens.

Results

Table 31.1 shows a large interindividual variation in CSF HVA concentration. HVA and 5-HIAA decrease significantly with increase in plasma phenylalanine on the free diet ($p < 0.01$ and $p < 0.0005$, respectively; exponential regression). Tyrosine supplementation was found to increase the CSF HVA level in four of six cases and consistently raised the HVA/5-HIAA ratio (Table 31.2).

It is apparent from Table 31.1 that the five patients in whom dietary control was achieved late (at 3 months of age or later) had long and fluctuating reaction times. From Tables 31.1 and 31.2, it is seen that the six patients in whom dietary treatment was initiated at or before 1 month of age all had normal reaction times (±2 SD) even without diet and in spite of the fact that the RT test was carried out at plasma phenylalanine levels up to 1877 μmol/l. Tables 31.1 and 31.2 show that both dietary phenylalanine restriction and tyrosine supplementation have a normalizing effect on reaction times. This effect is particularly conspicuous in patients with very long reaction times. Consequently, the interindividual variations in reaction time were much larger without treatment, regardless of the order of the regimens.

The data from the double-blind crossover test, on the effect of addition of a tyrosine supplement to a free diet, are shown in Table 31.2 (patients 1–4, 9, and 10). The long reaction times are significantly shortened ($p < 0.02$, paired Student's t test), whereas normal reaction time values (below the normal 90th percentile) were unaffected. Also the pure motor task of finger tapping is unaffected. Hence, tyrosine supplementation

TABLE 31.1. Effect of cessation of dietary phenylalanine (Phe) restriction on plasma Phe, reaction time (RT), and CSF neurotransmitter metabolites in adolescents with PKU.[a]

Pt.	Age/ sex	Age at dietary control	IQ	EEG	Neurol exam.	RT[a] (1/100 s) 10%	50%	90%	Plasma Phe (mol/l)	CSF HVA (nmol/l)	CSF 5-HIAA (nmol/l)
1	19/M	3 months	71	Unprovoked normal	Slight dyscoor-	27	27	37	1011	189	31
				FS 1 1/2–3 Hz + polyspikes	dination	22	27	44	1664	149	18
2	23/M	22 months	85	Normal	Slight dyscoord.	25	28	32	1008	264	48
					+ pyr. signs	25	30	38	2050	165	29
3	24/F	4 years	70	Normal	Slight	22	25	32	576	171	69
					dyscoord.	22	27	34	1613	93	28
4	15/F	5 months	96	Normal	Normal	20	23	30	606	174	61
						20	24	32	1431	126	25
5	16/M	14 months	100	3–5 Hz	Slight	17	22	28	830	114	36
				+ sharpwaves	dyscoord.	17	22	32	1260	120	24
6	20/F	1 1/2 months	57	4–7 Hz	Normal	24	28	32	456	144	116
				+ sharpwaves		24	26	31	1255	68	37
7	17/M	11 days	102	Unprov. normal	Normal	22	25	30	605	118	39
				FS:5–7 Hz		23	28	31	1160	119	29
8	16/M	3 days	85	3–5 Hz	Normal	19	23	29	675	178	47
				+ sharpwaves		18	22	28	980	216	58
15	16/M	1 month	82	2 1/2–4 Hz	Slight	20	23	27	1052	77	18
				+ sharpwaves	dyscoord.	20	23	26	1877	84	14
Normal controls (age range 15–24, $n = 14$)						18.1 ± 2.8	21.5 ± 3.4	26.4 ± 3.6			

[a]For each patient, first line is with Phe restriction; second line, no Phe restriction.

TABLE 31.2. Effect of addition of tyrosine (Tyr) (160 mg/kg) to free diet on plasma Phe, Tyr, reaction time (RT), and CSF HVA and HVA/5-HIAA.[a]

Pt.	Age/ sex	Age at dietary control	IQ	EEG abn.	Neurol exam. abn.	RT[a] (1/100 s) 10%	50%	90%	Plasma Phe (μmol/l)	Plasma Tyr (μmol/l)	CSF HVA (nmol/l)	CSF 5-HIAA (nmol/l)	CSF HVA/5-HIAA (nmol/l)
1	19/M	3 months	71	+	+	22	26	36	1320	50			
						23	27	34	1360	72			
2	23/M	22 months	85	–	+	35	41	46	1760	46			
						32	38	44	1790	63			
3	24/F	4 years	70	–	+	23	25	30	1205	44			
						23	25	28	1300	51			
4	15/F	5 months	96	–	–	22	24	31	1431	49			
						21	24	31	1504	169			
9	17/M	10 months	59	+	+	26	31	46	1500	44			
						27	31	41	1375	68			
10	15/M	1 year	101	–	–	23	25	29	925	72			
						20	23	28	940	121			
11	21/M	3 months	83	–	+	22	27	46	1371	33	78	17	4.6
						22	26	32	1355	88	131	23	5.7
12	16/F	16 days	107	+	–	21	27	33	1385	23	116	27	4.3
						21	22	28	1220		154	30	5.2
13	16/M	25 days	100	+	+	17	22	32	1260	40	—		—
						16	22	32	1240	83			—
7	17/M	11 days	102	+	–	23	28	31	1492	51	119[b]	30	4.0
						21	22	27	1450	143	70	13	5.4
6	20/F	1 1/2 months	57	+	–	24	26	29	1170	39	—		—
						21	26	28	1410	200	—		—
14	15/M	1 month	105	+	+	16	22	27	1205	35	455[b]	78	5.8
						16	22	27	1245	55	164	24	6.8
15	16/M	1 month	82	+	+	20	23	26	1877	60	84	14	6.0
						19	23	27	1657	123	104	14	7.4
16	19/F	18 months	77	–	–	20	22	26	989	38	85	16	5.2
						20	22	27	1242	187	166	18	9.4

appears to shorten the abnormally long reaction times in PKU by facilitating attention and/or the initiation of motor activity as a response to the sensory stimulus. The effect is *not* due to increased speed of the execution of the finger tapping motor task.

The effect of tyrosine on vigilance is confirmed by the results shown in Table 31.2 for eight additional patients. In the reaction time test, a total of 12 patients out of 14 have 90th percentile values above the 90th percentile (mean) of 14 normal controls (Table 31.1). These abnormally long reaction times were reduced in all 12 patients by tyrosine ($p < 0.001$, paired Student's t test).

Discussion

The present study confirms an early case report suggesting that dietary discontinuation in patients with PKU may result in decreased neurotransmitter synthesis and impaired mental function, even in adolescene and young adulthood.

Increased plasma phenylalanine may, in theory, affect CNS synthesis of dopamine and serotonin through interference with the transport mechanisms of the transmitter precursors tyrosine and tryptophan (Herrero et al. 1983). However, in an earlier study we did not find tyrosine and tryptophan consistently reduced in CSF (Lou 1985). It is therefore unlikely that decreased tyrosine and tryptophan concentrations in the interstitial fluid in the CNS is responsible for decreased neurotransmitter synthesis. If transport inhibition is involved in reduced neurotransmitter synthesis, it is more likely due to impaired transport across the neuronal cell membrane. Another possible mechanism is competitive inhibition of tyrosine-3-hydroxylase and tryptophan-5-hydroxylase by phenylalanine. Such an inhibition has been demonstrated both in vivo and in vitro (Curtius et al. 1982, Ikeda et al. 1967). However, the consistent increase in HVA/5-HIAA ratio after tyrosine administration suggests that dopamine synthesis may be enhanced by the intake of large amounts of its amino acid precursor tyrosine.

The finding that tyrosine administration consistently reduces prolonged reaction times without affecting the speed of the motor component (finger tapping) indicates that dopamine synthesis is of central importance in the regulation of attention and/or the initiation of motor activity in man.

It is our opinion that monitoring of HVA and 5-HIAA levels in CSF and tests of cerebral vigilance may assist in defining a group of young adult PKU patients who will benefit from continued treatment. However, in many young patients, a satisfactory compliance with a strict dietary regimen becomes increasingly difficult. It is therefore encouraging that administration of a naturally occurring neurotransmitter precursor such as tyrosine seems very effective in correcting the deficient synthesis of dopamine as well as the decreased vigilance. Supplementing the diet with a large dose of tyrosine may therefore be a new and important therapeutic alternative in cases where strict dietary control is impractical. The long-term effects of such a regimen are not yet known and will require further investigations.

References

Curtius, H.C., Niederwieser, A., Visconti, M. et al. (1982). Serotonin and dopamine synthesis in phenylketonuria. *In* Haber, B., et al. (eds.). Serotonin: current aspects of neurochemistry and function. New York: Plenum, pp. 227–89.

Elsass, P., Christensen, S.E., Ranek, L., Theilgaard, A., and Tygstrup, N. (1985). Continuous reaction time in patients with hepatic encephalopathy. A quantitative measure of changes in consciousness. Scand. J. Gastroenterol. **16:**441–47.

Güttler, F. (1980). Hyperphenylalaninemia. Acta Paediatr. Scand. Suppl. **280:**1–80.

Herrero, E., Aragon, M.C., Gumenez, C., and Valdivieso, F. (1983). Inhibition by L-phenylalanine of tryptophan transport by synaptosomal plasma membrane vesicles: implications in the pathogenesis of phenylketonuria. J. Inher. Metab. Dis. **6:**32–35.

Ikeda, M., Levitt, M., and Udenfriend, S. (1967). Phenylalanine as substrate and inhibitor of tyrosine hydroxylase. Arch. Biochem. Biophys. **102:**420–27.

Lou, H.C. (1985). Large doses of tryptophan and tyrosine as potential therapeutical alternative to dietary phenylalanine restriction in phenylketonuria. Lancet **ii:**150–51.

Lou, H.C., Güttler, F., Lykkelund, C., Bruhn, P., and Niederwieser, A. (1985). Decreased vigilance and neurotransmitter synthesis after discontinuation of dietary treatment for phenylketonuria (PKU) in adolescents. Eur. J. Pediatr. **144:**17–20.

CHAPTER 32

Effect of Dietary Tryptophan Supplement on Neurotransmitter Metabolism in Phenylketonuria

Jytte Bieber Nielsen*

Eleven young adults with treated phenylketonuria received a normal diet supplemented with the serotonin precursor tryptophan. This supplementation prevented the decrease in the cerebrospinal fluid concentration of the serotonin metabolite 5-hydroxyindoleacetic acid that would otherwise have been observed after termination of the PHE-restricted diet.

Termination of dietary treatment in phenylketonuria (PKU) causes disturbances in neurotransmitter metabolism (Nielsen et al. 1988), especially in the synthesis of serotonin. The present study has examined whether supplementing a normal diet with serotonin's precursor, tryptophan, can stimulate serotonin synthesis in spite of the presence of high concentrations of phenylalanine in plasma and cerebrospinal fluid (CSF).

Patients and Methods

Patients

The patients were 11 young adults with PKU, aged 14 to 25 years. They were on a normal diet for at least three weeks and consumed this diet supplemented with tryptophan (4.5 g, in tablets) for at least one week, in random order.

Controls

Controls were 18 young adults without PKU, aged 18 to 31 years, undergoing radiculography on the suspicion of prolapsed lumbar disk.

This paper reviews, in somewhat different form, a portion of the material covered more extensively in a forthcoming paper in the journal Brain Dysfunction (Karger).

*Neuropediatric Department, The John F. Kennedy Institute, DK-2600 Glostrup, Denmark.

Lumbar Puncture

Lumbar puncture was performed at the end of each treatment period. The liquor was collected in seven 0.5-ml fractions, which were stored at −70°C until assay for homovanillic acid (HVA), 5-hydroxyindoleacetic acid (5-HIAAA), phenylalanine, tyrosine, and tryptophan by high-pressure liquid chromatography (HPLC), as described earlier (Lykkelund et al. 1987). Immediately before the lumbar puncture, venous blood was drawn. Phenylalanine, tyrosine, and tryptophan in plasma were determined (Lykkelund et al. 1987).

Statistical Methods

Pratt's rank sum test for matched pairs was applied for comparing the two sets of data for the patients; the Mann-Whitney (Wilcoxon) rank sum test was used for comparing the patient group to the normal controls. In all instances, a 5% level of significance was used.

Results

Patients consuming the normal diet had 40-fold higher CSF phenylalanine concentrations than the normal controls. Their median tyrosine concentrations were 67% higher ($p < 0.005$) and median tryptophan concentrations were 50% higher ($p < 0.001$). Their CSF HVA concentrations did not differ from those of the controls, but their CSF 5-HIAA concentration was only about one-third that of the controls ($p < 0.001$).

When the diet given the patients was supplemented with tryptophan, a fivefold increase in its plasma concentration was observed, while the concentrations of phenylalanine and tyrosine in plasma or CSF were unaltered. The median tryptophan concentrations in CSF were three times higher ($p < 0.001$) than in subjects not receiving the supplement. Consistent with this change, CSF 5-HIAA increased a little less than three times ($p < 0.001$) following tryptophan supplementation. HVA did not change.

Discussion

Inasmuch as the high plasma phenylalanine levels of PKU are thought to inhibit the transport of other large neutral amino acids across the blood-brain barrier (Pratt 1983), the high CSF concentrations of tyrosine and tryptophan found in our subjects were not anticipated. We also previously found these CSF concentrations to be increased in PKU patients on a phenylalanine-restricted diet (Nielsen et al. 1988). The results may reflect impaired amino acid transport across the neuronal cell membranes. This hypothesis is supported by the finding of a decrease in the contents of tyrosine and tryptophan in brain cortex from PKU patients, obtained postmortem (McKean 1972). Hence, CSF concentrations of these amino acids may be a poor index of the cerebral content in PKU.

It is not readily understood why serotonin metabolism seems to be more disturbed than dopamine metabolism in our patients. The concentration of CSF 5-HIAA was significantly lower in the PKU patients than in the normal controls. It was increased somewhat on a low-phenylalanine diet but remained lower than in the controls (Nielsen et al. 1988). The high intracerebral concentration of phenylalanine inhibits the transport of tyrosine into the neurons and causes a competitive inhibition of the tryptophan hydroxylase. The other enzyme in serotonin biosynthesis, L-aromatic amino acid decarboxylase, may also be inhibited by phenylalanine metabolites (Sandler 1982). Dopamine synthesis should be inhibited in parallel, as well as by the expected decrease in the tyrosine available for its synthesis. Nevertheless, the concentration of HVA was normal in our PKU subjects, on and off the phenylalanine-restricted diet (Nielsen et al. 1988). Contrary to the present findings, an earlier study (Lou et al. 1985) demonstrated decreased vigilance and neurotransmitter synthesis after discontinuation of the low-phenylalanine diet, and another showed that both vigilance and dopamine synthesis increased when the normal diet was supplemented with tyrosine (Lou et al. 1987).

The present study shows that supplementation of the normal diet with tryptophan causes a threefold increase in CSF 5-HIAA concentrations in PKU, elevating them to normal levels. It thus seems to be possible to prevent the inhibition of serotonin synthesis caused by the high phenylalanine concentration. The clinical effect of the tryptophan is less clear: we could not find any unambiguous effects on vigilance or on performance, as tested by neuropsychological examination (results not published). To ensure optimal neurotransmitter synthesis and clinical improvement in PKU, a dietary supplement with both tryptophan and tyrosine may be an alternative to the standard treatment of PKU in adolescence, where compliance with the diet often is poor.

References

Lou, H.C., Güttler, F., Lykkelund, C., Bruhn, P., and Niederwieser, A. (1985). Decreased vigilance and neurotransmitter synthesis after discontinuation of dietary treatment for phenylketonuria in adolescents. Eur. J. Pediatr. **144:**17–20.

Lou, H.C., Lykkelund, C., Gerdes, A-M., Udesen, H., and Bruhn, P. (1987). Increased vigilance and dopamine synthesis by large doses of tyrosine or phenylalanine restriction in phenylketonuria. Acta Paediatr. Scan. **76:**560–565.

Lykkelund, C., Nielsen, J.B., Lou, H.C., Rasmussen, V., Gerdes, A-M., Christensen, E., and Guttler, F. (1987). Increased neurotransmitter biosynthesis in phenylketonuria induced by phenylalanine restriction or by supplementation of unrestricted diet with large amounts of tyrosine. Eur. J. Pediatr. (submitted for publication.)

McKean, C.M. (1972). The effects of high phenylalanine concentrations on serotonin and catecholamine metabolism in the human brain. Brain Res. **47**:469–76.

Nielsen, J.B., Lou, H.C., and Güttler, F. (1988). Effects of diet discontinuation and dietary tryptophan supplement on neurotransmitter metabolism in phenylketonuria. *In*. Brain Dysfunction (in press).

Pratt, O.E. (1983). Amino acid transport across the blood-brain barrier in conditions in which amino acid metabolism is disturbed. *In* Kleinberg, G., and Deutsch, E. (eds.), New Aspects of Clinical Nutrition. Vienna: Espen, pp. 453–63.

Sandler, M. (1982). Inborn errors and disturbances of central neurotransmission (with special references to phenylketonuria). J. Inher. Metab. Dis. **5**(suppl. 2):65–70.

CHAPTER 33

Effect of Aspartame in Diabetic Rats

V.M. Sardesai, J.F. Holliday, G.K. Kumar, and J.C. Dunbar*

The effect of acute and chronic ingestion of aspartame was studied in normal and streptozotocin-induced diabetic rats. In both groups, a single dose of aspartame caused an increase of approximately 12% in liver tryptophan oxygenase (TO) activity. In diabetic animals, the sweetener caused a decrease in blood and brain tryptophan and an increase in serum glucose and glucagon. Chronic ingestion of aspartame also caused increase in TO activity in both groups. It caused a decrease in fasting serum insulin in normal animals and an increase in serum glucose in diabetic animals. Aspartame may adversely affect the capacity to control glucose metabolism in the diabetic.

Introduction

The most restricted items in the diet of diabetic individuals are refined carbohydrates because these sugars aggravate the postprandial glucose levels. The search for low-calorie, noncarbohydrate sweeteners has been intense. For the last few years, the use of the synthetic non-nutritive sweetener saccharin has been questioned because of its potential carcinogenic effect (Kalthoff and Levin 1978). The most recently approved sweetener is aspartame which is a dipeptide, L-aspartyl-L-phenylalanine methyl ester. It is approximately 180 times sweeter than glucose, and its use can help reduce the intake of calories and refined carbohydrates.

Several studies have demonstrated the safety of aspartame in normal individuals. However, some have raised concerns because of the products of its digestion—phenylalanine (Wurtman 1983), aspartic acid, and methanol. Diabetics are likely to be frequent users of aspartame-sweetened foods and beverages, but there are only minimal data available on the metabolic effects of this sweetener in diabetes, especially from chronic studies in insulin-dependent diabetics (Nehrling et al. 1985).

* Robert S. Marx Surgical Laboratories of the Department of Surgery and Department of Physiology, Wayne State University School of Medicine, Detroit, MI 48201, USA.

A study of the effect of aspartame on tryptophan metabolism is of interest because of its phenylalanine content. It has been reported that administration of phenylalanine to rats causes a 3 to 4-fold increase in their liver tryptophan oxygenase (TO) activity (Knox and Mehler 1951). Shibata et al. (1975) have shown that excess phenylalanine in the diet caused increased urinary excretion of xanthurenic acid, a tryptophan metabolite, which has been claimed to have diabetogenic action. We have reported (Sardesai et al. 1986) that aspartame does affect tryptophan metabolism.

This study was undertaken to determine the effect of acute and chronic ingestion of aspartame in diabetic animals.

Methods

Sprague-Dawley male rats weighing between 200 and 225 g were used. Animals were made diabetic by intravenous injection of streptozotocin (60 mg/kg body weight) in 1 ml of 0.1 *M* citrate buffer, pH 4.5.

For acute studies, animals were administered either 200 mg of aspartame per kg body weight in 1 ml of distilled water or plain distilled water (controls) by gavage. After 5 hours, they were sacrificed, and the liver, brain, and blood were quickly removed for analysis of biochemical parameters.

For chronic studies, the animals were either maintained on aspartame-sweetened water (1200 mg/l) or plain water as the drinking fluid and laboratory chow for 22 days. They were sacrificed, and the tissues were used for biochemical analysis.

Liver TO activity was determined according to the procedure previously described (Sardesai and Provido 1972). Serum and brain tryptophan were assayed by the method of Bloxan and Warren (1974), and serum glucose was evaluated by the glucose oxidase method (Sigma Kits). Serum insulin was measured by using Micromedic Autopack Kits and glucagon by the method of Foa et al. (1977).

Results and Discussion

As can be seen in Table 33.1, a single dose of aspartame (200 mg/kg body weight) caused liver TO activity to increase by approximately 12% in both normal and diabetic animals. This was accompanied by a drop in blood and brain tryptophan and a rise in blood glucose in diabetic animals. As expected, the level of blood insulin was lower in diabetics, but was not affected by the sweetener. Glucagon was elevated (286 vs. 58 pg/ml) in the diabetics but was not changed in controls.

In normal and diabetic animals, aspartame ingestion for 3 weeks caused approximately a 15% increase in liver TO activity (Table 33.2). The sweetener caused a decrease in blood and brain tryptophan in diabetic animals. The ingestion of aspartame caused a decrease in blood insulin in

TABLE 33.1. Acute effects of aspartame.[a]

Group	Tryptophan oxygenase (μmol/g)	Blood tryptophan (μmol/dl)	Brain tryptophan (nmol/g)	Blood glucose (mg/dl)	Insulin (μu/ml)
Normal, water	3.88	10.05	21.59	53	114
Normal, aspartame	4.93	8.90	20.37	75	120
Diabetic, water	6.44	6.22	19.09	326	60
Diabetic, aspartame	7.47	6.52	16.08	490	58

[a]Values are averages of 10 experiments.

TABLE 33.2. Chronic effects of aspartame.[a]

Group	Tryptophan oxygenase (μmol/g)	Blood tryptophan (μmol/dl)	Brain tryptophan (nmol/g)	Blood glucose (mg/dl)	Insulin (μu/ml)
Normal, water	3.64	9.48	19.06	125	87
Normal, aspartame	4.21	10.26	19.38	123	38
Diabetic, water	6.52	9.10	16.52	313	28
Diabetic, aspartame	7.58	8.08	14.51	470	28

[a]Values are averages of 12 experiments.

normal fasting animals and an increase in fasting blood glucose in diabetic animals.

The induction of TO after aspartame ingestion could be due to the presence of phenylalanine in the sweetener. Knox and Mehler (1951) and Shibata et al. (1975) reported alteration in tryptophan metabolism in rats administered phenylalanine. Conditions that induce TO, such as pregnancy and use of oral contraceptives, tend to cause a diversion of tryptophan from the serotonin to the kynurenine pathway. Many of the above states are also associated with decreased glucose tolerance. The rise in serum glucose in diabetic rats and decrease in serum insulin in normal rats fed aspartame could be due to the accumulation of tryptophan metabolites, some of which may be diabetogenic (Kotake and Ineda 1953). Also, it may be possible that elevated serum glucose is due to increased glucagon secretion.

These findings suggest that aspartame may alter the capacity to control glucose metabolism, which may be a significant factor in the already compromised diabetic.

References

Bloxan, D.L., and Warren, W.H. (1974). Error in the determination of tryptophan by the method of Denkla and Dewey. A revised procedure. Anal. Biochem. **60:**621–625.

Foa, P.P., Matsumyana, T., and Foa, N.L. (1977). *In* Abraham, G.E. (ed.), Handbook of radioimmunoassay. New York: Marcel Dekker, pp. 299–314.

Kalthoff, R.K., and Levin, M.E. (1978). The saccharin controversy. Diabetes Care **1**:211–222.

Knox, W.E., and Mehler, A.H. (1951). The adaptive increase of the tryptophan peroxidase-oxidase system of the liver. Science **113**:237–238.

Kotake, Y., and Ineda, T. (1953). Studies on xanthurenic acid. II. Preliminary report on xanthurenic acid and its relation to pyridoxine. J. Biochem. **40**:291–294.

Nehrling, J.K., Kobe, P., McLane, M.P., Olson, R.E., Kamath, S., and Horowitz, D.L. (1985). Aspartame use by persons with diabetes. Diabetes Care **8**:415–417.

Sardesai, V.M., Holliday, J.F., Kumar, G.K., and Dunbar, J.C. (1986). Effect of aspartame in normal and diabetic rats. Biochem. Arch. **2**:237–243.

Sardesai, V.M., and Provido, H. (1972). The effect of ethyl alcohol on rat liver tryptophan oxygenase. Life Sci. **11**(2):1023–1028.

Shibata, Y., Nishimoto, Y., Takeuchi, F., and Tatsuma, Y. (1975). Tryptophan metabolism in various nutritive conditions. Acta. Vit. Enzymol. **29**:190–193.

Wurtman, R.J. (1983). Neurochemical changes following high-dose aspartame with dietary carbohydrates. N. Engl. J. Med. **309**:429–430.

CHAPTER 34

Maternal Phenylketonuria Collaborative Study (MPKUCS): USA and Canada

R. Koch,* E.G. Friedman,* C. Azen,* F. dela Cruz,† H. Levy,‡ R. Matalon,§ B. Rouse,‖ and W.B. Hanley#

The fact that pregnancy in untreated phenylketonuria women results in microcephaly in 92% of the offspring has motivated a study to determine if this remarkably high occurrence of fetal morbidity can be obviated. All fifty states and the provinces of Canada are collaborating in this study sponsored by the National Institute of Child Health and Human Development. Each pregnancy will be treated according to a set protocol, and control data are also being collected. Results will be available in a preliminary form in 1989 and a final report will be issued in 1991.

It is clear that maternal phenylketonuria (PKU) poses a serious risk to fetal development in the absence of dietary restriction of phenylalanine. The degree of fetal risk with therapy is as yet unclear, but most believe that phenylalanine restriction is beneficial. To assess the degree of risk, the National Institute of Child Health and Human Development has initiated a study involving the United States and Canada. The Medical Genetics Division at Childrens Hospital of Los Angeles was selected as the coordinating center for the project, with four regional contributing centers encompassing 50 states and the District of Columbia.

The research design is prospective, longitudinal, and observational in nature and will attempt to include 200 hyperphenylalaninemic (HPA) pregnancies and their offspring. Women with blood phenylalanine levels persistently greater than 240 μmol/l, on an unrestricted diet, would be eligible.

The treatment plan will consist of:

1. provision of adequate nutrition during pregnancy;
2. offering the phenylalanine-restricted diet to HPA women with blood

*Childrens Hospital of Los Angeles, 4650 Sunset Boulevard, Los Angeles, CA 90027, USA.
†National Institute Child Health and Human Development, Washington, D.C. 20892, USA.
‡Boston Childrens Hospital, Boston, MA, USA.
§University of Illinois Medical Center, Urbana, IL 61801, USA.
‖University of Texas Medical Branch, Galveston, TX 77550, USA.
#Hospital for Sick Children, Toronto, Ontario, Canada.

phenylalanine concentrations consistently greater than or equal to 600 μmol/l;
3. aiming to maintain blood phenylalanine concentrations between 120 and 600 μmol/l;
4. supplementation with tyrosine and trace elements as medically indicated.

The research questions that the study is designed to answer are the following:

1. Does the phenylalanine-restricted diet reduce the frequency of mental retardation, spontaneous abortion, low birth weight, congenital malformations, and neurological and behavioral impairment reported in pregnancies of HPA mothers who were on unrestricted phenylalanine intake during pregnancy?
2. Is pregnancy outcome in HPA women who restrict phenylalanine intake during pregnancy comparable to that of non-HPA women?
3. Is pregnancy outcome in HPA women related to maternal phenylalanine levels during pregnancy?
4. Is gestational age, at the onset of intervention, predictive of fetal outcome?
5. Are there beneficial effects of starting diet prior to conception?
6. What are the levels of tyrosine and trace elements during pregnancy and what are the effects on pregnancy outcome of supplementation if levels are found to be reduced?

In order to have demographically and genetically similar groups of non-HPA women to whom subjects may be compared, the study will follow, prospectively, (1) normal controls matched for race, age, gestation, and parity, and (2) familial controls obtained from the sibship of the HPA woman or her mate, or first cousins of the HPA woman or her mate.

Additional comparison groups will include (1) prospective evaluation of the pregnancies of the mates of HPA males, (2) prospective evaluation of offspring from previous pregnancies of HPA subjects, and (3) historical data on untreated HPA pregnancies from the 1980 Lenke-Levy survey.

In an observational study such as the present one, many extraneous factors could contribute to observed differences between subjects and comparison groups. Some are accounted for in matching criteria for controls. Others will need to be routinely assessed and controlled for at the time of data analysis. Among this latter group are parental intelligence and head cricumference, socioeconomic status of the family, nutritional status apart from phenylalanine level prior to and during pregnancy, and maternal exposure to teratogens, such as drugs, tobacco, and alcohol.

The very nature of long-term collaborative studies calls for dedication and self-sacrifice for the research effort, by the families and many individuals contributing to the data collection and analysis. The present study began on 1 May 1984, and has completed its organizational phase. Subject enrollment commenced on 1 November 1984. To date, 49 pregnancies have been followed and 101 more PKU women are enrolled in the study.

It is hoped that the study will allow us to determine what phenylalanine level during pregnancy will maintain normal fetal development, whether preconceptual phenylalanine restriction is necessary, and whether supplementation with tyrosine and various trace elements such as zinc are necessary for normal pregnancy outcome.

In this early phase of the study, most of the enrolled women are mildly retarded or borderline in intelligence. This problem hopefully will resolve as the study progresses and more women are enrolled with normal intellectual ability.

Acknowledgment. This study is supported by National Institute Child Health and Human Development Contract N01-HD-4-3807.

Part VI Aspartame and Food Intake

CHAPTER 35

Effects of Aspartame on Appetite and Food Intake

John E. Blundell, Andrew J. Hill, and Peter J. Rogers*

Sweetness is a powerful psychobiological phenomenon which has physiological significance and exerts strong effects on the pattern of food consumption. A number of mechanisms play a role in mediating these effects. In natural foods, sweetness is frequently associated with high energy. Together, these elements influence the effect of foods on appetite. Uncoupling sweet taste and calories, by means of a high-intensity sweetener, is therefore likely to induce some changes in the expression of appetite. Aspartame—in addition to its role as an agent which disengages sweet taste and energy value—may influence eating via hormonal or neurochemical mechanisms. Is aspartame inert when consumed by humans under laboratory circumstances or in natural conditions? Evidence on the action of aspartame is ambiguous. Some studies have indicated a paradoxical separation of indices of motivation to eat. Given to experimental subjects in aqueous solutions, aspartame may facilitate hunger or leave a residue of hunger. Other studies have suggested a suppressive effect of aspartame, or little effect when given in larger doses by-passing sweet taste receptors. The field is at present greatly under-researched.

Introduction

Aspartame is a dipeptide compound which has the potential to influence the expression of appetite by means of its sweetening power or by means of a possible action upon hormonal or neurochemical mechanisms involved in the control of food intake. Some of these mechanisms will be common to other intense sweeteners.

Artificial sweeteners constitute an important class of food additives which are being used in progressively increasing quantities. However, intake of these substances is not evenly distributed throughout the population. Owing to the inclusion of artificial sweeteners in particular types of products and to the perceived value of sweeteners in helping to treat obesity or to prevent weight gain, this class of food additives is likely to be

*Biopsychology Group, Psychology Department, University of Leeds, Leeds LS2 9JT, UK.

consumed in very large amounts by certain subgroups of the population. Despite the widespread ingestion of artificial sweeteners and the likely effect that sweet substances will exert on eating and appetite, only a handful of studies have investigated the relationship between consumption of artificial sweeteners and mechanisms of hunger and food consumption. The most substantial studies have been conducted by Porikos, Van Itallie, and colleagues (Porikos 1981, Porikos et al. 1977, Porikos et al. 1982, Porikos and Van Itallie 1984), but the particular design chosen together with certain other methodological constraints leave the final outcome open to question, as the authors themselves point out. Oddly, despite the importance of the issue, there has been no series of experiments carried out to investigate the action of artificial sweeteners under varying environmental or physiological circumstances. There is also a lack of basic information concerning sweeteners and appetite under standard laboratory conditions. Although considerable attention has been directed to the possible toxicity of artificial sweeteners, effects on appetite are grossly under-researched.

Role of Sweeteners in Eating

Sweetness is a powerful psychobiological phenomenon which has physiological significance and exerts strong effects on the pattern of food consumption (Naim and Kare 1982). A preference for sweet-tasting substances is probably an innate disposition of many animals including man (Beauchamp and Cowart 1987). Sweet substances are inherently pleasurable to most individuals and can act as potent rewards. Consequently, on the basis of sensory qualities alone, the sweetness of a food will tend to promote its ingestion.

There are two particular ways in which the adjustment of sweetness could influence food consumption and appetite. First, in nature, the sweet taste is almost invariably linked to high-calorie, nutritionally useful commodities. This association between the taste of food and its metabolic value is important in the control of energy intake. One way in which organisms (including man) control episodes of eating is by predicting caloric value on the basis of taste (e.g., Booth 1980). There is good evidence that the biological system is innately programmed to establish links between cues (taste sensations) and consequences (metabolic effects of foods), and this constitutes a part of the biological wisdom of the body (e.g., Garcia et al. 1974). The operation of this system allows organisms to anticipate the effects of food ingestion in advance of the digestion and absorption of the products. It follows that the use of artificial sweeteners in foods will effectively uncouple sweet taste from calorific value. Since normally sweetened foods are also likely to be consumed along with artificially sweetened items, the sweet taste of food no longer provides a reliable cue for predicting metabolic effects. Consequently, one mechanism for regulating energy intake would be undermined.

Second, the cephalic phase reflex is the term given to anticipatory responses which prepare the gastrointestinal system for the arrival of food. Such responses can be detected at various levels of the system including salivary secretion, gastric outflow, and pancreatic secretion. In man, the sight and smell of food are sufficiently potent to trigger cephalic phase insulin responses (Sjostrom et al. 1980, Simon et al. 1986), and the response in obese subjects is four times that in lean people. In rats, the taste of saccharin can provoke a cephalic phase insulin response (Berthoud et al. 1981). In turn, the size of the cephalic phase response is related to the degree of hyperphagia on a good-tasting diet. The stimulus-induced cephalic phase responses are accompanied by a raised tendency to consume food. Consequently, it may be expected that in humans, sweet substances will enhance cephalic phase reflexes and promote ingestion.

Artificial Sweeteners and Mechanisms Controlling Eating

The potent rewarding capacity of the sweet taste means that the acceptability of foods can be increased by making them sweet. Palatability is one of the most important dimensions controlling short-term intake in man, and this is one way in which artificial sweeteners will influence ingestion. It is also likely that artificial sweeteners will trigger cephalic phase responses, and owing to the disengagement of sweet taste and calories, they may disrupt a natural biological regulatory system. However, not all intense sweeteners have the same composition. Of the sweeteners currently in use—saccharin, acesulfame-K, cyclamates, and aspartame—one has attracted special interest because of its dipeptide structure.

Aspartame and Appetite

Aspartame is composed of two amino acids—phenylalanine and aspartic acid. This particular composition means that aspartame may influence feeding control mechanisms in a variety of ways. For example, phenylalanine is known to release cholecystokinin (Gibbs et al. 1976), an anorexigenic hormone released from the duodenum following food consumption. It has been proposed that cholecystokinin is an endogenous satiety agent, and it has been demonstrated to inhibit food intake in animals (Smith and Gibbs 1979) and man (Kissileff et al. 1981). In addition, phenylalanine may trigger amino acid receptors within the alimentary canal which are involved in regulatory processes concerning digestive functions, salivation, and food intake (Mei 1985). A further effect of ingested phenylalanine could occur via changes in brain synthesis of the catecholamine neurotransmitters dopamine and noradrenaline (Wurtman 1983; see also this volume), which have been shown to influence the control of feeding patterns and food selection in the rat (e.g., Leibowitz and Shor-Posner 1986). Although the dynamics of phenylalanine-induced changes in catecholamine synthesis are different in man, similar feeding mechanisms could well be affected.

It follows that aspartame may influence the expression of appetite and eating by means of a number of mechanisms:

1. cephalic phase reflexes,
2. cue-consequence conditioning (uncoupling of sweetness-energy),
3. triggering of endogenous satiety factors,
4. action on brain mechanisms of hunger.

It follows, of course, that these mechanisms are independent of each other and that some or all may be brought into play at any particular time. Consequently, the ingestion of aspartame may have a number of possible actions:

1. no effect on hunger-satiation,
2. leave a residual hunger,
3. facilitate hunger,
4. intensify satiation (induce anorexia),
5. combinations of the above—simultaneous or sequential.

In addition, it should be considered that there may be sizable individual differences in the sensitivity of particular mechanisms. For example, people who are dieting severely may be particularly vulnerable to the triggering of cephalic phase reflexes. It was noted earlier that the size of the cephalic phase insulin response is four times greater in obese than in lean subjects (Sjostrom et al. 1980).

On the basis of the comments set out above, it seems likely that artificial sweeteners will exert some action on mechanisms controlling appetite and eating. The questions are whether the effects will be sufficiently prominent to be readily measured, whether experimental tests can be conducted so as to reveal the nature of the effects, and whether these actions are likely to be invoked by natural day-to-day ingestion of artificial sweeteners.

Experimental Studies on Aspartame and Hunger/Satiety

The above arguments provide a rationale for a systematic investigation of the effects of artificial sweeteners on appetite—as experimental tools which exquisitely disengage the sweet taste from calories, and as components of food products widely consumed. Some possibilities have been tested in a series of experiments in which the effects of aspartame and other artificial sweeteners (saccharin, acesulfame-K) have been compared with the effects produced by energy-rich sweeteners such as glucose or sucrose.

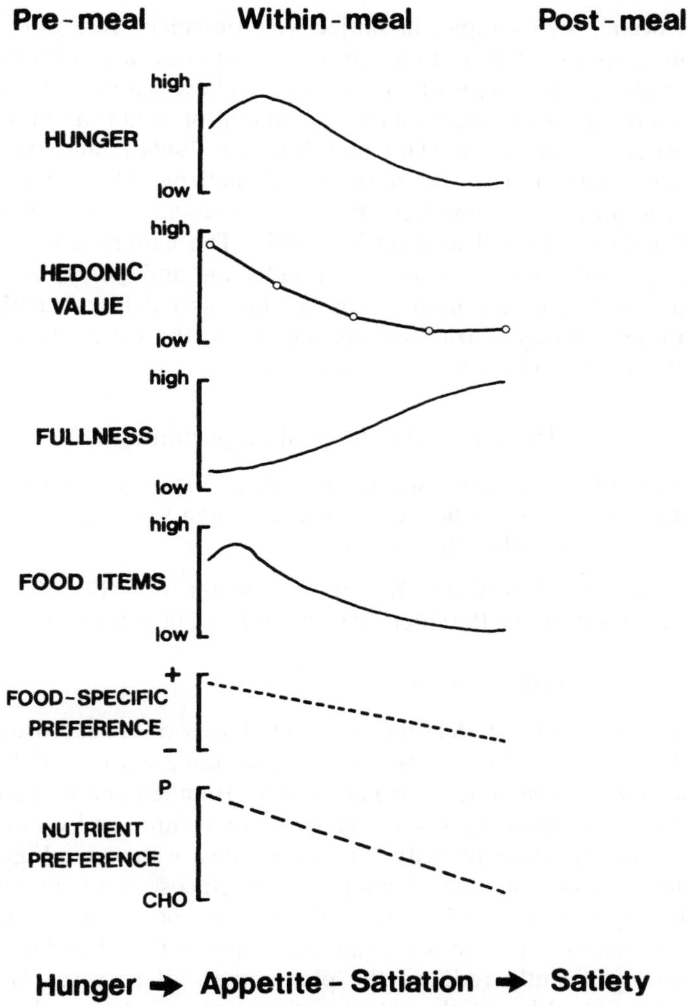

FIGURE 35.1. Fluctuations in motivational parameters which accompany eating during the course of a meal.

Monitoring Hunger and Satiation

It follows from the possible mechanisms outlined above that if a number of actions occurred together, even weakly, then artificial sweeteners could generate a rather unusual profile of changes in the processes of hunger and satiation which normally control patterns of eating. Consequently, monitoring techniques are required which permit the disclosure of different types of effects. Figure 35.1 illustrates the fluctuations which occur in various parameters during the ingestion of a meal or following a preload.

Changes occur, for example, in subjectively perceived ratings of hunger and fullness, in the perceived pleasantness of food or taste stimuli, and in expressed choices for foods of various types. These parameters are sensitive to manipulations in the total energy content of food (Hill et al. 1987), macronutrient composition (Hill and Blundell 1986a), and palatability (Hill et al. 1984), and to pharmacological challenge (Hill and Blundell 1986b). The procedures have been described in detail elsewhere (Blundell and Hill 1988b, Blundell and Burley 1987). The experiments described below employed various measures of behavior and perception widely regarded as reflecting the intensity of satiation and satiety, together with certain other techniques traditionally applied to the experimental investigation of sweet substances (Cabanac 1971).

Paradoxical Effects of Aspartame

In the first study, 50 young adults (35 females, 15 males) of normal weight for height were tested. A between-subjects design was used and subjects received one of the following three loads:

A. 50 g of glucose dissolved in 200 ml of tap water ($N = 17$);
B. 162 mg of aspartame (9 tablets of Canderel) in 200 ml of tap water ($N = 17$);
C. 200 ml of tap water ($N = 16$).

Subjects made ratings of pleasantness and intensity of the sucrose solution on 100 mm visual analogue rating scales (providing a score of 0 to 100). The procedure has been described elsewhere (Blundell and Hill 1988a).

The result of consuming loads of glucose, aspartame, and water on perceived pleasantness and intensity of a sweet taste are shown in Figure 35.2. There was a highly significant effect of the glucose load on ratings of pleasantness [$F(6,96) = 5.952, p < 0.01$]. On every occasion after the load, the pleasantness of sucrose was significantly lower than baseline. Aspartame also significantly reduced the pleasantness of sucrose [$F(6,96) = 2.741, p < 0.05$], although the extent of this change was about half of that experienced by the subjects in the glucose condition. Consuming an equal volume of water did not alter subjects' perception of pleasantness. There were no effects of load upon ratings of intensity, i.e., altered perception of pleasantness was not due to changes in the perception of intensity.

Regarding the changes in motivation, the treatments did not produce massive effects. This is probably because subjects had recently eaten, and motivation to eat was relatively low. The data are shown in Figure 35.3. It can be seen that the changes produced by aspartame and glucose were invariably in opposite directions. The clearest separation occurred in ratings of prospective consumption, where glucose reduced subjects' estimates of how much food they could eat (significant at 40 minutes). Overall, there was a clear tendency for glucose to reduce ratings of hunger and

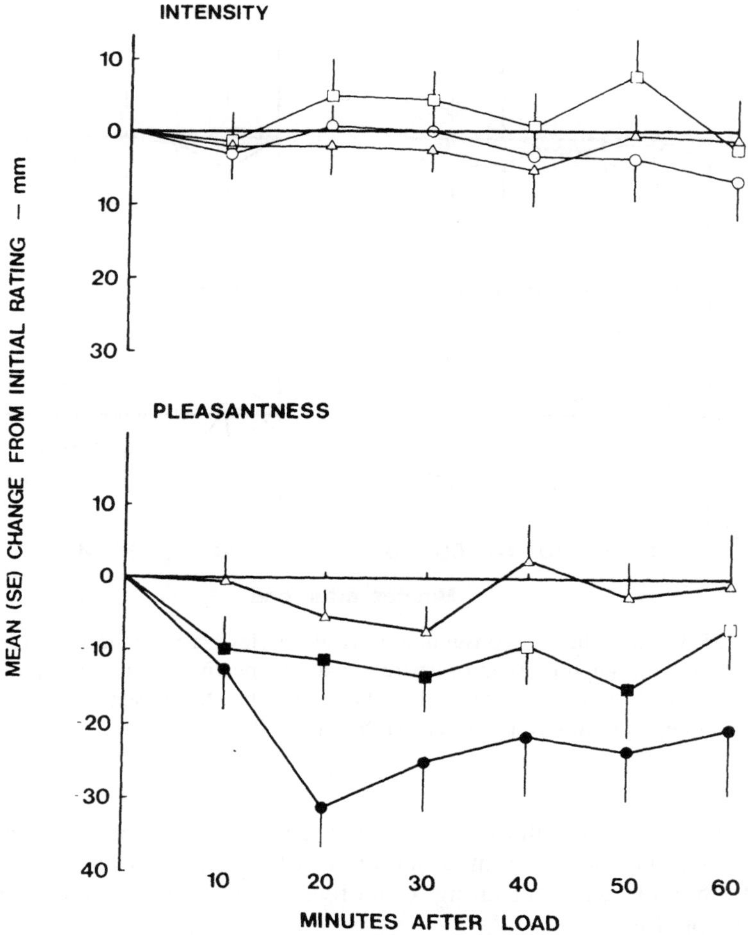

FIGURE 35.2. The effect of consuming glucose (circles), aspartame (squares), or water (triangles) upon the perceived intensity and pleasantness of sucrose at 10-minute intervals after the loads. Closed symbols indicate a significant change from baseline ($p < 0.05$, two-tailed test).

appetite and to increase fullness, whereas aspartame acted in the opposite direction.

In other words, in this study the oral glucose load, which combined sweet taste with calorific properties, produced a consistent effect upon the measures of satiation. Glucose markedly reduced the perceived pleasantness of sweet solutions, decreased measures of hunger and appetite, and increased ratings of fullness. In contrast, aspartame gave rise to an ambiguous profile of effects. On one hand, the aspartame load produced a

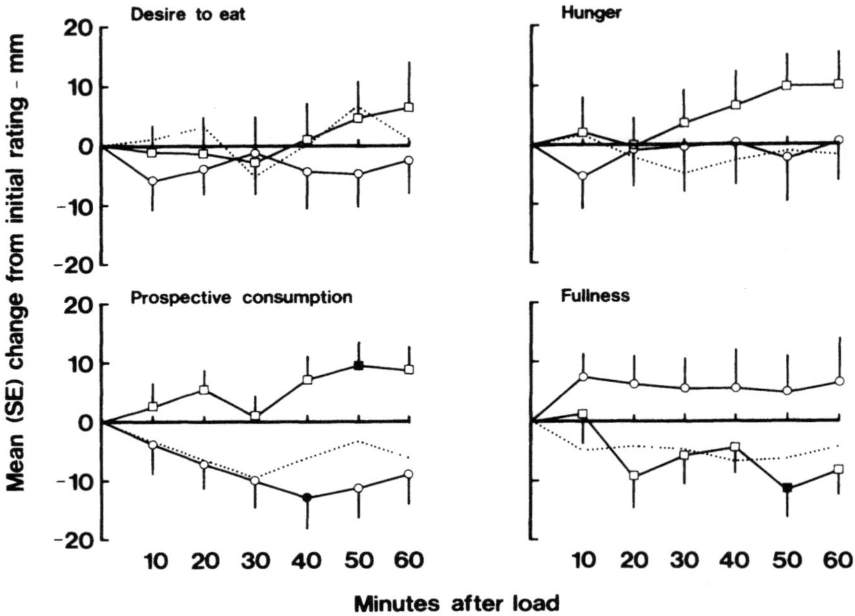

FIGURE 35.3. The effect of consuming glucose (circles), aspartame (squares), or water (dotted line) upon ratings of desire to eat, prospective consumption, hunger, and fullness at 10-minute intervals after the loads. Statistically significant changes from baseline are indicated by closed symbols ($p < 0.05$).

moderate negative alliesthesia (decreased perceived pleasantness of sweetness), indicative of mild satiation, while on the other hand this agent tended to augment ratings of motivation (indicating a mild stimulation of appetite).

Effects of Aspartame and Sucrose on Appetite

In a further study in this series, we have compared the effects of aspartame with those produced by the natural sweetening agent against which all artificial sweeteners are compared, namely, sucrose. Here a within-subjects design was used and all subjects were female ($N = 10$). They were tested following a two-hour period of not eating. The four treatment conditions were as follows:

A. 50 g of sucrose dissolved in 200 ml of tap water;
B. 288 mg of aspartame (16 tablets of Canderel) in 200 ml of tap water, calibrated so as to be of equal sweetness to the sucrose load;
C. 200 ml of tap water;
D. no ingestion.

These treatments therefore had the following properties:

	Volume	Sweetness	Energy
A	Yes	Yes	Yes (188 kcal)
B	Yes	Yes	No (5 kcal)
C	Yes	No	No
D	No	No	No

The results of this experiment indicate that the sucrose load produced a consistent and synchronized effect on all measures of satiation. As predicted, the sucrose load brought about a consistent decrease in ratings of pleasantness (i.e., moderate negative alliesthesia) together with a marked suppression of hunger and desire to eat and an increase in fullness. The effects of aspartame were quite different. Interestingly, this high dose of aspartame (288 mg, nearly twice the concentration used in our previous experiments) did not produce a stronger decrease in ratings of pleasantness—in fact, only a rather weak negative alliesthesia was observed. This may indicate a rather unusual dose-response function for aspartame. (Indeed, in an unpublished study, we have noted that this high concentration of aspartame produced more variable responses than the lower-strength solution—162 mg in 200 ml.) For measures of motivation, the effects of aspartame were similar to those for the water load and control condition. Over the course of the 60-minute test period, aspartame gave rise to a significant increase (from baseline) in hunger and a decrease in fullness. An exception was that aspartame produced a significant decrease in ratings of desire to eat at the 10-minute test point. This probably reflects a cognitive effect of consuming the sweet-tasting solution (see below).

An interesting profile of changes emerged from the analysis of food checklist responses (see Blundell and Hill 1987). In keeping with the decrease in ratings of hunger and appetite, the sucrose load gave rise to a small negative shift in the checking of food items (Figure 35.4). In contrast, all other treatments gave rise to increases, with the largest effect produced by aspartame. In other studies, we have reported strong positive correlations between hunger, actual food consumption, and checklist scores. Therefore, these data suggest that this dose of aspartame is producing a mild facilitative effect on appetite.

Taken together, the effects of sucrose demonstrated in this experiment are similar to those noted previously for glucose. A sweet and calorific load gave rise to a constellation of effects on measures of satiation which are consistent. In contrast, aspartame, in this experiment, produced an ambiguous profile. Surprisingly, this high concentration of aspartame produced only a weak negative alliesthesia, indicating a very mild satiating effect. The pattern of effects for the motivational ratings showed the opposite effect—a slight facilitation of appetite. The checklist scores indicated a tendency of aspartame to promote appetite.

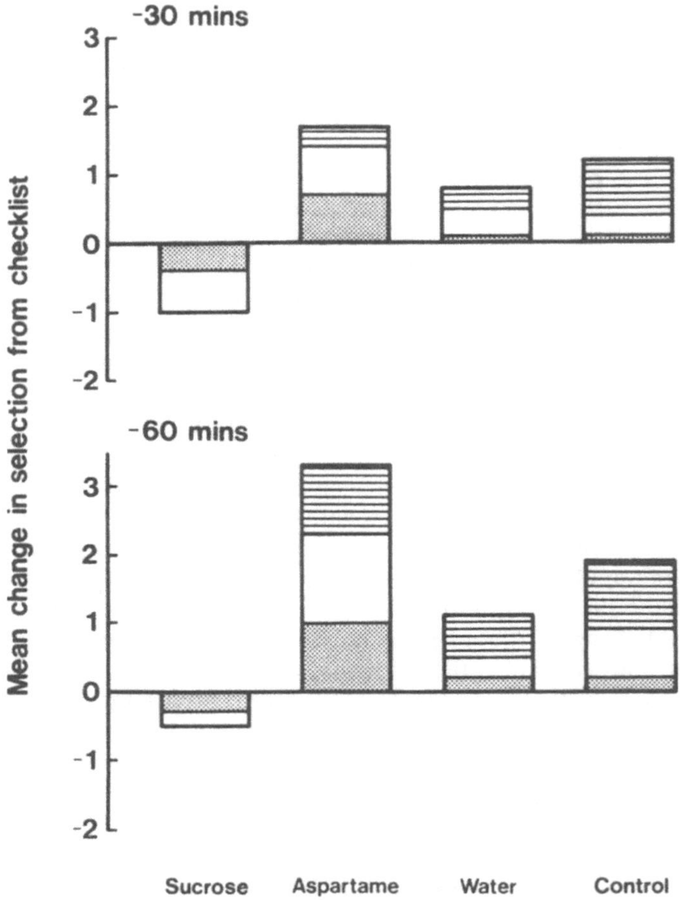

FIGURE 35.4. The effect of treatments on the selection of high-carbohydrate (stippled columns), high-fat (plain columns), and high-protein (graded columns) food items from the food preference checklist at 30 and 60 minutes after ingestion. The increase in selected items following aspartame is significant at 60 minutes ($p < 0.05$, two-tailed test).

Comparison of Aspartame and Other Sweeteners

In order to establish whether the effects revealed above were specific to aspartame or were more general features of all artificial sweeteners, we have compared the effects of preloads of aspartame, saccharin, and acesulfame-K, matched for sweetness and volume with a 25% glucose solution. Twelve subjects were tested in a repeated-measures design with each subject receiving each treatment. After ingesting the solutions at 12:00 M., subjects filled in rating scales periodically during the following 60 minutes

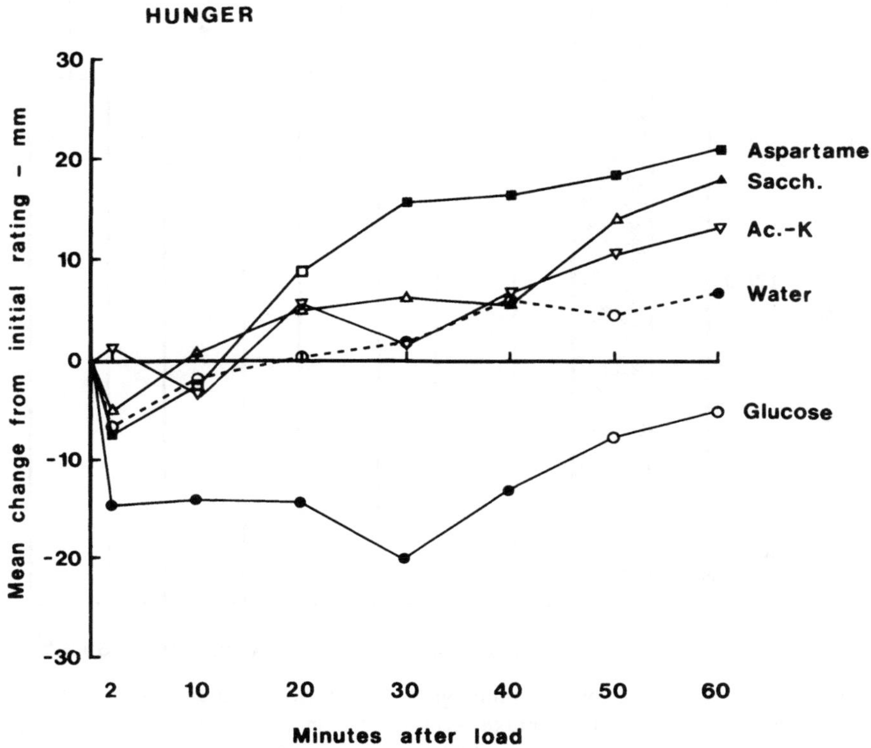

FIGURE 35.5. The effect of consuming glucose (circles), aspartame (squares), saccharin (triangles), acesulfame-K (inverted triangles), and water (dotted line) upon hunger ratings at 10-minute intervals after the loads. Closed symbols indicate a significant change from baseline ($p < 0.05$).

in order to track the changes in hunger and other perceived sensations. At the end of this time, subjects were given a lunchtime test meal (at 1:00 P.M.) comprising sandwiches, cake, biscuits, and yoghurt. The amount of each product consumed together with total calories and intakes of specific macronutrients (protein, carbohydrate, and fat) were measured.

The effects of the preloads upon ratings of perceived hunger are shown in Figure 35.5. Although the subjects were mildly deprived at the time of the experiment (no food eaten since 9:00 A.M.), the artificial sweeteners gave rise to a facilitation of perceived hunger, in contrast to the glucose load, which markedly suppressed hunger. Statistical analysis revealed a significant effect of time [$F(6,60) = 16.02$, $p < 0.01$] and a significant preload by time interaction [$F(24,240) = 2.27$, $p < 0.01$]. The food intake data were equally interesting and are shown in Figure 35.6. The effect of preload was marginally significant [$F(4,44) = 2.45$, $p < 0.1$, two-tailed test).

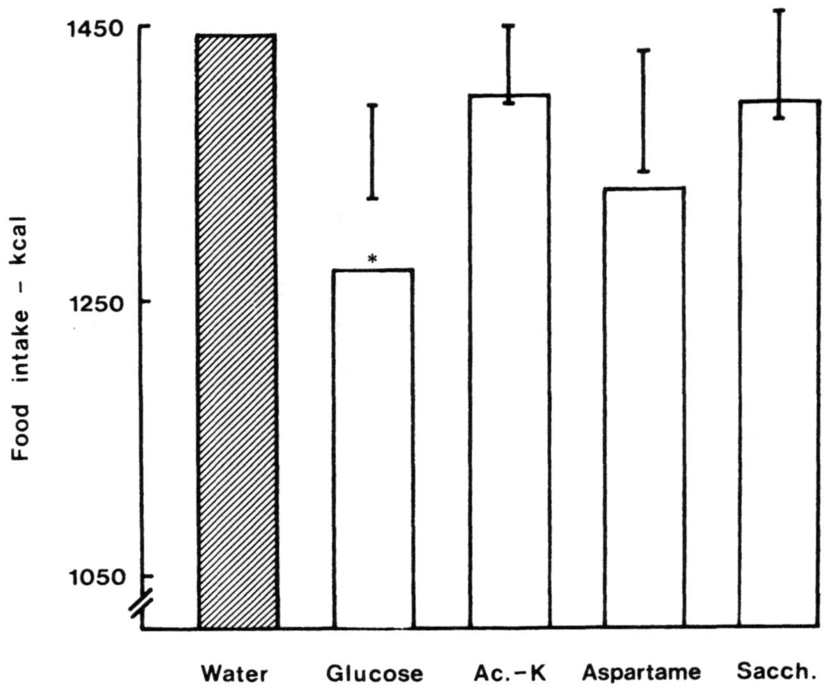

FIGURE 35.6. The effect of consuming glucose, aspartame, saccharin, acesulfame K, and water upon mean food intake (kcal) in a test meal one hour later. Error bar indicate standard error of the mean difference between water and preload. *Signif icant reduction compared to water ($p < 0.05$).

Post-hoc two-sample comparisons (Newman-Keuls) revealed that compared to water, the glucose load significantly suppressed the amount of food consumed in the test meal ($p < 0.05$). Indeed, the energy reduction brought about by glucose almost precisely matched the energy contained in the preload. In other words, subjects compensated very well for this caloric preload. The artificial sweeteners did not significantly suppress food intake, indicating that the residual hunger was functional and maintained consumption. However, it is noticeable that the artificial sweeteners did suppress intake by small amounts ranging from 2% (acesulfame-K) to 8% (aspartame). The low intake for aspartame is a little misleading since with this treatment six subjects increased their intake (compared to water load) and six showed reductions. Of these last six, two subjects displayed substantial reductions of 52% and 27%, respectively. In these subjects, for some reason, the aspartame load produced an anorexic response.

One interesting feature of the study is the relationship between the effects of the treatments on hunger and on food consumption. For the glucose load, the effects are consistent—a marked suppression of hunger together with a reduction in food consumption. However, the artificial sweeteners—and aspartame in particular—showed elevations in hunger but weak suppressive effects on intake. This dissociation is very unusual in our experience. However, it is not surprising that increases in food intake were not observed. The subjects were hungry at the time of the test meal and ate considerable amounts of food (average of 1450 kcal). Therefore, intake was probably maximal for this time of the day, and it would have been extremely difficult for any manipulation to further improve on this degree of energy consumption. In addition, it should be kept in mind that subjects were eating from a limited range of foods (those appropriate for the nutritional analysis employed), and especially favored foods which may have been sought under conditions of enhanced hunger may not have been present. These explanations probably account for the lack of an increase in food intake following artificial sweetener preloads, but why should a mild decrease have occurred? This was probably engendered by a carry-over of the sensory and cognitive processes mediating satiety (see later) which normally help to suppress intake in advance of postingestive and postabsorptive effects.

Implications of Uncoupling Sweetness and Calories with Aspartame

In all of the experiments described above, subjects given oral loads of solutions high in sweetness and in energy displayed a consistent profile of responses. Measures of satiation including hedonic ratings, hunger motivation scales, and the food item checklist all revealed a concerted pattern of effects. Consistent profiles were also displayed by treatments which were devoid of energy and sweetness. These conditions (water load or no ingestion) did not produce a decline in hedonic ratings (of the sucrose test solutions) and during the course of the testing period (60 minutes) appetitive motivation increased—subjects began to feel more hungry and less full. Consequently, when sweetness and calories covary, consistent effects on indices of satiation and on food consumption are demonstrated. What happens when sweetness and energy are experimentally disengaged through the agency of artificial sweeteners?

In the first series of studies described here, the intense sweetener aspartame has been compared with glucose, sucrose, and control treatments. In contrast to those treatments in which sweetness and energy were linked, aspartame generally gave rise to a mixed profile of effects—some measures of satiation showing a moderate downward response and others showing no effect or being facilitated. Usually, aspartame produced a decline in hedonic ratings but did not reduce, and sometimes enhanced, other motivational measures. Interestingly, a high concentration of aspartame did not strengthen the alliesthesia effect—indeed, it appeared weaker. This feature requires further investigation, but suggests that the effects of aspartame may vary in an unpredictable way with the dose applied. This may be particularly important in view of the additional effects on appetite control mechanisms which may be brought into play by the particular chemical composition of aspartame.

Consequently, in experimental investigations in the laboratory, aspartame has been demonstrated to display irregular effects on measures of appetite control. What are the implications of this for artificial sweeteners in general use? First, for people consuming dietary products containing artificial sweeteners, it can be predicted that the lowering of energy intake would fail to bring about a normal depression of motivation. This would leave people with a residual hunger, i.e., a tendency to eat if given the opportunity. This was demonstrated to be the case in the last series of experiments where aspartame (and the other artificial sweeteners) failed to significantly depress food consumption in the test meal. Second, the intense sweeteners appear to display ambiguous effects on factors reflecting the operation of satiation (the process which stops eating). This suggests that the appetite control mechanisms will be activated in possibly conflicting ways, with some systems being turned down while others remain active. This confusion of psychobiological information cannot help the development of good control over eating. Third, there is evidence that under some circumstances, aspartame actually stimulated eating. This response appears very variable and probably depends on the dose, time after administration, and the characteristics of a particular subject. However, it does mean that certain individuals could respond to aspartame (and other artificial sweeteners) by becoming more rather than less hungry. The circumstances under which this facilitation of hunger leads to increases in food consumption remain to be further investigated.

Can Aspartame-Sweetened Products Provide Satiating Power?

At the present time, only a handful of studies have explored the effects of aspartame (and other artificial sweeteners) on appetite and food consumption. Apart from the investigations mentioned above, these studies include the three experiments on the substitution of aspartame for sucrose in foods (see Porikos et al. 1977, 1982, Porikos and Van Itallie 1984, Porikos, 1981), an experiment on aspartame in Jell-o (Rolls et al. 1986), a comparison of aspartame and sucrose in milk shake loads (Brala and Hager 1983), and the administration of large doses of aspartame and phenylalanine in capsules, bypassing the taste receptors (Ryan-Harshman et al. 1987). These studies have produced markedly different effects ranging from an apparent aspartame-induced anorexia (Rolls et al. 1986) to aspartame-induced increases in intake—9% more than placebo, 26% more than

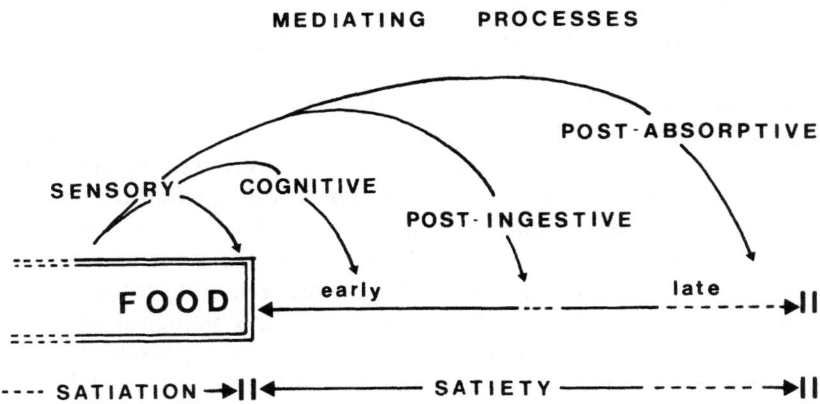

FIGURE 35.7. A conceptualization of the contributions of sensory, cognitive, post-ingestive, and postabsorptive stimuli to satiation and the time course of satiety.

sucrose (Brala and Hager 1983). Interestingly, in a staunch defence of artificial sweeteners, Booth (1987) has categorically stated that "all the evidence is that aspartame does not stimulate eating" without even referring to the work of Brala and Hager.

One of the key questions is whether artificially sweetened products (i.e., sweet but low in energy) can provide the satiating power of high-energy sweeteners. That is, do they provide a strong suppression of hunger and an inhibition of later food consumption? The studies reported above suggest that artificial sweeteners do not possess strong satiating power. However, a certain degree of satiety may be present under some circumstances. Thinking on this issue may be clarified by considering the processes set out in Figure 35.7. This diagram illustrates various ways in which the ingestion of foods or preloads may inhibit later consumption. It follows that a sweet-tasting solution or food (low in calories) will provide certain sensory and cognitive suppression of intake in the immediate postingestive period, but will lack the later postabsorptive actions. Consequently, the belief about a sweet-tasting product or the triggering of established cue-consequence associations (derived from the consumption of high-calorie sweeteners) may serve to provide short-lived satiety under some circumstances. These mechanisms probably account for the weak inhibition of food consumption by artificial sweeteners in the test meals described above. In experiments on the satiating effect of sweeteners, the precise nature of the preload or the sweetened food will be crucial; some products/solutions will impose potent sensory/cognitive effects while others will be much weaker. However, artificial sweeteners would not be expected to provide late-phase satiety.

The issue of satiating power is clearly relevant to the three studies of

Porikos, Van Itallie, and colleagues in which subjects (a total of 13 obese and 14 lean) consumed food from a menu in which the sucrose content of sweet items had been replaced by aspartame. These elegant experiments fall between clinical trials of a product (aspartame) and studies on energy regulation. Since the designs used were not completely balanced (consumption of sucrose-containing foods always preceded consumption of aspartame-loaded items), the studies cannot readily be used to throw light on the body's capacity to monitor energy intake. The results indicated that when subjects were switched from the sucrose to the aspartame products, overall energy intake fell. However, despite the technical accomplishments in the use of calorically-disguised sweet items, the biological system was not totally deceived and subjects showed a caloric compensation of about 40%. Therefore, on the full aspartame regimen, the subjects did overeat but not by a large enough margin to achieve full compensation for the lost calories.

However, a number of comments can be made about these very ingenious studies. First, given the elaborate deception, the degree of caloric compensation is quite impressive. In order to regain the lost calories through consumption of the sweet items alone, the subjects would have been obliged to consume massive and unacceptable amounts. Since they were obliged by the methodology to sustain a certain energy deficit through the soft drinks (by mandatory consumption), which are calorically dense (calories per weight), the degree of overeating required through the other food items to obtain perfect compensation was unrealistic. Second, there is evidence that the normal subjects used were an unrepresentative sample who may have been nutritionally imbalanced before entering the study. Certainly, the first baseline intake of these subjects was remarkably high and the quality and palatability of the food offered probably caused an artificial elevation of intake over the first few days. In our view, these considerations suggest that a compensation of 40% is rather remarkable. Indeed, taking account of the elevated baselines, the compensation may well be nearer 100% (Booth 1987).

The satiating power of artificial sweeteners is also questioned by the survey of use of these products carried out by the American Cancer Society. In a preliminary report on more than 78,000 women aged 50 to 69 "the rate of weight gain (over a one-year period) in AS users was significantly greater than in nonusers irrespective of initial relative weight" (Stellman and Garfinkel 1986, p. 197). In addition, "the proportion of AS users who gained 10 pounds or more was significantly greater than the proportion of nonusers who gained 10 pounds or more at each weight level." These data should be treated cautiously since all the information was collected retrospectively by questionnaire. Moreover, the users of artificial sweeteners may well have been dieters, and it should be considered whether it is the dieting strategy rather than artificial sweetener use (as part of this strategy) which is counterproductive in the attempt to control body weight.

Although these data do not prove that artificial sweetener use causes weight gain (although the data are certainly consistent with this view), they certainly demonstrate that artificial sweetener use does not prevent weight gain. Similar findings have been revealed for other dietary products. For example, in a retrospective study of dieters, it was found that "use of slimmers' meal replacements . . . was associated with failure to lose weight during the diet and with failure to maintain lost weight" (Lewis and Booth 1985, p. 199). The authors further assert that "there is evidence that such products are abused and/or the regulation of claims is inadequate." Since this study, like that of Stellman and Garfinkel, also relied upon retrospective supply of information, the data should be considered with some caution. However, the similarity of the two findings is interesting.

It is noticeable that in the Stellman and Garfinkel analysis, the authors present a deliberately balanced view of the outcome, in particular drawing attention to one category of women (the most obese) in which more artificial sweetener users than nonusers lost 10 pounds or more. The authors question how these women may differ from the large numbers who gained weight. The presence of different types of responders gives credibility to the analysis. The major detected effect of artificial sweetener use, being increased weight gain, is consistent with the idea that artificial sweetener use may either facilitate appetite or lead to a buildup of residual hunger (Blundell and Hill 1986, 1987). Other dieting strategies may have similar effects.

In the discussion presented above, the major characteristic of artificial sweeteners under consideration has been the capacity to uncouple sweetness from calories. However, in the case of aspartame, certain postingestive mechanisms could be brought into play by way of triggering of amino acid receptors in the alimentary canal, a release of endogenous satiety hormones, or the adjustment of plasma amino acid profiles and ultimately brain neurotransmitter synthesis. In an attempt to investigate this last issue, one study has examined the effects of the administration of very large doses (up to 10.0 g) of phenylalanine and aspartame (Ryan-Harshman et al. 1987). Since subjects ingested these amounts in capsules (swallowing as many as 24 capsules in some conditions), the substances by-passed the generation of sweet sensations in the mouth and therefore would avoid triggering cephalic phase reflexes or disrupting learned associations between sweetness and calories. Consequently, this study is not related to the issues investigated in our research presented previously. Nevertheless, it is worth examining. The results apparently failed to support any action of phenylalanine or aspartame on hunger or food consumption. However, no data were shown for the effects of amino acid loads on hunger and no statistical analysis was presented. In the case of food intake (in a test meal), no F values or other statistical data were presented and no post-hoc tests were applied, even though one dose of phenylalanine (10.08 g) produced an approximate 16% suppression of intake. It is worth noting that this degree of suppression is greater than that normally achieved by 50 g of glucose (Blundell and Hill 1988a) or by a potent anorexic drug (Blundell and Rogers 1980, Hill and Blundell 1986b). Lastly, it is worth pointing out that the study was carried out on men who were unrestrained (i.e., not dieting) and hungry at the time of the test meal. As noted previously, this design is insensitive to the detection of increases in food intake, and it would be particularly interesting to compare the changes in hunger ratings with food intake (see Figures 35.5 and 35.6). Consequently, although this study provides some fascinating data, the report still leaves unanswered a number of critical questions. Evidence from our own studies using much smaller doses of aspartame in capsule form (therefore bypassing the sweet sensation) indicates that this manipulation is not inert (Rogers and Blundell 1987). We feel that it is important to consider the possibility that aspartame may exert effects on more than one mechanism implicated in appetite and eating control.

The Methodology of Future Research with Aspartame

Artificial sweeteners, and aspartame in particular, constitute exquisite tools for researchers examining mechanisms controlling appetite and eating: they are also agents of considerable commercial importance. This means that investigations of artificial sweeteners are not just a scientific enterprise but acquire political significance if research outcomes are perceived to have an influence on commercial prospects. It is therefore difficult to carry out research which, while throwing light on the action of sweeteners, is not perceived as being advantageous or detrimental to one or another commercial interest. The tendency to view a research outcome in terms of marketing advantage is a great handicap to independent researchers seeking to disclose valid information about important psychobiological manipulations. In a recent report (Blundell and Hill 1986), we drew attention to certain paradoxical effects of artificial sweeteners. However, reports in the media were largely focused on one single aspect of this study—namely, the possible facilitation of appetite—with a commensurate degree of misrepresentation of the outcome of the research. This action of journalists linked to a particular commercial lobby is undesirable but may be expected. Rather alarmingly, at least one scientific investigator has engaged in a heroic attempt to undermine the essence of the outcome of the above-mentioned report. Although one may expect journalists to get facts wrong or fail to properly understand subtle issues, it is disturbing when scientists display similar weaknesses. In the report alluded to (seen by the present writers in draft form but citation withheld), the distinguished commentator made eight errors of perception including the misreading of sucrose for glucose.

It is worth repeating that knowledge concerning the effects of artificial sweeteners on appetite is still extremely limited: the field is under-researched, and wordy polemic is no substitute for sound experimentation. Artificial sweeteners, in general, are likely to influence psychobiological systems, and effects on appetite and eating should be expected. Research on this issue should be encouraged, not stifled. The nature of the effects on appetite remain to be precisely defined; effects may vary with amount ingested, with environmental or physiological circumstances, and with the type of individual. Our research indicates that these sweeteners may give rise to unexpected effects or to multiple and paradoxical effects. When used as an aid to weight control, it should be kept in mind that artificial sweeteners do not per se perform a weight-reducing or weight-regulating function. Individuals who consume artificial sweeteners and who assign responsibility to these products may fail to check facilitative effects or the existence of residual hunger. It would seem judicious to point out that the ingestion of artificial sweeteners for weight control should be accompanied by a greater vigilance and greater degree of self-monitoring, not less. As pointed out by other researchers, a number of questions must be answered before the efficacy of low-calorie sweeteners in the management of human obesity can be established.

> First, studies in which people know they are using low calorie products need to be carried out. It is possible that people would use the known energy savings from low calorie products as an excuse to increase intake of other foods. (p. 284) . . . [The second] issue needing direct study is that of actual weight loss with low-calorie products. A carefully controlled long-term study is needed to look at the effects of using low calorie food analogs on body weight. (p. 285) . . . Given the sizeable role that low-calorie foods have in the American diet, it is surprising how little information exists concerning their effectiveness in promoting weight loss. (p. 282) (Porikos and Van Itallie 1984)

In addition to these recommendations, there is a need for comparisons between aspartame and other intense sweeteners. The study described in this chapter which involved a comparison between glucose, aspartame, saccharin, and acesulfame-K appears to be the only comparative examination of the effects of intense sweeteners in man. Since all intense sweeteners have certain common features, comparisons among these compounds are necessary in order to determine whether aspartame possesses any specific appetite-influencing properties which are not shared by other compounds. There is no doubt that aspartame is a fascinating compound that has attracted a vast amount of research interest. However, at the present time, there is still much to learn about its effects on human appetite.

References

Beauchamp, G.K., and Cowart, B.J. (1987). Development of sweet taste. *In* Dobbing, J. (ed.), Sweetness. London: Springer-Verlag, pp. 127–138.

Berthoud, H.R., Bereiter, D.A., Trimble, E.R., Siegel, E.G., and Jeanrenaud,

B. (1981). Cephalic phase, reflex insulin secretion. Diabetologia **20**:393–401.
Blundell, J.E., and Burley, V.J. (1987). Satiation, satiety and the action of fibre on food intake. Int. J. Obesity **11** (Suppl. 1):9–25.
Blundell, J.E., and Hill, A.J. (1986). Paradoxical effects of an intense sweetener (aspartame) on appetite. Lancet **i**:1092–1093.
Blundell, J.E., and Hill, A.J. (1987). Artificial sweeteners and the control of appetite: implications for the eating disorders. *In* Worden, A., Parke, D., and Marks, J. (eds.), The future of predictive safety evaluation, Vol. 2. Lancaster: MTP Press, pp. 263–282.
Blundell, J.E., and Hill, A.J. (1988a). On the mechanisms of action of dexfenfluramine: effect on alliesthesia and appetite motivation in lean and obese subjects. Clin. Neuropharmacol. (in press).
Blundell, J.E., and Hill, A.J. (1988b). Descriptive and operational study of eating in man. *In* Blinder, B.J., Chaitin, B.F., and Goldstein, R. (eds.), Modern concepts of the eating disorders: research, diagnosis, treatment. New York: Spectrum (in press).
Blundell, J.E., and Rogers, P.J. (1980). Effects of anorexic drugs on food intake, food selection and preferences and hunger motivation and subjective experiences. Appetite **1**:151–165.
Booth, D.A. (1980). Acquired behaviour controlling energy intake and output. *In* Stunkard, A.J. (ed.), Obesity. Philadelphia: Saunders, pp. 101–143.
Booth, D.A. (1987). Evaluation of the usefulness of low-calorie sweeteners in weight control. *In* Grenby, T.H. (ed.), Developments in sweeteners, Vol. 3. Elsevier (in press).
Brala, P.M., and Hager, R.L. (1983). Effects of sweetness perception and caloric value of a preload on short term satiety. Physiol. Behav. **30**:1–9.
Cabanac, M. (1971). The physiological role of pleasure. Science **173**:1103–1107.
Garcia, J., Hawkins, W.G., and Rusiniak, K.W. (1974). Behavioural regulation of the milieu interne in man and rat. Science **185**:824–831.
Gibbs, J., Falasco, J.D., and McHugh, P.R. (1976). Cholecystokinin-decreased food intake in rhesus monkeys. Am. J. Physiol. **230**:15–18.
Hill, A.J., and Blundell, J.E. (1986a). Macronutrients and satiety: the effects of a high protein or a high carbohydrate meal on subjective motivation to eat and food preferences. Nutrition and Behaviour **3**:133–144.
Hill, A.J., and Blundell, J.E. (1986b). Model system for investigating the actions of anorectic drugs: effect of *d*-fenfluramine on food intake, nutrient selection, food preferences, meal patterns, hunger and satiety in healthy human subjects. *In* Ferrari, E., and Brambilla, F. (eds.), Advances in the BioSciences, Vol. 60. Oxford: Pergamon Press, pp. 377–389.
Hill, A.J., Magson, L.D.,and Blundell, J.E. (1984). Hunger and palatability: tracking ratings of subjective experience before, during and after the consumption of preferred and less preferred food. Appetite **5**:361–371.
Hill, A.J., Leathwood, P.J., and Blundell, J.E. (1987). Some evidence for short term caloric compensation in normal weight human subjects: the effects of high and low energy meals on hunger, food preference and food intake. Hum. Nutr.: Appl. Nutr. **41A**:244–257.
Kissileff, H.R., Pi-Sunyer, F.X., Thornton, J., and Smith, G.P. (1981). C-Terminal octapeptide of cholecystokin decreases food intake in man. Am. J. Clin. Nutr. **34**:154–160.

Leibowitz, S.F., and Shor-Posner, G. (1986). Monoamine meal patterns in the rat. *In* Carruba, M.O., and Blundell, J.E. (eds.), Psychopharmacology of eating disorders: theoretical and clinical advances. New York: Raven Press, pp. 29–50.

Lewis, V.J., and Booth, D.A. (1985). Causal determinants within an individual's dieting thoughts, feelings and behaviour. *In* Diehl, J.M., and Leitzman, C. (eds.), Measurement and determinants of food habits and food preferences. EC Workshop Report, Gienen, May 1–4.

Mei, N. (1985). Intestinal chemosensitivity. Physiol. Rev. **65**:211–237.

Naim, M., and Kare, M.R. (1982). Nutritional significance of sweetness. *In* Birch, G.G., and Parker, K.J. (eds.), Nutritive Sweeteners. London: Applied Science Publishers, pp. 171–193.

Porikos, K.P. (1981). Control of food intake in man: response to covert caloric dilution of a conventional, and palatable diet. *In* Cioffi, L.A., James, W.P.T., and Van Itallie, T.B. (eds.), The body weight regulatory system: normal and disturbed mechanisms. New York: Raven Press, pp. 83–87.

Porikos, K.P., and Van Itallie, T.B. (1984). Efficacy of low-calorie sweeteners in reducing food intake: studies with aspartame. *In* Lewis, D., Stegink, L.J., and Filer, J. (eds.), Aspartame: physiology and biochemistry. New York: Marcel Dekker, pp. 273–286.

Porikos, K.P., Booth, G., and Van Itallie, T.B. (1977). Effect of covert nutritional dilution on the spontaneous food intake of obese individuals: a pilot study. Am. J. Clin. Nutr. **30**:1638–1644.

Porikos, K.P., Hessner, M.F., and Van Itallie, T.B. (1982). Caloric regulation in normal-weight men maintained on a palatable diet of conventional foods. Physiol. Behav. **29**:293–300.

Rogers, P.J., and Blundell, J.E. (1987). Effect of low doses of aspartame, with and without sweet taste, on hunger and food intake. Unpublished report.

Rolls, B.J., Hetherington, M., Burley, V.J., and Duijvenvoorde, P.M. (1986). Changing hedonic responses to foods during and after a meal. *In* Kare, M.R., and Brand, J.G. (eds.), Interaction of the chemical senses with nutrition. New York: Academic Press, pp. 247–268.

Ryan-Harshman, M., Leiter, L.A., and Anderson, G.H. (1987). Phenylalanine and aspartame fail to alter feeding behaviour, mood and arousal in men. Physiol. Behav. **39**:247–253.

Simon, C., Schlienger, J.L., Sapin, R., and Imler, M. (1986). Cephalic phase insulin secretion in relation to food presentation in normal and overweight subjects. Physiol. Behav. **36**:465–469.

Sjostrom, L., Garellick, G., Krotkiewski, M., and Luyckx, A. (1980). Peripheral insulin in response to the sight and smell of food. Metabolism **29**:901–909.

Smith, G.P., and Gibbs, J. (1979). Postprandial satiety. Progress in Psychobiology and Physiological Psychology **8**:179–242.

Stellman, S.D., and Garfinkel, L. (1986). Artificial sweetener use and one-year weight change among women. Preventative Medicine **15**:195–202.

Wurtman, R.J. (1983): Neurochemical changes following high-dose aspartame with dietary carbohydrates. New England J. Medicine **309**:429–430.

Chapter 36

Effects of Phenylalanine and Aspartame on Mealtime Food Intake and Behavior in Adult Males

Lawrence A. Leiter, Milly Ryan-Harshman, and G. Harvey Anderson*

Two experiments were conducted to investigate the neurobehavioral effects of phenylalanine (Phe) and aspartame (APM) on energy and macronutrient selection and on subjective feelings of hunger, mood, and arousal in normal-weight males. Neither mean energy intakes nor macronutrient selection at mealtime were altered by the administration of up to 10 g of Phe or APM. In both experiments, visual analogue scale scores were unaffected by the treatments, but time increased ($p < 0.05$) scores for emptiness, rumbling, weakness, degree of hunger, and urge to eat. Absolute and relative concentrations of plasma Phe and tyrosine (Tyr) were increased in some instances by capsule administration; yet food intake and behavior were unaffected. In a third experiment, the administration of 10 g of APM in conjunction with a 118 g carbohydrate (CHO) breakfast resulted in a 0.09 m*M* rise in plasma Phe, in contrast to a 0.19 m*M* rise when given alone in the second experiment. Once again, there was no treatment effect on hunger, mood, and arousal. We conclude that Phe, when given as the free amino acid or as APM, in doses up to 10 g, does not affect feeding behavior in normal-weight adult males.

Introduction

Elevations in plasma phenylalanine (Phe) and tyrosine (Tyr) may influence brain catecholamine synthesis which, in turn, may be involved in the regulation of energy balance and food choice (Anderson et al. 1984). Although brain catecholamine synthesis seems to respond to the availability of the precursor, Tyr (Gibson and Wurtman 1978), the specific effect of bolus doses of Phe on short-term food intake and behavior in normal-weight adult males has received little attention. In addition, aspartame (APM), which contains approximately 50% by weight of Phe, might also be expected to have effects. Finally, the effect of Phe on behavior when APM is administered concurrently with a high-carbohydrate meal is also of interest. Yokogoshi et al. (1984) suggested that alterations in brain catecholamine levels might occur when APM is consumed with carbohy-

*Departments of Nutritional Sciences and Medicine, University of Toronto, Ontario, Canada M5S 1A8.

drate, thus influencing the physiological and behavioral mechanisms mediated by catecholamines.

The objectives of this research were to measure the effects of Phe and APM on short-term energy and macronutrient intakes at lunch time and on subjective feelings of hunger, mood, and arousal in normal-weight adult males. Plasma amino acid levels were also measured to determine the effectiveness of the supplements in altering the large neutral amino acid (LNAA) profiles of the subjects.

Methods

Three experiments were conducted with men between the ages of 20 and 35 years, and within 90 and 110% of ideal body weight. In experiment 1 (n = 13), four feeding trials were conducted following administration of capsules containing Phe in doses of 0 (5.04 g of alanine (Ala) placebo), 0.84 g, 2.52 g, and 5.04 g. Visual analogue scales (VAS) were administered at 11:00 A.M. (capsules taken at 11:15 A.M.), 12:00 M. (buffet-style luncheon served at 12:15 P.M.), and 12:45 P.M. to assess 17 subjective feelings of hunger, mood, and arousal.

In experiment 2 (n = 13), conditions were similar to experiment 1 except that doses were different and additional time (a duration of 90 min) was allowed between administration of Phe and APM by capsule and the second VAS test. Doses administered were Ala placebo (10.08 g), 5.04 g APM plus 5.04 g Ala, 10.08 g APM, and 10.08 g Phe, and VAS tests were administered at 10:10 A.M. (capsules taken at 10:15 A.M.), 11:45 A.M. (lunch served at noon), and 12:30 P.M.

Venous blood was drawn from a subset of subjects (n = 5, experiment 1; n = 7, experiment 2) on days separate from the feeding experiments for plasma amino acid analysis. Blood samples were obtained prior to capsule administration and 45 min (90 min in experiment 2) after the capsules were ingested. The methods have been described more fully by Ryan-Harshman et al. (1987).

In experiment 3, six subjects were given either 10.08 g Ala placebo of 10.08 g APM by capsule in conjunction with a high-carbohydrate (118 g) breakfast at 8:00 A.M. VAS tests were administered at 8:00 A.M., 10:00 A.M., 12:00 M., and 2:00 P.M. Blood samples for amino acid levels were drawn at baseline (7:30 A.M.) and 90 min (9:30 A.M.) after the capsule administration.

Results

Mean (± SD) energy intakes (kcal) did not differ significantly in either experiment 1 (1330 ± 311, 5.04 g Ala placebo; 1543 ± 471, 0.84 g Phe; 1334 ± 479, 2.52 g Phe; 1215 ± 366, 5.04 g Phe) or experiment 2 (1230 ± 358, 10.08 g Ala placebo; 1103 ± 355, 5.04 g APM; 1124 ± 387, 10.08 g

APM; 1070 ± 385, 10.08 g Phe). Macronutrient selection was not significantly different between feeding trials, but remained relatively constant across both experiments at about 16% protein, 36% carbohydrate, and 48% fat.

Analysis of the VAS indicated that there were no treatment effects of Phe or APM in both experiments, although VAS scores were significantly altered ($p < 0.05$) over time. Higher values for emptiness, stomach rumbling, weakness, degree of hunger, and urge to eat were reported by subjects prior to lunch; these effects of time were eliminated by eating lunch.

In experiment 1, the 5.04 g dose of Phe resulted in a rise ($p < 0.05$) in plasma Phe from 5.4 μmol/dl to 27.7 μmol/dl after 45 min. The plasma Phe/LNAA ratio was significantly higher ($p < 0.05$) after all Phe doses, but the plasma Tyr/LNAA ratio was significantly increased ($p < 0.05$) only after doses of 2.52 g Phe and 5.04 g Phe. In experiment 2, the plasma Tyr/LNAA ratio increased significantly ($p < 0.05$) at a dose of 10.08 g APM, but the high plasma Phe levels at 10.08 g Phe caused the plasma Tyr/LNAA to remain unchanged. The absolute and relative plasma Phe concentrations rose significantly ($p < 0.05$) at all doses given (placebo excepted).

In experiment 3, the administration of 10.08 g of APM in conjunction with the high-carbohydrate meal once again did not exhibit any treatment effect on hunger, mood, and arousal although a similar effect of time was once again seen at 4 hours (emptiness, degree of hunger, and urge to eat). A 0.09-m*M* rise in plasma Phe was seen after 90 min (6.4 ± 1.1 to 15.3 ± 5.3 μmol/dl). This compares to a 0.19-m*M* rise (6.9 ± 1.2 to 26.2 ± 7.8 μmol/dl) observed in experiment 2 when APM was given alone.

Discussion

The lack of effect on VAS of large doses of APM seen in this study is of particular interest because of the Centers for Disease Control (CDC) report evaluating consumer complaints regarding APM (Bradstock et al. 1986). The VAS used in our experiments contained descriptive words such as headache, dizziness, stomachache, nausea, drowsiness, alertness, and depression which are similar to those symptoms reported by complainants after APM consumption in the CDC report. Because our subjects reported no unusual reactions to APM consumption, the symptoms noted by complainants in the CDC report may have been idiosyncratic and not representative of the general population. These responses may also be normal responses to caloric deprivation which is self-inflicted, and the reason many consumers choose APM-containing beverages or foods (Ryan-Harshman et al. 1987).

Current views on precursor control of neurotransmitter synthesis suggest that alterations in brain neurotransmitter synthesis would be expected in these experiments from the increased brain Phe and Tyr (Sved 1983). Clearly, if this occurred, no behavioral consequences were detected. Perhaps the approaches used in the experiments reported herein may not qualitatively assess the interactions among neurotransmitter precursors and the multiple neurochemical systems regulating feeding behavior. That Phe or APM might enhance brain catecholamine synthesis and subsequently influence feeding behavior when administered concurrently with carbohydrate has been suggested (Yokogoshi et al. 1984), but our data suggest that plasma Phe levels are lower following aspartame and carbohydrate administration than when aspartame is given alone. Although it is conceivable that brain uptake of Phe was enhanced, it is also likely that increased oxidation of Phe occurred following the carbohydrate meal.

Conclusions

Phe and APM in doses up to 10 g failed to alter food intake or hunger, mood, and arousal in normal-weight adult males, even though significant changes in the absolute and relative concentrations of plasma Phe and Tyr were observed. Even the administration of 10 g of APM in conjunction with a high-carbohydrate meal had no effect on hunger, mood, and arousal.

Acknowledgment. These studies were supported by the International Life Sciences Institute-Nutrition Foundation.

References

Anderson, G.H., Li, E.T.S., and Glanville, N.T. (1984). Brain mechanisms and the quantitative and qualitative aspects of food intake. Brain Res. Bull. **12:**167–173.

Bradstock, M.K., Serdula, M.K., Marks, J.S., Barnard, R.J., Crane, N.T., Remington, P.L., and Trowbridge, F.L. (1986). Evaluation of reactions to food additives: the aspartame experience. Am. J. Clin. Nutr. **43:**464–469.

Gibson, C.J., and Wurtman, R.J. (1978). Physiological control of brain norepinephrine synthesis by brain tyrosine concentration. Life Sci. **22:**1399–1406.

Ryan-Harshman, M., Leiter, L.A., and Anderson, G.H. (1987). Phenylalanine and aspartame fail to alter feeding behavior, mood and arousal in men. Physiol. Behav. **39:**247–253.

Sved, A.F. (1983). Precursor control of the function of monoaminergic neurons. *In* Wurtman, R.J., and Wurtman, J.J. (eds.), Nutrition and the brain, Vol. 6. New York: Raven Press, pp. 223–275.

Yokogoshi, H., Roberts, C.H., Caballero, B., and Wurtman, R.J. (1984). Effects of aspartame and glucose administration on brain and plasma levels of large neutral amino acids and brain 5-hydroxyindoles. Am. J. Clin. Nutr. **40:**1–7.

Part VII Aspartame and Headache

CHAPTER 37

Aspartame and Headache

Donald R. Johns*

There is evidence that aspartame consumption can provoke adverse neurological reactions, most notably headache, in a susceptible subset of the population. The available experimental and clinical data on human aspartame consumption and adverse reactions, the concept of chemical headache and its relevance to aspartame-provoked headache, the methodological issues in the assessment of adverse reactions, and the studies of aspartame-provoked neurological symptoms currently in progress are reviewed. Some future implications for the study of aspartame-provoked headache are formulated.

Introduction

The possibility that the consumption of aspartame-containing products could result in adverse reactions has been raised by both consumer complaints to the Food and Drug Administration (FDA) (Centers for Disease Control 1984, Bradstock et al. 1986, Tollefson and Barnard 1987) and clinical reports in the medical literature (Ferguson 1985, Novick 1985, Wurtman 1985, Blundell and Hill 1986, Drake 1986, Johns 1986a, Kulczycki 1986, Walton 1986). A disproportionate number of these adverse reactions have been neurological in nature, most notably headache. The thesis to be explored is that aspartame may act as a provoking factor in the pathogenesis of headache in a susceptible subset of the general population.

Chemical Headache

Headache is one of the most common disorders which afflicts mankind and by its very nature is a uniquely human subjective experience. Migraine is a type of vascular headache with a number of special features and diverse

*Department of Neurology, Massachusetts General Hospital, Boston, MA 02114, USA. Current address: Department of Neurology, Johns Hopkins University, Baltimore, MD 21205, USA.

neurological accompaniments (Ad Hoc Committee 1962). The pathophysiology of migraine is poorly understood but is postulated to represent a complex, multifactorial process including genetic predisposition, interactive precipitating factors [including chemical substances and sexual activity (Johns 1986b)], and paroxysmal central nervous system disorders, notably headache and autonomic nervous system dysfunction (Selby and Lance 1960, Dalessio 1972, Leviton 1984). A diverse array of mechanistic theories have been proposed, invoking psychological, vascular reactivity, platelet aggregability, neuronal, and electrical activity factors. Most recently, the possible role of the trigeminovascular system has been demonstrated by Moskowitz et al. (1986). The lack of an animal model of headache has greatly hindered our understanding of this disorder.

Chemical headache, of which dietary migraine is the most common example, may be defined as a headache disorder which occurs in susceptible individuals with an increased probability after exposure to a precipitating chemical stimulus. Dietary migraine occurs in approximately 10% to 25% of migraine patients, with certain foods most frequently implicated (including alcohol, chocolate, cheese, citrus fruits) (Selby and Lance 1960, Peatfield et al. 1984). Chemical substances appear to play the major role in dietary migraine (Dalessio 1972, Sandler et al. 1974, Littlewood et al. 1982, Peatfield et al. 1984, Raskin 1986), but immunologic mechanisms (Monro et al. 1984) and psychological factors (e.g., taste aversion) (Perkin and Hartje 1983) may also be involved. Implicated factors in chemical headache (Raskin 1986), many of which have been verified in controlled studies, include tyramine (Smith et al. 1970, Moffett et al. 1972, Kohlenberg 1982, Littlewood et al. 1982, Peatfield et al. 1983), phenylethylamine ("chocolate headache"; Moffett et al. 1974, Sandler et al. 1974), nitrates in medication and industrial exposure ("dynamite headache"), nitrites in cured meats ("hot-dog headache"; Henderson and Raskin 1972), alcoholic beverages (Seltzer 1982, Peatfield et al. 1984), monosodium glutamate (MSG) ("Chinese restaurant syndrome"; Schaumberg et al. 1969), and hypoglycemia (Blau and Cumings 1966). In addition, the rapid consumption of very cold substances can provoke severe headache via direct physical stimulation of receptors in the oropharynx ("ice cream headache"; Raskin and Knittle 1976).

These varied substances are thought to act via a chemical mechanism in sensitive patients who have a lowered biologic threshold for that particular precipitant, in concert with other triggering factors, to produce clinically evident headache on some occasions (Leviton 1984, Peatfield et al. 1984, Raskin 1986). The probability of overt headache occurrence on any given exposure depends on a number of factors, including dosage and duration of exposure and other environmental factors.

Human Exposure to Aspartame

Aspartame received extensive preclinical laboratory study in experimental animals, especially rodents, but as stated in a review of these studies:

> The data from preclinical studies . . . cannot by themselves assure that a food additive is completely safe for all segments of the consuming population . . . Only use, concomitant with regulatory surveillance, can lead to a final determination of whether aspartame is safe for use by all segments of the population. (Molinary 1984, p. 299)

In contrast, only a limited number of human studies of aspartame's clinical effects have been carried out, involving fewer than 400 total subjects. Populations studied include healthy infants and children (Frey 1976, Koch et al. 1976a, Filer et al. 1984), diabetics (Stern et al. 1976, Nehrling et al. 1985), phenylketonuric (PKU) heterozygotes (Koch et al. 1976b), and young persons attempting weight reduction (Knopp et al. 1976). No adverse reactions attributable to aspartame were noted in these general studies or in a small focused study of glutamate-susceptible individuals (Stegink et al. 1981). It is unlikely that low-frequency (e.g., on the order of 1/1000 exposed individuals), idiosyncratic adverse reactions would be recognized until a much larger population was exposed to aspartame under a variety of dosages and other circumstances. This phenomenon is reflected in the marked increase in aspartame-related consumer complaints to the FDA in the last half of 1983, following its approval for use in carbonated beverages in July 1983 and the commensurate increase in aspartame consumption (Centers for Disease Control 1984).

Estimates for human consumption of aspartame were based at least in part on the assumption that aspartame would replace dietary sucrose, with the 99th percentile for projected daily intake set at 34 mg/kg and the allowable/acceptable daily intake (ADI) set by the FDA at 50 mg/kg (Sturtevant 1985). Aspartame consumption may exceed these limits in a proportion of patients, particularly in the form of diet beverages in hot weather. Ironically, the very features that make aspartame attractive as a substitute sweetener (noncarcinogenic, low caloric content, non-dental cariogenic, no adverse effects on glucose intolerance) make it liable to dietary abuse, since it seemingly can be consumed without the adverse consequences of equivalent quantities of sucrose-sweetened foods and beverages.

Assessment of Adverse Reactions

Methodological principles to evaluate the validity of possible adverse reactions have been applied most extensively to pharmacologic agents, but have been modified to apply to food additives. Criteria used include the temporal association, effect of cessation and rechallenge, and biologic plausibility, while the principles of dose-response relationship and relative risk have proven more difficult to apply (Bradstock et al. 1986).

Many different methods are available to study these adverse reactions, each with its own strengths, weaknesses, and limitations. The most extensive data available are in the form of passive surveillance of consumer-generated complaints to the FDA (Centers for Disease Control 1984, Bradstock et al. 1986), the Adverse Reaction Monitoring System (ARMS) (Tollefson and Barnard 1987). Such a system has limitations and is most useful in assessing the likelihood of an adverse reaction being due to a particular food additive and identifying areas for more intense, scientifically rigorous study (Bradstock et al. 1986). These issues are explored in greater detail in Chapter 39 of this volume. It should be mentioned that there may be significant under-reporting of adverse reactions to this passive surveillance system, in that many patients simply discontinue and avoid a possible offending agent without reporting their experience to a physician or regulatory agency.

A suitable control group has been difficult to identify for an epidemiological cohort study. Focused clinical studies employing appropriate scientific design are useful in the assessment of a particular adverse reaction in a defined population, but their generalizability is rather limited, and the adverse reaction must first be identified by other methods. An alternative experimental design is the single-patient randomized clinical trial ("*N* of one study"; Guyatt et al. 1986, McLeod et al. 1986) which involves randomization, double-blind placebo control, multiple crossovers, and the use of a relevant target outcome.

Adverse Reactions to Aspartame in Humans

The majority of currently available reports in the medical literature on adverse reactions to aspartame are at the anecdotal or single-case report level. The possibility of a fortuitous coincidental occurrence must be acknowledged (i.e., temporal association does not necessarily imply causality), particularly with a common adverse reaction and a commonly consumed substance. While these reports have obvious limitations, many adverse reactions to medications or other substances were first recognized and published in such a format. These reports thus serve to alert us to potential adverse reactions and to identify areas for more intense study.

Table 37.1 is a summary of reported adverse reactions to aspartame in individual patients including the demographic characteristics, methodology employed, and the aspartame product implicated. Note the preponderance of neurological adverse reactions (nos. 1 to 5) and the occurrence of adverse reactions which have an apparent immunologic basis (nos. 6 and 7). Most of these adverse reactions are now under more intense scientific investigation (see below).

TABLE 37.1. Adverse reactions to aspartame in man.

Source	Demographics[a]	Adverse reaction	Methodology	Aspartame product
Ferguson 1985	22 yo F	Headache	Elimination	Sweetener + MAOI[c]
Wurtman 1985	42 yo F	Headache, seizure	Elim/Rechall[b]	Diet soda
	27 yo M	Headache, seizure	Elim/Rechall	Diet beverage
	36 yo M	Seizure	Elimination	Diet beverage
Johns 1986a	31 yo F	Migraine headache	Single-blind	500 mg (pure)
Walton 1986	54 yo F	Seizure, mania	Elimination	Diet beverage
Drake 1986	33 yo F	Panic attack	Elim-Rechall	Diet soda
Kulczycki 1986	23 yo F	Urticaria	Double-blind	25–50 mg (pure)
	42 yo F	Urticaria/ angioedema	Double-blind	75 mg (pure)
Novick 1985	22 yo F	Granulomatous panniculitis	Single-blind	200 mg (pure)
Nehrling et al. 1985	Diabetic	Diarrhea	Double-blind	900 mg (pure)

[a]yo = years old, M = male, F = female.
[b]Elim/Rechall = elimination and rechallenge.
[c]MAOI = monoamine oxidase inhibitor.

Preliminary studies of the effect of dietary aspartame on migraine patients revealed a significant decrease in headache incidence upon restriction of aspartame-containing products and a significant increase in headache incidence upon addition of these products to the diet (S. Koehler, personal communication).

The data pertaining to neurological adverse reactions generated by the FDA's fourth quarter of 1986 Adverse Reaction Monitoring System (ARMS) are presented in Table 37.2. Neurological symptoms constituted more than half of 3133 complaints, with headache the most frequent single symptom by nearly a factor of three over the next most common nonneurological symptom [878 complaints (19.1%), versus 311 (6.8%) for nausea and vomiting]. For all complainants (including those with nonneurological symptoms), 78% were female, 74% were between 20 and 59 years of age, and 41% identified diet soft drinks and 22% identified tabletop sweetener as the associated product type. Alleged reactions were also classified according to severity (Type classification) and strength of association (Group classification). The Type I classification (Severe, including seizures, severe migraine headache, coma, and severe behavioral or mood disturbance) was assigned to 9% and 66% were classified as Group A or B (multiple episodes involving one or more products containing aspartame) (Tollefson and Barnard 1987).

TABLE 37.2. Summary of neurological complaints to the adverse reaction monitoring system.

		No. of complaints	Percent of total complaints
I. Symptom category			
Neurological		1045	23.6
Headache		983	22.2
Behavioral problem		263	5.9
Seizures		175	4.0
	Total	2466	55.7
II. Reported symptoms			
Headache		878	19.1
Dizziness/imbalance		402	8.7
Mood quality/level		332	7.2
Vision change		145	3.2
Fatigue/weakness		137	3.0
Other neurological		129	2.8
Sleep problems		121	2.6
Seizures/convulsions		121	2.6
Memory loss		117	2.5
Numbness/tingling		81	1.8
	Total	2453	53.5

In formulating a possible mechanism whereby aspartame might provoke headache, the principles discussed above for the pathogenesis of migraine should be recalled. In that schema, aspartame would serve as one of the interactive precipitating factors which could act in the susceptible individual to increase the probability of the occurrence of clinical headache. Among the possible mechanisms are an immunologic/allergic reaction (Novick 1985, Kulczycki 1986), modifications in central nervous system neurochemical levels by aspartame or its constituent amino acids (Fernstrom et al. 1983, Wurtman 1983, Yokogoshi et al. 1984, Coulombe and Sharma 1986, Yokogoshi and Wurtman 1986), a toxic reaction to aspartame or its diketopiperazine metabolite (Boehm and Bada 1984), a direct interaction with pharyngeal receptors (analogous to "ice cream headache"; Raskin 1986), a psychological taste aversion reaction (Perkin and Hartje 1983), an effect on brain electroencephalographic activity, or an adverse reaction to other chemical substances in the aspartame-containing products (e.g., fluid or electrolyte imbalance induced by ingestion of large volumes of diet soda).

On the basis of the clinical data reviewed above, a number of scientific clinical studies have been initiated. These include studies of: seizures (Yale, B.A. Shaywitz; Mount Sinai, A.J. Rowan; Massachusetts Institute of Technology and Beth Israel Hospital, R.J. Wurtman and D.L. Schomer): urticaria (Washington University, A. Kulczycki); and other adverse reactions and their mechanisms (Emory, L.J. Elsas; UCLA, W.M. Pardridge).

The possible effect of aspartame on headache is under investigation at several institutions using studies focused on patients who are suspected of having aspartame-provoked headache. At Duke University, S. Schiffman has performed an in-patient, randomized, double-blind, placebo-controlled, crossover design study with aspartame (30 mg/kg) administered in capsules in a single daily challenge to 40 subjects. S. Solomon at Montefiore Hospital is employing an ambulatory double-blind challenge of aspartame versus placebo administered in a liquid vehicle. Such a liquid challenge more closely approximates the physiological condition under which most consumers ingest aspartame, but is less reliably blinded. At the Massachusetts General Hospital, M. Barry, S. Skates, and I have designed a series of ambulatory randomized, double-blind, placebo-controlled, multiple crossover trials of aspartame capsules (12 to 18 mg/kg) using an "*N* of One Study" paradigm (Guyatt et al. 1986, McLeod et al. 1986). We are also performing a questionnaire study of suspected aspartame-provoked headache patients to define more completely the clinical and epidemiologic features of these patients, particularly those of neurological interest.

Conclusions

In summary, the currently available data are consistent with the thesis that there is a susceptible subset of the population which has aspartame-provoked adverse neurological reactions, particularly headache. Aspartame should therefore be added to the list of substances which are suspected of provoking "chemical headache." Obviously, further data are needed to verify or refute this purported association and to clarify its epidemiological, pathophysiological, and clinical features. Focused scientific clinical studies will likely play the most important role, but the value of other methodological approaches (including anecdotal reports of previously unrecognized reactions) should not be ignored.

Recognition of a possible association between aspartame consumption and headache can serve to heighten awareness of environmental factors in the pathogenesis of neurological disease. Prompt identification of aspartame as a headache provoker can obviate the need for extensive diagnostic and therapeutic interventions, and allow expeditious limitation or elimination of the offending agent. This point illustrates the need for more extensive labeling of products as to the presence and quantity of aspartame, so that rational decisions can be made on the relative risk of aspartame consumption. Aspartame-provoked headache could potentially serve as an experimental model of human headache.

References

Ad Hoc Committee on Classification of Headache. (1962). Classification of headache. Arch. Neurol. **6:**173–176.

Blau, J.N., and Cumings, J.N. (1966). Method of precipitating and preventing some migraine attacks. Br. Med. J. **2:**1242–1243.

Blundell, J.F., and Hill, A.J. (1986). Paradoxical effects of an intense sweetener (aspartame) on appetite. Lancet **ii:**1092–1093.

Boehm, M.F., and Bada, J.L. (1984). Racemization of aspartic acid and phenylalanine in the sweetener aspartame at 100°C. Proc. Natl. Acad. Sci. USA. **81:**5263–5266.

Bradstock, M.K., Serdula, M.K., Marks, J.S., Barnard, R.J., Crane, N.T., Remington, P.L., and Trowbridge, F.L. (1986). Evaluation of reactions to food additives: the aspartame experience. Am. J. Clin. Nutr. **43:**464–469.

Centers for Disease Control. (1984). Evaluation of consumer complaints related to aspartame use. Morb. Mortal. Weekly Rep. **33:**605–607.

Coulombe, R.A., and Sharma, R.P. (1986). Neurobiochemical alterations induced by the artificial sweetener aspartame (NutraSweet). Toxicol. Appl. Pharmacol. **83:**79–85.

Dalessio, D.J. (1972). Dietary migraine. Am. Family Physician **6:**60–65.

Drake, M.E. (1986). Panic attacks and excessive aspartame ingestion. Lancet **ii:**631.

Ferguson J.M. (1985). Interaction of aspartame and carbohydrates in an eating-disordered patient. Am. J. Psychiatr. **142:**271.

Fernstrom, J.D., Fernstrom, M.H., and Gillis M.A. (1983). Acute effects of aspartame on large neutral amino acids and monoamines in rat brain. Life Sci. **32:**1651–1658.

Filer, L.J., Baker, G.L., and Stegink, L.D. (1984). Aspartame ingestion by human infants. *In* Stegink, L.D., and Filer, L.J. (eds.), Aspartame: physiology and biochemistry. New York: Marcel Dekker, pp. 579–591.

Frey, G.H. (1976). Use of aspartame by apparently healthy children and adolescents. J. Toxicol. Environ. Health **2:**401–415.

Guyatt, G., Sackett, D., Taylor, D.W., Chong, J., Roberts, R., and Pugsley, S. (1986). Determining optimal therapy-randomized trials in individual patients. N. Engl. J. Med. **314:**889–892.

Henderson, W.R., and Raskin, N.H. (1972). "Hot-dog" headache: individual susceptibility to nitrite. Lancet **ii:**1162–1163.

Johns, D.R. (1986a). Migraine provoked by aspartame. N. Engl. J. Med. **315:**456.

Johns, D.R. (1986b). Benign sexual headache within a family. Arch Neurol. **43:**1158–1161.

Knopp, R.H., Brandt, K., and Arky, R.A. (1976). Effects of aspartame in young persons during weight reduction. J. Toxicol. Environ. Health **2:**417–428.

Koch, R., Schaeffler, G., and Shaw, K.N.F. (1976a). Results of loading doses of aspartame by two phenylketonuric (PKU) children compared with two normal children. J. Toxicol. Environ. Health **2:**459–469.

Koch, R., Shaw, K.N.F., Williamson, M., and Haber, M. (1976b). Use of aspartame in phenylketonuric heterozygous adults. J. Toxicol. Environ. Health **2:**453–457.

Kohlenberg, R.J. (1982). Tyramine sensitivity in dietary migraine: a critical review. Headache **22:**30–34.

Kulczycki, A. (1986): Aspartame-induced urticaria. Ann. Intern. Med. **104:**207–208.

Leviton, A. (1984). To what extent does food sensitivity contribute to headache recurrence? Dev. Med. Child Neurol. **26:**534–545.

Littlewood, J.,Glover, V., and Sandler, M. (1982). Platelet phenolsulphotransferase deficiency in dietary migraine. Lancet **i:**983–986.

McLeod, R.S., Taylor, D.W., Cohen, Z., and Cullen, J.B. (1986). Single-patient randomised clinical trial. Use in determining optimum treatment for patient with inflammation of Koch continent ileostomy reservoir. Lancet **i:**726–728.

Moffett, A., Swash, M., and Scott, D.F. (1972). Effect of tyramine in migraine: a double-blind study. J. Neurol. Neurosurg. Psych. **35:**496–499.

Moffett, A., Swash, M., and Scott, D.F. (1974). Effect of chocolate in migraine: a double-blind study. J. Neurol. Neurosurg. Psych. **37:**445–448.

Molinary, S.V. (1984). Preclinical studies of aspartame in nonprimate animals. *In* Stegink, L.D., and Filer, L.J. (eds.), Aspartame: physiology and biochemistry. New York: Marcel Dekker, pp. 289–306.

Monro, J., Carini, C., and Brostoff, J. (1984). Migraine is a food-allergic disease. Lancet **ii:**719–721.

Moskowitz, M.A., Henrikson, B.M., and Beyerl, B.D. (1986). Trigeminovascular connections and mechanisms of vascular headache. *In* Rose, F.C. (ed.), Handbook of clinical neurology, Vol. 4. New York: Elsevier, pp. 107–115.

Nehrling, J.K., Kobe, P., McLane, M.P., Olson, R.E., Kameth, S., and Horwitz, D.L. (1985). Aspartame use by persons with diabetes. Diabetes Care **8:**415–417.

Novick, N.L. (1985). Aspartame-induced granulomatous panniculitis. Ann. Intern. Med. **102:**206–207.

Peatfield, R., Littlewood, J.T., Glover, V., Sandler, M., and Rose, F.C. (1983). Pressor sensitivity to tyramine in patients with headache: relationship to platelet monoamine oxidase and to dietary provocation. J. Neurol. Neurosurg. Psych. **46:**827–831.

Peatfield, R.C., Glover, V., Littlewood, J.T., Sandler, M., and Rose, F.C. (1984). The prevalence of diet-induced migraine. Cephalagia **4:**179–183.

Perkin, J.E., and Hartje, J. (1983). Diet and migraine: a review of the literature. J. Am. Diet. Assoc. **83:**459–463.

Raskin, N.H. (1986). Ice cream, ice pick and chemical headache. *In* Rose, F.C. (ed.), Handbook of Clinical Neurology, Vol. 4. New York: Elsevier, pp. 441–448.

Raskin, N.H., and Knittle, S.C. (1976). Ice cream headache and orthostatic symptoms in patients with migraine. Headache **16:**222–225.

Sandler, M., Youdim, M.B.H., and Hannington, E. (1974). A phenylethylamine oxidising defect in migraine. Nature **250:**335–337.

Schaumberg, H.H., Byck, R., Gerstl, R., and Mashman, J.H. (1969). Monosodium L-glutamate: its pharmacology and role in the Chinese restaurant syndrome. Science **163:**826–828.

Selby, G., and Lance, J.W. (1960). Observations on 500 cases of migraine and allied vascular headache. J. Neurol. Neurosurg. Psych. **23:**23–32.

Seltzer, S. (1982). Foods, and food and drug combinations, responsible for head and neck pain. Cephalalgia **2:**111–124.

Smith, I., Kellow, A.H., and Hannington, E. (1970). A clinical and biochemical correlation between tyramine and migraine headache. Headache **10:**43–52.

Stegink, L.D., Filer, L.J., and Baker, G.L. (1981). Effect of aspartame and sucrose loading in glutamate-susceptible subjects. Am. J. Clin. Nutr. **34:**1899–905.

Stern, S.B., Bleicher, S.J., Flores, A., Gombos, G., Recitas, D., and Shu, J. (1976). Administration of aspartame in non-insulin-dependent diabetes. J. Toxicol. Environ. Health **2:**429–439.

Sturtevant, F.M. (1985). Use of aspartame in pregnancy. Int. J. Fertil. **30:**85–87.

Tollefson, L., and Barnard, R.J. (1987). Quarterly Report on adverse reactions associated with aspartame ingestion. Health and Injury Related Surveillance Subprogram Postmarketing Surveillance System. Department of Health and Human Services, Washington, D.C.

Walton, R.G. (1986). Seizure and mania after high intake of aspartame. Psychosomatics **27:**218–220.

Wurtman, R.J. (1983). Neurochemical changes following high-dose aspartame with dietary carbohydrates. N. Engl. J. Med. **309**:429–430.

Wurtman, R.J. (1985). Aspartame: possible effect on seizure susceptibility. Lancet **ii:**1060.

Yokogoshi, H., Roberts, C.H., Caballero, B., and Wurtman, R.J. (1984). Effects of aspartame and glucose administration on brain and plasma levels of large neutral amino acids and brain 5-hydroxyindoles. Am. J. Clin. Nutr. **40:**1–7.

Yokogoshi, H., and Wurtman, R.J. (1986). Acute effects of oral or parenteral aspartame on catecholamine metabolism in various regions of rat brain. J. Nutr. **116:**356–364.

CHAPTER 38

The Effect of Aspartame Consumption on Migraine Headache: Preliminary Results

Shirley M. Koehler*

Almost all dietetic products now contain aspartame, a sweetener consisting of phenylalanine and aspartic acid. Recently, three pilot studies have proposed that the addition of products containing aspartame to the diets of migraine headache sufferers may produce a significant increase in the frequency of their migraines. The present study was a controlled thirteen-week, double-blind, randomized cross-over study comparing the effect of aspartame to that of a matched placebo on the frequency and intensity of migraine headache. The results of this study indicated that the ingestion of aspartame by migraineurs caused a significant increase in headache frequency for some subjects. The possible biochemical bases of these findings and their implications for research are discussed.

Recently, three studies have proposed that the addition of products containing aspartame (NutraSweet) to the diets of migraineurs may produce a significant increase in the frequency of their headaches (Koehler and Hartje 1985, Koehler 1987, Johns 1986). The present study was a controlled nine-week, double-blind, randomized crossover study comparing the effect of aspartame to that of a matched placebo on the frequency and intensity of migraine headache. It was hypothesized that the ingestion of aspartame would increase the frequency and intensity of migraine headache in comparison to the ingestion of placebo.

Aspartame is now found in almost all dietetic foods. It is being marketed extensively with no warnings as to any adverse side effects except a warning to those suffering from phenylketonuria (PKU) of the presence of phenylalanine. Aspartame is composed of approximately 55% phenylalanine and 45% aspartic acid (Stegink and Filer 1983). Phenylalanine, an amino acid, is implicated in the migraine process by its involvement in the production of serotonin. Serotonin may influence the cerebrovascular changes associated with migraine headache pain (Adams et al. 1980, Andrasik et al. 1982, Blanchard and Andrasik 1982, Dexter 1983, Kewman and Roberts 1980,

*Department of Clinical Psychology, University of Florida, Gainesville, FL 32601, USA.

Murray 1981). The production of serotonin depends on the availability of L-tryptophan which is derived from proteins in the diet (Wurtman and Wurtman 1979). Phenylalanine competes with L-tryptophan for limited entry to the brain, and this competition may be the main cause of a decrease of serotonin in the brain (Boullin 1978, Gibson et al. 1982, Harper 1983). The resulting reduction in serotonin causes the vasodilation which is thought to produce the pain of migraine (Adams et al. 1980, Murray 1981).

Subjects and Procedures

Subjects

Subjects were persons who met the criteria for migraine established by the 1962 Ad Hoc Committee on Classification of Headache, i.e., recurrence of attacks with varied intensity, frequency, and duration; unilaterality in onset; anorexia, sometimes with nausea and vomiting; heredity factors; and in some, the presence of sensory, motor, and mood disturbances (the prodrome of "classical migraine"). Participants are also selected based on these criteria: a) medical diagnosis of migraine; b) currently taking no medication other than analgesics such as aspirin, acetaminophen, etc.; c) meet the Ad Hoc Committee's criteria; d) recurrent attacks of headache separated by headache-free periods; e) at least one to three migraines in a four-week period; and f) a belief that aspartame has an effect on their migraines.

Procedure

Twenty-five subjects were requested to keep records of their diets and headaches for a baseline period of 4 weeks. Following this baseline period, subjects were randomly assigned to receive either capsules containing 300 mg of aspartame or matching placebo. The participants each received a diet chart on which to record their daily diets and were required to mail these in weekly. Each subject took one capsule 4 times daily (at meals and at bedtime) for a period of 4 weeks. This dosage is approximately equivalent to consuming six diet colas daily. Following this 4-week period, a 1-week "wash-out" period was observed in which participants did not take any capsules. This wash-out period served to decrease possible treatment interference. At the end of this week, a crossover design was employed in which the placebo group received the aspartame capsules and the aspartame group received the placebo. Subjects were instructed to deal with their migraine headaches in their customary manner (e.g., go to the doctor or take pain medication). They were requested to record any medications taken at these times.

During the course of the study, eight subjects dropped out, with the majority citing lack of sufficient time for record keeping as their reason. Four subjects were omitted from the data analyses when it was found that they had begun medication (e.g., Inderal) during the study, and two subjects' data was incomplete. The remaining subject pool consisted of 2 males and 9 females.

Data Analyses

Eleven subjects' records of the frequency, intensity, duration, and associated symptoms of migraine headaches and amount of medications taken during their migraine-headache episodes were analyzed using Wilcoxon's Matched-Pairs Signed-Ranks Test for the placebo and treatment periods. As predicted, the results of these analyses revealed a significant increase in the frequency of migraine headaches ($p < 0.05$) from the placebo to the treatment phases. The mean number of migraines increased from 1.55 to 3.55 with the consumption of aspartame capsules. Analyses of changes in the frequency, intensity, duration, and associated symptoms of tension and unknown headaches during the placebo and aspartame phases failed to achieve significance.

Discussion

The results of the present study corroborate the hypothesis that the consumption of aspartame by migraineurs may cause a significant increase in headache activity for some subjects. The hypothesis that aspartame would increase the intensity of the migraine headache was not supported.

It is proposed that it is the phenylalanine in aspartame that affects the onset of migraine by causing a decrease of serotonin in the brain. The autonomic nervous system of migraine patients is thought to be more unstable than other people's. This instability results in migraineurs, even in remission, having exaggerated cranial arterial responsiveness and cranial vascular variability (Yates 1980). A biochemical defect in the metabolism of serotonin is thought to be a causative factor in producing this vascular lability. Therefore, it may be that some migraineurs are more sensitive to environmental influences on serotonin (such as diets containing high amounts of phenylalanine). This difference in sensitivity might be thought of as an individual's threshold and could account for the lack of a positive reaction to the consumption of aspartame by some of this study's migraineurs.

Acknowledgments. This study was supported by the Center for Brain Sciences and Metabolism Charitable Trust. This study was completed as a partial requirement of the University of Florida for the degree of Master of Science. The author wishes to acknowledge the numerous contributions and encouragement of her committee members.

References

Adams, H.E., Feuerstein, M., and Fowler, J.L. (1980). Migraine headache: review of parameters, etiology, and intervention. Psychol. Bull. **87**:2.a.

Ad Hoc Committee on Classification of Headache (1962). Classification of headache. J. Am. Med. Assoc. **179**:127–128.

Andrasik, F., Blanchard, E.B., Arena, J.G., Saunders, N.L., and Barron, K.D. (1982). Psychophysiology of recurrent headache. Behav. Ther. **13**:407–429.

Blanchard, E.B., and Andrasik, F. (1982). Psychological assessment and treatment of headache: recent developments and emerging issues. J. Consult. and Clin. Psychol. **50**:859–879.

Boullin, D.J. (ed.). (1978). Serotonin in mental abnormalities. New York: John Wiley & Sons.

Dexter, J.D. (1983). Biochemical pathophysiology of migraine. Neurol. Clin. **1**:527–531.

Gibson, C.J., Deikel, S.M., Young, S.N., and Binik, Y.M. (1982). Behavioral and biochemical effects of tryptophan, tyrosine, and phenylalanine. Psychopharmacology **76**:118–121.

Harper, A.E. (1983). Phenylalanine metabolism. *In* Stegink, L.O., and Filer, L.J. (eds.), Aspartame: physiology and biochemistry. New York: Marcel Dekker, pp. 77–102.

Johns, D.R. (1986). Migraine provoked by aspartame. N. Engl. J. Med. **315**:456.

Kewman, D., and Roberts, A.H. (1980). Skin temperature biofeedback and migraine headaches: A double-blind study. Biofeedback and Self-Regulation, **48**:327–344.

Koehler, S.M. (1987). The effect of dietetic products containing phenylalanine on migraine headache. Paper presented at the Florida Psychological Association, Orlando, June.

Koehler, S.M., and Hartje, J.C. (1985). Migraine headache and phenylalanine. Paper presented at the Society of Behavioral Medicine, New Orleans, March 30.

Murray, J.B. (1981). Psychophysiological aspects of migraine headaches. Psychol. Rep. **48**: 139–162.

Stegink, L.D., and Filer, L.J. (eds.). (1983). Aspartame: physiology and biochemistry. New York: Marcel Dekker.

Wurtman, R.J., and Wurtman, J.J. (eds.). (1979). Nutrition and the brain, Vol. 3. New York: Raven Press.

Yates, A.J. (1980). Biofeedback and the modification of behavior. New York: Plenum.

CHAPTER 39

Monitoring of Adverse Reactions to Aspartame Reported to the U.S. Food and Drug Administration

Linda Tollefson, Robert J. Barnard, and Walter H. Glinsmann*

The Food and Drug Administration's (FDA's) Center for Food Safety and Applied Nutrition (the Center) recently developed the Adverse Reaction Monitoring System (ARMS) which is concerned with the retrieval, processing, and analysis of data related to adverse health effects associated with specific food products and additives. The ARMS is a passive surveillance system that is designed to identify specific areas for focused clinical investigations on potentially causal associations. An overview of the Center's monitoring system is presented, including the mechanics of the system and its limitations.

Reports of adverse reactions to aspartame received by the FDA are monitored in the ARMS and evaluated by specifically defined criteria in order to accurately assess the safety issues involved. The criteria that are used in evaluating reports of adverse effects associated with food products are discussed, and a summary of the adverse reactions associated with the consumption of aspartame, both those evaluated by the Centers for Disease Control (CDC) and those received by the FDA since the CDC report, is presented.

Introduction

An adverse reaction to a food product is defined as any clinically abnormal response which can be attributed to a food or food additive (Anderson 1986). Adverse reactions to foods are a recognized public health problem; however, these vary widely in both extent and severity. Currently, there is no reliable information on the prevalence of most types of these reactions in the general population (Nichaman and McPherson 1986).

The U.S. Food and Drug Administration's (FDAs) Center for Food Safety and Applied Nutrition (the Center) recently developed the Adverse Reaction Monitoring System (ARMS), which is concerned with the retrieval, processing, and analysis of data related to adverse health effects

* Office of Nutrition and Food Sciences, Center for Food Safety and Applied Nutrition, U.S. Food and Drug Administration, Washington, DC, 20204, USA.

associated with specific foods, food and color additives, and vitamin/mineral supplements. The ARMS was established as a data monitoring system to assess complex food/illness/injury relationships to help assure a safe food supply. Alleged adverse reactions to food products or additives experienced by consumers and reported to the FDA are monitored in the ARMS by routing them to the proper personnel for response, investigation, and analysis. The ARMS is concerned with spontaneous reports from consumers regarding alleged adverse effects from food products, and, as such, is a form of passive surveillance. Used as a sentinel system, the ARMS can alert FDA officials to any hazardous effects associated with an approved food additive.

The underlying issues associated with reports of adverse effects that allegedly have occurred or may theoretically occur as a result of the consumption of aspartame-containing products are diverse and complex. Adverse reactions to aspartame received by the FDA are monitored in the ARMS, thoroughly reviewed, and evaluated for any potentially causal relationship between the reported adverse effect and the ingestion of aspartame.

Methods

All consumer complaints reporting a food-associated reaction are entered into a computerized tracking system. The initial information abstracted from the complaint is primarily geographic. The complaints are then sent to the Center's epidemiology staff for review. In order to establish a basis for the association of the adverse reaction to the implicated product, clearly defined and mutually exclusive case definitions are formulated for every product monitored in the ARMS.

The reports are classified by severity of the reaction (described by the Type classification referenced from the reported symptoms to an affected organ or physiological system) and by the frequency and consistency of the association with ingestion of the product of interest (described by the Group classification). A Type I, or severe, reaction includes but is not limited to serious respiratory distress or chest pain; cardiac arrhythmias; anaphylactic or hypotensive episodes; severe gastrointestinal distress such as protracted vomiting or diarrhea leading to dehydration; severe neurological distress such as extreme dizziness, fainting, or seizures; or any reaction requiring emergency medical treatment. All reports of cancer and congenital anomalies are always considered serious. A Type II, or moderate reaction, includes mild respiratory symptoms not requiring medical treatment; mild gastrointestinal and urinary tract symptoms such as nausea, vomiting, diarrhea, or increased urination; generalized urticaria without severe accompanying reactions such as laryngospasm, respiratory distress, or hypotension; localized rash, pruritis, or edema; and mild neurological symptoms such as mild anxiety, insomnia, headache, or fatigue.

The Group classification is used to determine the feasibility that the reported symptoms are a result of ingesting the incriminated product. This was modified from an earlier classification scheme used in reports on adverse reactions associated with aspartame (Centers for Disease Control 1984, Bradstock et al. 1986). Complainants are placed in Group A if symptoms recurred each time they consumed different products containing the item of interest; in Group B if symptoms recurred each time they consumed the same product; in Group C if symptoms were associated with the ingestion of a product or products containing the item of interest, but the complainant did not rechallenge himself or herself; and in Group D if symptoms did not recur every time products were consumed containing the item of interest or if the complainant consulted a physician who stated that the symptoms were unlikely to be due to the substance being monitored.

Reports describing severe reactions are investigated by FDA field personnel. This involves interviewing the complainant or a close family member; interviewing the attending physician, if possible, and obtaining medical records; and if applicable, analyzing samples of the food suspected to have caused the reaction. Specialized questions or questionnaires may be used to elicit specific information required by a case definition. Appendix A contains the questionnaire administered during the investigation of an adverse reaction associated with aspartame.

The adverse reaction reports are reviewed, and data from these reports and any follow-up investigations are extracted, coded, and entered into a computerized tracking system. The data entered include geographic information, descriptive epidemiology such as sex, age, race, date of episode, and time from ingestion of the product until symptoms appeared. In addition, the Type and Group classification, the product type, and the symptoms experienced by the complainant are coded and entered into the computer. The product type can be the type, specific brand, or specific item description which adequately identifies the product used. Symptom complexes are divided into two categories, descriptive and anatomic. Descriptive symptoms are nondiagnostic categories used to capture the complainant's reported physical symptoms. Anatomic symptoms consist of a secondary list used only for the severe (Type I) reactions. The anatomic list was abstracted from FDA's Center for Drugs and Biologics COSTART (Coding Symbols for Thesaurus of Adverse Reaction Terms) manual (FDA). It relates the reported symptom to the anatomic system involved.

Summary of Adverse Reactions to Aspartame Received by the FDA

Aspartame was initially approved by the FDA for use as a food additive in dry form in July 1981. A major increase in exposure developed in the latter portion of 1983 and early 1984 following expanded approval for use in carbonated beverages. Acceptance was high and market penetration was rapid. At that time, considerable public media attention was focused on potential adverse effects that might accompany the increased exposure. The FDA, G.D. Searle Company, and others began to receive complaints alleging multiple adverse reactions associated with the use of aspartame-containing products.

In early 1984, a decision was made to obtain aspartame-related adverse reaction reports from all sources and request assistance from the Center for Health Promotion and Education, a component of the Centers for Disease Control (CDC) in Atlanta, Georgia. A standardized questionnaire was developed to interview all complainants in a focused and comparable fashion. FDA's field staff then interviewed all complainants, completed the questionnaire, and obtained all available medical records. A total of 517 cases were able to be reviewed, and 231 were analyzed extensively for possible correlations using a computer program. These cases provided the basis for a detailed evaluation (Bradstock et al. 1986). The analyses by CDC found "no clear symptom complex that suggested a widespread public health hazard associated with aspartame use"; however, there were "some case reports in which the symptoms may be attributable to aspartame in commonly-consumed amounts" (Bradstock et al. 1986).

Since the CDC report, the FDA has received 3326 consumer complaints describing adverse reactions allegedly due to the consumption of aspartame. These include adverse reactions reported directly to the FDA, the G.D. Searle Company, and those received by the Center from Senator Metzenbaum and Dr. Richard Wurtman.

Complainants are predominately female (77%) between 20 and 59 years of age (76%); 10% are less than 20 years of age. Eight percent of reactions are classified as severe, using the criteria described in the Methods section.

We were unable to classify 28% of the reactions by the strength of the association with aspartame (the Group classification) because of insufficient information. Of those we were able to classify, 31% were placed into Group A in that they were associated with multiple episodes and multiple products containing aspartame. The majority (34%) of the reactions were classified as Group B in that they were associated with multiple episodes involving one particular aspartame-containing product, such as diet soda. Approximately 17% were classified as Group C, those associated with a single episode involving an aspartame-containing product or products. The remaining 18% were classified as Group D; the adverse reaction did not occur every time the complainant consumed a product containing aspartame, or a physician stated that the reaction was not likely to be associated with aspartame.

Table 39.1 shows the distribution of consumer reports by the product type mentioned by the complainant to be associated with the adverse reaction. Approximately 40% of the consumers reporting an aspartame-

TABLE 39.1. Aspartame consumer complaints by associated product type.[a]

Product type	No. of complaints	Percent of total complaints
Diet soft drinks	1697	40.5
Table-top sweetener	911	21.8
Puddings, gelatins	290	6.9
Lemonade	245	5.9
Kool-Aid	239	5.7
Hot chocolate	212	5.1
Iced tea	194	4.6
Other	156	3.7
Chewing gum	115	2.7
Cereal	56	1.3
Punch mix	37	0.9
Sugar substitute tablets	31	0.7
Nondairy toppings	4	0.1
Chewable multivitamins	3	0.1

[a]Multiple product types may be associated with each consumer complaint.

TABLE 39.2. Aspartame consumer complaints by reported symptoms.[a]

Reported symptom	No. of complaints	Percent of total complaints
Headache	951	19.3
Dizziness or problems with balance	419	8.5
Change in mood quality or level	349	7.1
Vomiting and nausea	329	6.7
Abdominal pain and cramps	254	5.2
Diarrhea	178	3.6
Change in vision	162	3.3
Fatigue, weakness	141	2.9
Seizures and convulsions	137	2.8
Sleep problems	127	2.6
Memory loss	125	2.6
Rash	111	2.3
Change in sensation (numbness, tingling)	91	1.8
Hives	80	1.6
Other	1464	29.7

[a]Multiple symptoms may be associated with each consumer complaint.

associated adverse reaction mention diet soft drinks as the source of exposure, and 22% mention table-top sweeteners as the exposure source. No other single product type is mentioned frequently in the consumer complaints.

Table 39.2 shows the distribution of the aspartame adverse reaction reports by the symptoms most frequently reported by the complainant. Over

70 different symptoms were described; 19% of the complainants mentioned headache, and 8% mentioned problems with dizziness or balance. Most of the symptoms reported are mild in nature and commonly found in the general population, as noted in the CDC report (Bradstock et al. 1986).

There are three symptom categories reportedly associated with aspartame which have received much scrutiny as possible hazards to the public health. These include headaches, seizures, and urticaria. Large numbers of consumers have reported experiencing headaches with the ingestion of aspartame. Based on the data in the ARMS, however, we have been unable to find a consistent link between any particular variety of headache and the consumption of aspartame. Furthermore, among the reports received by the FDA, it has not been possible to eliminate factors other than aspartame consumption as causing the headache.

Seizures have been a focus of concern because of their severity. Again, however, the ARMS data do not indicate a causal link between aspartame consumption and the occurrence of seizures. We have been able to identify underlying health problems as the probable cause of the seizure in most of the reports received by the FDA. In other cases, there were a number of circumstances that could not be eliminated as possibly causing the seizure, such as concurrent use of medications known to induce seizures, head trauma, electrolyte imbalances, and dietary factors. Seizure susceptibility can be increased by a number of factors such as estrogenic activity, insulin deficiency, hydration, hyponatremia, hypoglycemia, fever, and starvation. Epilepsies are common neurological disorders, estimated to affect between 0.5% and 2% of the population, and can occur at any age (Dichter 1983). Under these circumstances and because aspartame is frequently consumed by large numbers of people, it is not surprising that there may be a chance occurrence of seizure activity following ingestion of aspartame by seizure-prone people. Such a phenomenon, however, is not indicative of a causal relationship between consumption of aspartame and the onset of seizure activity. Only controlled studies such as well-designed clinical trials can determine whether aspartame induces seizures.

A number of reliable investigators have reported adverse symptoms from a variety of aspartame-containing products; these effects occurred following exposures as low as 1 to 2 mg/kg. There is possible evidence for a hypersensitivity reaction to aspartame, mediated by a Type I immune response. Dr. Anthony Kulczycki of the Washington School of Medicine at St. Louis, Missouri, has reported on patients with chronic urticaria who improve following cessation of all products containing aspartame and who develop hives when they are again exposed to small-dose (50-mg) aspartame challenges (Kulczycki 1986, Kulczycki and Danker 1986). A case of aspartame-induced granulomatous panniculitis has also been reported (Novick 1985). Thus, aspartame hypersensitivity, similar to other food sensitivities, may provide a causal basis for some of the reactions reported.

Discussion

Data generated from the ARMS are used to assess the likelihood that a problem exists with a particular food product or additive and allow us to define the possible problem by uncovering patterns in the reported reactions. The system functions as a sentinel by providing an early warning of unanticipated adverse effects which can provide important clues for further study by scientifically controlled methods.

The ARMS is based on spontaneous reports from consumers and health professionals, and so is a form of passive surveillance. Such a system has a number of limitations, the most important being that no causal relationship can be established definitely between the ingestion of the incriminated product and the occurrence of the reported symptom(s). Therefore, it is not only difficult to determine if the reaction is truly associated with the product under investigation since many other foods and drugs can cause similar reactions, but also because the segment of the population reporting the adverse reactions cannot be considered representative of the entire population ingesting the product. A reaction may be caused by any number of foods the complainant has consumed, by concurrent use of medication, or by underlying disease which may be present. Confounding by other variables is particularly likely to be present in situations such as aspartame consumption since both the food source and the reported outcomes are common.

In addition, it is not possible to determine the incidence or severity of reactions to a suspect food product in the general population from information gained through a passive surveillance system. We have no way of knowing the number of adverse reactions that occur due to a particular product, only those that come to our attention by passive surveillance. Therefore, reliable risk estimates should not be made from reported adverse reactions, and reporting rates should never be confused with occurrence rates (Faich 1986).

For all the reasons discussed here, an adverse reaction monitoring system is best used to uncover potential problems. Rigorous statistical inference on adverse effects should be drawn only from objectively conducted clinical trials. Spontaneous reporting systems can only generate suspicions and indicate where further clinical studies would be most productive.

The anecdotal reports of adverse reactions to aspartame should be used to identify specific areas for focused clinical investigations on potentially causal associations. Unfortunately, case reports similar to the ones reported to the FDA and described here are being accepted with little or no skepticism. Many reactions are incompletely documented, yet have served as the basis for proposing mechanisms of action. The results of adequately designed prospective epidemiology studies such as clinical trials will clarify the validity of the claims against aspartame, and enable clinicians, investigators, the food industry, and the FDA to make informed decisions about any potential public health hazards associated with its use.

In summary, the adverse reactions that have been reported to the FDA regarding aspartame, and which have been extensively reviewed, do not establish reasonable evidence of possible public health harm. There is currently no consistent or unique pattern of symptoms reported with respect to aspartame that can be causally linked to its use. Because the information reported to the FDA is anecdotal and often not accompanied by complete medical records, the agency has been unable to eliminate factors other than aspartame consumption as possible causes for the reported adverse effects.

References

Anderson, J.A. (1986). The establishment of common language concerning adverse reactions to foods and food additives. Symposium Proceedings on Adverse Reactions to Foods and Food Additives. J. Allergy Clin. Immunol. **78:**140–144.

Bradstock, M.K., Serdula, M.K., Marks, J.S., Barnard, R.J., Crane, N.T., Remingtom, P.L., and Trowbridge, F.L. (1986). Evaluation of reactions to food additives: the aspartame experience. Am. J. Clin. Nutr. **43:**464–469.

Centers for Disease Control (1984). Evaluation of consumer complaints related to aspartame use. Morb. Mortal. Weekly Rep. **33:**605–607.

Dichter, M.A. (1983). The epilepsies and convulsive disorders. *In* Petersdorf, R.G., et al. (eds.), Harrison's principles of internal medicine, 10th ed., New York: McGraw-Hill, pp. 2018–2028.

Faich, G.A. (1986). Adverse-drug-reaction monitoring. N. Eng. J. Med. **314:**1589–1592.

Food and Drug Administration (FDA). (1984). Rockville, Md. National adverse drug reaction directory "COSTART" (Coding Symbols for Thesaurus of Adverse Reaction Terms). Center for Drugs and Biologics, FDA.

Kulczycki, A., Jr. (1986). Aspartame-induced urticaria. Ann. Int. Med. **104:**207–208.

Kulczycki, A., Jr., and Danker, R.E. (1986). Aspartame-induced urticaria: double-blind challenges in two patients (abstract). J. Allergy Clin. Immunol. **77:**187.

Nichaman, M.Z., and McPherson, R.S. (1986). Estimating prevalence of adverse reactions to foods: principles and constraints. Symposium Proceedings on Adverse Reactions to Foods and Food Additives. J. Allergy Clin. Immunol. **78:**148–154.

Novick, N.L. (1985). Aspartame-induced panniculitis. Ann. Int. Med. **102:**206–207.

Appendix A: Aspartame Questionnaire Administered by CDC and FDA

ASPARTAME QUESTIONNAIRE

FINAL DISPOSITION OF INTERVIEW: Case Number ______________

Interview completed...____

Date completed.......................................__________

Interview not completed..________

Indicate Reason:

____

____

____

FINAL DISPOSITION OF MEDICAL RECORDS:

Medical Release Obtained...______

Medical Release Not Obtained.......................................______

Indicate reason for not obtaining records. If records are not required, so state.

____

____

____

Physician Contacted and Records Released.........................________

Medical Records Not Released.................................__________

Indicate Reason ____

____

____

WORKSHEET

Page 2

RECORD PHONE CALLS AND HOME VISITS:

Case Number ___________

Day of Week, Date, Time of Phone Call	Revisit/Recall Required:yes/no	Date of Interview or Appointment

Record Change of address or phone number if any:

____

____

Page 3

ASPARTAME CASE RECORD　　　　Case Number ________________

Name of Complainant ________________________________

Date(s) of episode(s) ________________________________

Date reported: ________________________________

Agency (individual) reported to:

____

CONSUMPTION INFORMATION ________________

Read the list of Aspartame-containing foods/beverages recorded below, to the complainant. Indicate any that were consumed in the year prior to the reported reaction:

(Indicate per day, per week, etc. Circle all products used. NOTE: the Use of Aspartame in carbonated soft-drinks was approved in July, 1983.)

Food/Beverage	Brand Name	Average Quantity	Average Frequency	Approx. Date First Used
Diet Soft Drink:				
Table-top sweetener:				
Kool-Aid type mix:				
Hot Chocolate:				
Iced Tea:				
Lemonade:				
Cereals:				
Puddings/Gelatins:				
Non-Dairy Toppings:				
Other (specify):				

Page 4

Aspartame Case Record
Consumption Information

Case Number ______________

Aspartame-containing food(s)/beverage(s) consumed within 24 hours prior to reported episode:

Product/Brand	Quantity Consumed	Date/Time of Consumption
______	______	______
______	______	______
______	______	______
______	______	______
______	______	______

Other food(s)/beverage(s) consumed with the Aspartame-containing products at the time of the reported episode:

Product/Brand	Quantity Consumed	Date/Time of Consumption
______	______	______
______	______	______
______	______	______
______	______	______
______	______	______

Were any abnormalities noted in the product at the time of use? (e.g., off-taste, off-odor, foreign object, container integrity, etc.,)

Yes ______

No ______

If yes, describe briefly: ____

____

____

Are you still using Aspartame?...yes/no

Page 5

Aspartame Case Record Case Number ____________

SYMPTOMS/DIAGNOSES:

List symptoms: Space for additional description is available below, if needed.

Symptoms	Date of onset	Time from most recent Aspartame Ingestion to onset	Duration* of symptoms

(*For each symptom, indicate if the person is still symptomatic.)

What was your activity at the time the symptom(s) first appeared? (e.g., exercise, sleeping, resting, etc.)

____

____

____

Page 6

Aspartame Case Record
Symptoms/Diagnosis

Case Number________ _____

Was anything done that made symptoms better or worse (such as resting, exercise, medications, eating other foods, etc.)?

____

____

____

____

Was a physician seen for this problem?
If yes, name, address, phone number of physician:

Yes ________

No ________

____

____

____

Was hospitalization requested?
If yes, name, address, phone number of hospital and dates of hospitalization:

Yes ________

No ________

____

____

____

____

Have similar symptoms been experienced prior to initial use of Aspartame-containing products?

Yes ________

No ________

If yes, describe briefly and indicate if the symptoms were associated with other factors (e.g., medications, physical activity, or use of other artificial sweeteners, etc):

____

____

Page 7

Aspartame Case Record
Symptoms/Diagnosis

Case Number________________

Have similar symptoms occurred after consuming an Aspartame-containing product at any other time?

Yes ________

No ________

If yes, describe number of times, date(s), time to onset of symptoms, and duration:

____

____

____

____

____

____

____

____

Have you eaten any other Aspartame-containing products and not had the same symptoms? If yes, list products:

______________________________ Yes ________

______________________________ No ________

______________________________ Do not know ________

Do you know if other family members had similar symptoms after using Aspartame containing products?

Yes ________
No ________
Do not know ________

If yes, describe briefly and obtain names and phone numbers for purpose of contacting.

____

____

Page 8

Aspartame Case Record
Symptoms/Diagnosis

Case Number________________

Do you know if other family members who do not use Aspartame-containing products had similar symptoms?

Yes ________
No ________
Do not know ________

If yes, describe briefly

____

____

Page 9

Aspartame Case Record Case Number ____________

MEDICAL HISTORY

Were you under care or hospitalized for any medical problems now or within the past 2 years?

____

____

____

____

____

____

____

____

____

____

Are you a diabetic? Yes ______ No ______

If yes, how long ______________________

Do you have any food, drug or insect bite allergies? Yes ______ No ______

If yes, describe and how long have you had the allergies: Do not know __________

____

____

____

____

Is there a family history of PKU (phenylketonuria)?

Yes ______
No ______

If yes, who?

____

Page 10

Aspartame Case Record   Case Number ____________

MEDICATION HISTORY:

Have you been on any medications now or in the past 2 years?
(Specifically inquire for diet pills, over the counter medications, medications for anxiety or depression, or herbal products taken at the time of the reported symptoms.)

Yes ______
No ______

If yes, list:

Medication	Approx. date begun	on currently?	On at time of symptoms

Page 11

Aspartame Case Record
Medications

Case Number ________

Were you following a weight reduction diet at the time of symptoms:
If yes, when did you start the diet:

________________ Yes ____ No ____

If yes:

____

At the time of the symptoms:

Approximately how many calories per day were you eating? ________

How much weight lost since start of diet? ________________

In what time period was the weight lost? ________________

What was the name or type of diet being followed? ________________

Were you engaging in active physical exercise as part of the diet program?

Yes ____
No ____

Do you use any other type of artificial sweetener (e.g., sorbitol, saccharin, etc.)

Yes ____
No ____

If yes, type ________________

At the time of symptoms, what was your height?____ft.______in.

and weight?__________lbs.

Why do you think your illness is related to Aspartame ingestion?

____

____

____

____

Page 12

Aspartame Case Record
Medications

Case Number ______________

Were you familiar with newspaper or news media advertising on Aspartame-containing products at the time of the symptoms?

Yes ________
No ________

Were you familiar with newspaper or news media publicity concerning possible reactions to the use of Aspartame-containing products at the time of the symptoms?

Yes ________
No ________

IDENTIFYING INFORMATION

Name of Complainant:

____

Age ________ years Sex __________ Race ____________

Address/phone
Home ____

____

____

Work ____

____

____

Occupation: ____

Parent/Guardian (minors) ____

Person interviewed (if not named above):

____

Indicate reason for not interviewing complainant:

____

____

____

If this case report is subsequent to the original complaint filed by another family member, record name of original complainant:

____

Page 13

Aspartame Case Record Case Number ____________

MEDICAL RECORDS

Physician Name: ____________________

Address: ____________________

Phone: ____________________

The following information should be obtained through contact with the physician:

Were medical records requested? Yes _______
No _______

Were medical records obtained? Yes _______
Approximate dates of complainant visit:
No _______

____

Were you able to interview the physician? Yes _____ No ________

Medical Diagnosis? ____________________

____

Did physician feel the diagnosis was related to Aspartame use?

Yes _______
No _______

Investigator ____________________

Address: ____________________

Telephone Number: ____________________

Date: ____________________

Part VIII Regulatory Status of Phenylalanine as a Nutritional Supplement

CHAPTER 40

Facts and Myths Related to the Use and Regulation of Phenylalanine and Other Amino Acids

Simon N. Young*

Clinical studies on the effect of L-phenylalanine in depression and on the use of D-phenylalanine in the treatment of depression and pain have failed to demonstrate any therapeutic effects. In a small placebo-controlled study, DL-phenylalanine did produce a modest improvement in adult hyperactive patients, but a follow-up study showed that all patients rapidly became tolerant of the therapeutic effect. In spite of the absence of any established clinical indication for phenylalanine, it is widely available in health-food stores in the USA. Such sales have recently been prohibited in Canada, but this regulation is at the moment being challenged legally. Books available to the general public make many unsupported claims about the use of phenylalanine, for example, that it can relieve or prevent depressions, is effective in many types of severe pain, is useful in restoring energy and vitality after stressful activities, and improves memory, and aids in weight loss.

Introduction

Studies carried out over the past 25 years have established that tryptophan influences brain metabolism and various aspects of brain function when given to humans. Work on phenylalanine is at a much earlier stage. The exact effect phenylalanine has on metabolism and function of the human brain is uncertain. The purpose of this paper is to review data relevant to the clinical use of L- or D-phenylalanine, to discuss briefly some of the problems related to regulation of amino acids, and to describe the folklore that exists in certain sections of the general population about the effects of phenylalanine.

Use of L-Phenylalanine as an Antidepressant

L-Phenylalanine is the precursor of tyrosine and can be converted to the catecholamines, dopamine and noradrenaline. Animal data indicate that phenylalanine administration can increase catecholamine synthesis under

*Department of Psychiatry, McGill University, Montreal, Quebec, Canada H3A 1A1.

some circumstances (Carlsson and Lindqvist 1978), but this has not been studied in humans. Phenylalanine can be decarboxylated to phenylethylamine, a compound with some chemical and behavioral similarities to amphetamine. Phenylalanine administration increases rat brain phenylethylamine threefold. However, this is small compared with the 50-fold increase seen with a monoamine oxidase inhibitor (Saavedra 1974).

Because of phenylalanine's dual role as a precursor of the catecholamines and of phenylethylamine, it has been tested as an antidepressant. In one open study, depressed patients were treated with a combination of the monoamine oxidase inhibitor deprenyl and phenylalanine. At least 80% of the 155 patients responded well (Birkmayer et al. 1984). However, as the dose of phenylalanine used (250 mg/day) was only about 10% of the normal daily dietary intake of phenylalanine, it is doubtful whether any of the therapeutic effect was due to phenylalanine. In a second open study, 31 of 40 depressed patients responded to doses of phenylalanine of up to 14 g/day (Sabelli et al. 1986). However, in view of the high response rate of depressed patients to placebo, this open study does no more than provide the rationale for a placebo-controlled study. The hypothesis that L-phenylalanine is an antidepressant remains unproven.

Use of D and DL-Phenylalanine

Metabolic Effects of D-Phenylalanine

Part of the interest in D-phenylalanine was prompted by a report that it was effective in raising phenylethylamine levels in mouse brain through a mechanism not involving conversion to L-phenylalanine (Borison et al. 1978). However, in this study the level of phenylethylamine in brain, which was measured by a gas-liquid chromatographic method, was considerably higher than that reported with a more specific mass-spectrometric method (Boulton 1979). Thus, the idea that D-phenylalanine is an effective precursor of phenylethylamine remains in doubt. Studies on experimental animals and in vitro have shown that D-phenylalanine definitely can inhibit degradation of met-enkephalin by carboxypeptidase A or aminopeptidase, the enzymes which normally degrade enkephalin in vivo (Ehrenpreis et al. 1983). Because of this, D-phenylalanine has an analgesic action in animals, although this effect is seen only at relatively high doses (250 mg/kg).

One important question is the metabolic fate of an unnatural D-amino acid. Limited studies in humans suggest that D-phenylalanine is absorbed rapidly after oral ingestion. About one-third of the amount ingested is excreted relatively rapidly in the urine (a negligible pathway for the L-isomer), while another third is converted to the L-isomer (Tokuhisa et al. 1981, Lehmann et al. 1983). Conversion probably occurs by deamination of D-phenylalanine to phenylpyruvic acid and subsequent reamination to the L-amino acid. Both D- and DL-isomers have been used clinically. In situations where DL-phenylalanine has been used clinically, the dose of the L-isomer is a small fraction of the daily dietary intake. In this situation, clinical effects, if any, are presumably due only to the D-isomer.

The Antidepressant Effect of D-Phenylalanine

In three open studies, a total of 46 depressed patients were treated with 200 mg/day of D- or DL-phenylalanine. Response rates varied from 50% to 91% (Yaryura-Tobias et al. 1974, Spatz etl al. 1975, Beckmann et al. 1977). In a double-blind controlled study, 200 mg/day of the DL-isomer was compared with imipramine. There were no significant differences between the two groups (Beckmann et al. 1979). However, the group size in this study (20 patients) would not necessarily have been large enough to reveal a difference between imipramine and placebo. Thus, the results in all four of these studies could be explained by a placebo effect. On the other hand, the fact that only 2 of 11 patients responded to D-phenylalanine at a dose of 600 mg/day, even though 8 of them subsequently responded to tricyclic antidepressants (Mann et al. 1980), suggests that D-phenylalanine is not an effective antidepressant at this dosage level.

D-Phenylalanine in the Treatment of Pain

Studies of the action of D-phenylalanine on pain have been varied in their design. In a large open study, 78 chronic-pain patients of varied etiology were treated with 0.75 to 1g of D-phenylalanine per day. By the end of the first week, half the patients had shown some response (Balagot et al. 1983). In a smaller open study, 10 outpatients suffering from lumbosacral arachno-epiduritis were given 1 g of D-phenylalanine every day for 2 weeks. There was a decrease in pain and intake of analgesics (Godfraind et al. 1984). In a double-blind placebo crossover study, 21 patients with long-standing intractable pain of varied etiology were given D-phenylalanine (250 mg/day) and placebo for 2 weeks each. Seven patients did better on phenylalanine and one did better on placebo. No analysis for the whole group was reported (Budd 1983). Two double-blind, placebo-controlled studies on the ability of D-phenylalanine to potentiate acupuncture analgesia have been carried out. A single dose of 2 g raised pain threshold more than placebo in three subjects (Kitade et al. 1981). In seven patients with chronic low back pain being treated with acupuncture, a response was seen in one patient after placebo and in four after 4 g of D-phenylalanine (Balagot et al. 1983).

Although none of the studies described above are adequate in their design, they provide some slight indications that D-phenylalanine might be clinically useful. However, a recent well-designed study does not support this idea. Thirty chronic-pain patients of varied etiology were treated with placebo and D-phenylalanine (1 g/day), each for 4 weeks. There was no significant analgesic effect of D-phenylalanine when compared to placebo (Walsh et al. 1986). Taking all these studies into consideration, the conclusion must be that a clinically useful analgesic action of D-phenylalanine has not been demonstrated.

D-Phenylalanine in the Treatment of Attention Deficit Disorder

Nineteen adults with attention deficit disorder, residual type (adult hyperactivity), were given a 2-week double-blind, crossover trial of DL-phenylalanine (up to 1.2 g/day) versus placebo (Wood et al. 1985). Phenylalanine caused significant improvements in the patients' mood and their mood lability. There was a strong trend towards significance on a measure of overall functioning. However, in an open trial at the end of the controlled trial, all patients became tolerant of the therapeutic effect after 2 to 3 months.

Regulation of the Use of Amino Acids

Different countries have different policies concerning the use of amino acids. In the USA, amino acids are freely available in the health-food stores and pharmacies. The same situation existed in Canada until 1985, when Health and Welfare designated as drugs products containing single amino acids or mixtures of amino acids which have demonstrated pharmacologic effects or for which drug claims are made or implied. This includes the D, L, or DL forms of phenylalanine as well as tryptophan, arginine, ornithine, methionine, tyrosine, and lysine. Single amino acids for which no significant pharmacologic effects are documented or for which no drug claims are made are considered to be foods. This ruling has been the subject of two legal challenges, the first from a company which formerly distributed amino acids and the second from a health-food store which is being prosecuted for continuing to sell amino acids. At the time of writing, the second case has not come to trial. The first case was won by the government.

Regulation of the type in force in Canada is capable of preventing abuse of amino acids which influence central nervous system function. However, it does not address so directly the problem of possible toxic effect of amino acids which are still classified as foods. Animal studies indicate that ingestion of disproportionate amounts of any essential amino acids can have long-term toxic effects, especially when protein intake is low (Harper et al. 1970). However, to treat all amino acids as drugs would be impractical because of the widespread use of monosodium glutamate, while a ban on all essential amino acids would leave tyrosine available.

Popular Misconceptions Concerning the Effects of Phenylalanine

The widespread availability of both L- and DL-phenylalanine in the USA, sometimes in capsules containing 500 mg (the amount of phenylalanine present in aspartame in about six 280-ml cans of diet cola), is obviously due to widespread beliefs about the effects of these substances. To make definite statements about the nature of such beliefs is obviously impossible. However, I have made brief informal surveys in health-food stores in various cities across Canada and in the northeastern part of the USA over the past few years. Questions about the possible uses of amino acids have almost invariably produced answers similar to those found in various books written for the general public. One important source seems to be *Life Extension: a Practical Scientific Approach* (Pearson and Shaw 1982). Statements such as "Phenylalanine . . . can relieve or prevent depressions" (p. 172) may be based on an uncritical assessment of the scientific literature. There are certainly no grounds for the suggestion that "Phenylalanine is also useful in restoring energy and vitality after stressful activities. It can provide a lift so that you hop out of bed in the morning instead of dragging yourself out. It is excellent for writer's block" (p. 172). The *Amino Acids Book* (Wade 1985) makes similar claims, stating that "Phenylalanine gives important energy, wipes away gloom, helps soothe pain. Creates inner strength so that you have more resistance to common and uncommon disorders" (p. 60). A catalogue from the Vitamins Specialities Company, Wyncote, Pennsylvania, is succinct about the uses for phenylalanine. The L-isomer is "suggested as an aid in depression, providing improved memory and increased mental alertness. Also believed to aid in weight loss" while the D-form "is effective in many types of severe pain." Erroneous or unsubstantiated information of this type is readily available to the general public, as are capsules or tablets of phenylalanine containing many times the amount of phenylalanine present in the aspartame in a can of diet cola. The widespread concern about possible adverse effects of the phenylalanine in aspartame is perhaps out of proportion in relation to the possible adverse effects of phenylalanine sold in health-food stores.

References

Balagot, R.C., Ehrenpreis, S., Greenberg, J., and Hyodo, M. (1983). D-Phenylalanine in human chronic pain. *In* Ehrenpreis, S., and Sicuteri, F. (eds.), Degradation of endogenous opioids: its relevance in human pathology and therapy. New York: Raven Press, pp. 207–215.

Beckmann, H., Athen, D., Olteanu, M., and Zimmer, R. (1979). DL-Phenylalanine versus imipramine: a double-blind controlled study. Arch. Psychiatr. Nervenkr. **227**:49–58.

Beckmann, H., Strauss, M.A., and Ludolph, E. (1977). DL-Phenylalanine in depressed patients: an open study. J. Neural Transm. **41**:123–134.

Birkmayer, W., Riederer, P., Linauer, W., and Knoll, J. (1984). L-Deprenyl plus L-phenylalanine in the treatment of depression. J. Neural Transm. **59**:81–87.

Borison, R.L., Maple, P.J., Havdala, H.S., and Diamond, B.I. (1978). Metabolism of an amino acid with antidepressant properties. Res. Commun. Chem. Path. Pharmacol. **21**:363–366.

Boulton, A.A. (1979). Trace amines in the central nervous system. *In* Tipton, K.F. (ed.), International review of biochemistry: physiological and pharmacological biochemistry. Baltimore: University Park Press, pp. 179–206.

Budd, K. (1983). Use of D-phenylalanine, an enkephalinase inhibitor, in the treatment of intractable pain. *In* Bonica, J.J., Lindblom, V., and Iggo, A. (eds.), Advances in pain research and therapy. New York: Raven Press, pp. 305–308.

Carlsson, A., and Lindqvist, M. (1978). Dependence of 5-HT and catecholamine synthesis on concentrations of precursor amino-acids in rat brain. Naunyn-Schmiedeberg's Arch. Pharmacol. **303**:157–164.

Ehrenpreis, S., Balagot, R.C., Greenberg, J., Myles, S., and Ellyin, F. (1983). Analgesic and other pharmacological properties of D-phenylalanine. *In* Ehrenpreis, S., and Sicuteri, F. (eds.), Degradation of endogenous opioids: its relevance in human pathology and therapy. New York: Raven Press, pp. 171–187.

Godfraind, J.M., Plaghki, L., and De Nayer, J. (1984). A comparative study on the effects of D-phenylalanine, lysozyme and zimelidine in human lumbosacral arachno-epiduritis. A chronic pain state. *In* Bromm, B. (ed.), Pain measurement in man: neurophysiological correlates of pain. New York: Elsevier, pp. 501–511.

Harper, A.E., Benevenga, N.J., and Wohlhueter, R.M. (1970). Effects of ingestion of disproportionate amounts of amino acids. Physiol. Rev. **50**:428–558.

Kitade, T., Minamikawa, M., Nawata, T., Shinohara, S., Hyodo, M., and Hosoya, E. (1981). An experimental study on the enhancing effects of phenylalanine on acupuncture analgesia. Am. J. Chinese Med. **9**:243–248.

Lehmann, W.D., Theobald, N., Fischer, R., and Heinrich, H.C. (1983). Stereospecificity of phenylalanine plasma kinetics and hydroxylation in man following oral application of a stable isotope-labelled pseudo-racemic mixture of L- and D-phenylalanine. Clin. Chim. Acta **128**:181–198.

Mann, J., Peselow, E.D., Snyderman, S., and Gershon, S. (1980). D-Phenylalanine in endogenous depression. Am J. Psychiatr. **137**: 1611–1612.

Pearson, D., and Shaw, S. (1982). Life extension: a practical scientific approach. New York: Warner Books.

Saavedra, J.M. (1974). Enzymatic isotopic assay for and presence of beta-phenylethylamine in brain. J. Neurochem. **22**:211–216.

Sabelli, H.C., Fawcett, J., Gusovsky, F., Javaid, J.I., Wynn, P., Edwards, J., Jeffries, H., and Kravitz, H. (1986). Clinical studies on the phenylethylamine hypothesis of affective disorder: urine and blood phenylacetic acid and phenylalanine dietary supplements. J. Clin. Psychiatr. **47**:66–70.

Spatz, H., Heller, B., Nachon, M., and Fischer, E. (1975). Effects of D-phenylalanine on clinical picture and phenethylaminuria in depression. Biol. Psychiatr. **10**:235–239.

Tokuhisa, S., Saisu, K., Naruse, K., Yoshikawa, H., and Baba, S. (1981). Studies on phenylalanine metabolism by tracer techniques. IV. Biotransformation of D- and L-phenylalanine in man. Chem. Pharm. Bull. **29**:514–518.

Wade, C. (1985). Amino acids book. New Canaan. Connecticut: Keats Publishing.
Walsh, N.E., Ramamurthy, S., Schoenfeld, L., and Hoffman, J. (1986). Analgesic effectiveness of D-phenylalanine in chronic pain patients. Arch. Phys. Med. Rehabil. **67**:436–439.
Wood, D.R., Reimherr, F.W., and Wender, P.H. (1985). Treatment of attention deficit disorder with DL-phenylalanine. Psychiatr. Res. **16**:21–26.
Yaryura-Tobias, J.A., Heller, B., Spatz, H., and Fischer, E. (1974). Phenylalanine for endogenous depression. J. Orthomol. Psychiatr. **3**:80–81.

CHAPTER 41

Perspectives of the Health-Food Industry on the Use of Pure Phenylalanine

Gary L. Yingling*

This chapter discusses the Federal Food, Drug, and Cosmetic Act definitions of the terms "food" and "drug;" and the possibility of making labeling claims for phenylalanine that would make phenylalanine a drug or food in light of FDA's regulations. The definition of the term "food additive" and the regulations governing phenylalanine's use as a nutrient are also explored. A brief history of the development of regulations concerning dietary supplements is given.

According to the Dorland's Medical Dictionary, phenylalanine is a naturally occurring amino acid [$C_6H_5 \cdot CH_2CH(NH_2)COOH$] which is essential for optimal growth in infants and for nitrogen equilibrium in adults.

This chapter will focus on the limited issue of the marketing of phenylalanine by the health-food industry and the regulatory issues and concerns that play a role in that marketing. It will not attempt to explore the issue of the use or marketing of phenylalanine by the drug industry. However, the reader should recognize that the marketing of a product as a food or drug is not controlled totally by a preapproval process, but in large part by the manufacturer, the intended uses, and labeling.

The Food and Drug Administration (FDA) derives its basic authority over foods from the 1938 Federal Food, Drug, and Cosmetic Act. A food is defined as:

(1) articles used for food or drink for man or other animals, . . . and (3) articles used for components of any other such articles. Section 201(f) of the Act [21 U.S.C. §321(f)]

The drug definition focuses on:

(B) articles intended for use in the diagnosis, cure, mitigation, treatment, or prevention of disease in man or other animals; and (C) articles (other than food) intended to affect the structure or any function of the body of man or other animals. Section 201(g) of the Act [21 U.S.C. §321(g)]

*Burditt, Bowles & Radzius, 1029 Vermont Avenue, N.W., Washington, DC 20005, USA.

Based on such a broad definition, it is possible to make labeling claims for phenylalanine that would make the amino acid a drug. Indeed, courts have interpreted the statute to mean that water, if labeled as a cure for a disease, can be a drug and would be so regulated.

Many of the products that you purchase are regulated as drugs, such as mouthwash, toothpaste, antiperspirants, and certain vitamins and minerals. Their regulation as drugs is dependent on the labeling claims, e.g., the product kills germs, prevents cavities, or cures disease conditions such as scurvy.

As phenylalanine is a naturally occurring amino acid, some would suggest that it can be marketed as a food, but the question is how and what are the labeling restrictions, if any.

For me to suggest that a product can be categorized as a food does not mean that it will have a market as a food or even that the FDA will seek to regulate the product as food; the FDA may still consider it and regulate it as a drug.

If you were to look for phenylalanine in your local food store, you would not find it in the dairy case, with the canned peaches or even in the spice and condiment section. In the large service food stores that we are exposed to today, phenylalanine would be found in the section with vitamins and minerals and most likely would be in a multi-ingredient product. I was reviewing a "diet" product label the other day that contained grapefruit extract, vitamins, and L-phenylalanine. I would expect that you would find phenylalanine being marketed as "phenylalanine" or as a dietary supplement.

In our regulatory system, a product that is used as a "dietary supplement" is treated the same as a food. There is no statutory difference between a dietary supplement and a food. In fact, you will not find, in the Food, Drug, and Cosmetic Act, a definition for a "dietary supplement." However, some may view Section 411 of the Act as obviating the need for such a definition. As the use of phenylalanine in a tablet form is seen by some not to be a food use, it is prudent to carefully review the FDA's regulations.

In the Code of Federal Regulations, under Subchapter B-Food for Human Consumption, in 21 CFR under Part 184—Direct Food Substances Affirmed as Generally Recognized as Safe (GRAS), there is a list of a number of substances such as dextrin, corn gluten, niacin, and sorbitol that are GRAS according to FDA. There is no listing for phenylalanine. Therefore, for human food use, phenylalanine is not considered by the FDA to be a direct food substance that is generally recognized as safe. If you turn to 21 CFR Part 582, which is the listing of substances generally recognized as safe for use in animal feeds, you will find, under Subpart F—Nutrients and/or Dietary Supplement, that phenylalanine is listed under 21 CFR 582.5590. The regulation also states that, as a condition of use, "this substance is generally recognized as safe when used in accordance with good manufacturing or feeding practice." As phenylalanine is safe for use in animal feed and is not affirmed as safe for use in human food, it is either not regulated or the FDA has assigned it to a regulatory classification other than as a food.

A quick review of the regulations shows that FDA has determined that phenylalanine is a food additive, the definition of which is a:

> substance the intended use of which results or may reasonably be expected to result, directly or indirectly, in its becoming a component or otherwise affecting the characteristics of any food, . . . if such substance is not generally recognized, among experts qualified by scientific training and experience to evaluate its safety, as having been adequately shown through scientific procedures, . . . to be safe under conditions of its intended use; . . . Section 201(s) of the Act [21 U.S.C. §321(s)]

As a food additive, by definition, is not generally recognized as safe, it may be used only in accordance with an approved food additive regulation. If the use of a food additive is not in accordance with an approved food additive regulation, then the food in which the substance is present will be an adulterated food in accordance with Section 402(a)(2)(C) of the Act. That provision states that "[a] food shall be deemed to be adulterated if it is, or it bears or contains, any food additive which is unsafe within the meaning of Section 409." Section 409(b)(1) provides that a "person may, with respect to any intended use of a food additive, file with the Secretary a petition proposing the issuance of a regulation prescribing the conditions under which such additive may be safely used."

The FDA, in 21 CFR Part 172, has listed those food additives permitted for direct addition to food for human consumption. Within subpart D of Part 172 is established a "Special Dietary and Nutritional Additives" section (21 CFR 172.320) which lists phenylalanine under the heading "amino acids." Under this food additive provision, an amino acid, e.g., phenylalanine, may be safely used as a nutrient if:

(a) The food additive consists of one or more of the . . . individual amino acids in the free, hydrated or anhydrous form or as the hydrochloride, sodium or potassium salts; and

(c) The additive(s) is used or intended for use to significantly improve the biological quality of the total protein in a food containing naturally occurring primarily-intact protein that is considered a significant dietary protein source, provided that:
   (1) A reasonable daily adult intake of the finished food furnishes at least 6.5 grams of naturally occurring primarily intact protein . . . ;
   (2) the additive(s) results in a protein efficiency ratio (PER) of protein in the finished ready-to-eat food equivalent to casein . . . ;
   (3) each amino acid . . . added results in a statistically significant increase in the PER . . . ; and
   (4) the amount of the additive for nutritive purposes plus the amount

> naturally present in free and combined form . . . does not exceed . . . 5.8 per cent by weight of the total protein of the finished food.

The limitations just discussed for phenylalanine do not apply if the amino acid is intended for use *solely under medical supervision to meet nutritional requirements in specific medical conditions* and the amino acid complies with 21 CFR Part 105, Foods for Special Dietary Use.

Before we turn to Part 105, I want to explore for a moment the exemption of a "special dietary food intended for use solely under medical supervision." The food additive provision, 21 CFR 172.320(f), exempts the amino acid, say phenylalanine in this case, from the requirement of improving total protein and the protein efficiency ratio (PER), but the exemption is a very narrow one because it says *solely* under medical supervision. The FDA views such foods to be those involved in enteral feeding situations because the regulation requires the use be one to meet nutritional requirements in specific medical conditions. From a regulatory viewpoint, the limitation that the FDA has so clearly stated in 21 CFR 172.320 is less than clear if one reads the language of Part 105. Under 21 CFR 105.3(a)(1), Definitions and Interpretations, Foods for Special Dietary Uses, the term "special dietary uses" as applied to food for man, means "particular . . . uses of food," as follows:

(i) Uses for supplying particular dietary needs . . . physiological, underweight and overweight
(ii) Uses for supplying particular dietary needs . . . by reason of age . . . infancy
(iii) Uses for supplementing or fortifying the ordinary or usual diet with any vitamin, mineral or other dietary property.

Therefore, under Part 105, the regulation suggests that foods for special dietary uses include products for underweight, overweight, or supplementing vitamins and minerals. The problem with using the definition in Part 105 for the amino acids discussed in 21 CFR 172.320 is that subpart B of Part 105 does not contain any labeling standards for the amino acid phenylalanine. Since there is no labeling standard for the amino acid under the special dietary use provision, the labeler of a phenylalanine product finds it difficult to comply with FDA regulations if, in fact, the FDA has any regulations for dietary supplements such as phenylalanine products.

To confuse the issue more, let us, for a moment, examine the tormented path that brought us to this point of clarity in government regulation.

The issue of dietary supplements, and especially vitamins and minerals, has been one that has historically given FDA problems. In 1936, prior to the passage of the Food, Drug, and Cosmetic Act of 1938, the writers of that time were criticizing the overzealous exploitation of vitamins. Shortly after the passage of the 1938 Act, the FDA established regulations for "special dietary uses" pursuant to Section 403(j) of the Act [5 Fed. Reg. 5921 (5 September 1940)]. Section 403 is the misbranding provision of the Act as to foods, and subsection (j) states that foods that are represented for "special dietary uses" shall comply with the regulations promulgated by the Secretary. The regulations remained basically unchanged until 1962 when the FDA determined that there was insufficient control of vitamin-mineral products and their claims. On 20 June 1962 (27 Fed. Reg. 5815), the FDA published a proposal which set the standard for labeling and stated that dietary supplements could declare "only those nutrients recognized by competent authorities as essential." In June 1966, the FDA issued its final revision of the special dietary food regulations which required the "minimum daily requirement" (MDR) information [31 Fed. Reg. 8521 (18 June 1966)]. The regulation resulted in numerous objections, and the agency became involved in a two-year evidentiary hearing in which there were 162 witnesses, over 2000 documents, and about 25,000 pages of transcript.

After the hearing, the agency once again sought, in 1973, to establish a regulation, but this time the regulation adopted the "recommended daily allowances" (RDAs) [38 Fed. Reg. 20708 and 20730 (2 August 1973)]. Even this more moderate approach led to litigation and finally legislation. The Food, Drug, and Cosmetic Act was amended by Public Law 94-278 in a fashion that significantly restricted the Secretary's authority to regulate the nutritional formulation and composition of vitamins, minerals, and other ingredients in conventional foods.

Section 411(c)(3) of the Act, which was enacted at that time, states that:

> "special dietary uses" . . . means a particular use for which a food purports or is represented to be used, including but not limited to the following:
>
> (A) Supplying a special dietary need that exists by reason of . . . underweight, overweight . . . ;
> (B) Supplying a vitamin, mineral or other ingredient for use by man to supplement his diet by increasing the total dietary intake.

When the legislation passed, the FDA recognized the congressional mandate and did not seriously attempt to further define or regulate closely the marketing of vitamins and minerals and "special dietary foods."

The FDA, it appears, has now adopted a strategy of charging a misbranding violation under Section 403(j) if a special dietary food violates that provision of the statute, but the FDA is not actively seeking to regulate the "special dietary food" market by regulation.

Therefore, the health-food industry today has to be concerned with the accuracy and truthfulness of the labeling of the products being marketed. Phenylalanine-containing products are in the marketplace, and the issue, if there is an issue at all, is what is the FDA going to do about controlling that marketing from the standpoint of product availability or labeling.

Part IX Regulatory Status of Aspartame and Other Artificial Sweeteners

CHAPTER 42

Regulation of Food Additives with Neurotoxic Potential

David G. Hattan*

One popular misconception regarding food safety is that the Food and Drug Administration demands a standard of absolute safety for food additives—in effect, the equivalent of saying there is no risk whatever from exposure to a food additive. This chapter attempts to dispel this misinterpretation of responsibility regarding food safety by discussing the interaction between the requirements of toxicological testing, interpretation of test data, and the actual food additive safety standard of "reasonable certainty of no harm."

This chapter will attempt to demonstrate how the legislative authority provided by the Federal Food, Drug, and Cosmetic Act (1980) (referred to hereafter as the Act) is applied by the Food and Drug Administration (FDA). The present provisions of the Act are the culmination of a series of amendments to the Act which was passed in 1938. One of the key provisions of the 1938 Act was Section 402(a)(1) which set forth means for determining whether food was adulterated because of the presence of poisonous or deleterious substances (including certain constitutents of raw agricultural commodities).

Sections 409 (food additives) and 706 (color additives) were added as amendments to the Act in 1958 and 1960, respectively. These sections established the concept of premarket testing and clearance for these types of products added to foods. In addition, other sections dealing with control and regulation of environmentally added contaminants to food (Section 406) and of pesticides utilized in the production of agricultural commodities (Section 408) were passed by the U.S. Congress.

Thus, the various sections of the Act are concerned with various *categories* of food-related substances. A given substance or foodlike material will fall into a specified category based on such characteristics as its chemical structure or properties, its intended use, and even the toxicological nature it possesses (e.g., pesticides). Many people believe that by appropriate use

* Office of Nutrition and Food Sciences, Center for Food Safety and Applied Nutrition, Food and Drug Administration, Washington, DC 20204, USA.

TABLE 42.1. Food-related regulatory categories.

Category	Examples	Safety standard
Food	Raw agricultural commodities (apples, meat)	Ordinarily render injurious[a]
Food additives Direct	Preservatives, artificial sweeteners (sodium nitrite, aspartame)	Reasonable certainty of no harm
Indirect	Packaging materials (acrylamide)	Reasonable certainty of no harm
Color additives	Natural and synthetic colors (carotene, FD&C Red 40)	Reasonable certainty of no harm
Pesticides (residues)	Insecticides (organophosphates)	Residues set by EPA under 408
Natural toxins	Aflatoxin	Unavoidability under 406

[a] Section 402(a) of the Food, Drug, and Cosmetic Act states "A food shall be deemed adulterated if it bears or contains any poisonous or deleterious substances which *may render it injurous* to health; but in the case the substance is not an added substance such food shall not be considered adulterated under this clause if the quantity of such substance in such food does not *ordinarily* render it injurious to health . . . "

of the legal discretion allowed by categorization, the FDA can prevent *any* amount of a toxic substance from entering the food supply and thus ensure absolute safety. In fact, there is no standard of absolute safety, only degrees of certainty of safety (or absence of harm) depending on the standard applied to the category into which a compound falls. These differing degrees of certainty result from the application of the differing standards of safety established by the Act. Below is a fuller consideration of the utility of and need for these various categories.

Table 42.1 is an effort to succinctly present a number of the different food-related regulatory categories that are authorized by the different amendments to the Act. This table is not a comprehensive list of all food-related regulatory categories, but will serve to illustrate the interaction between science and law which we are discussing in this paper.

General Food Safety Provision (Section 402)

The Act refers to two standards for determining whether a poisonous or deleterious substance makes that food adulterated (unsafe for consumption). One standard is the "may render" standard (for added substances) and the other is the "ordinarily render" standard (for naturally occurring substances). For the purpose of invoking these standards, it is up to the FDA to demonstrate with the proper amount of evidence that the substance under consideration is a source of actual harm (supports "ordinarily render" standard) or of possible harm to consumers (supports "may render" standard). The FDA may take legal action against any food in interstate commerce which violates these standards.

For single-ingredient food and certain substances falling into the GRAS (Generally Recognized as Safe) category, the decision concerning which standard to apply will depend upon the circumstances surrounding the substance's addition to food. If the material is present naturally in food, the more rigorous standard of "ordinarily render" must be supported by adequate evidence. If the substance is added to food, the FDA would apply the less rigorous standard of "may render." Less rigorous in this instance means less evidence is needed to adequately support the standard applied.

There are numerous examples of substances which are covered by other sections of the Act (e.g., food additives, color additives, pesticide residues, environmental contaminants, and drugs for food-producing animals, respectively covered by Sections 409, 706, 408, 406, and 512).

Food Additives (Section 409)

Following the addition of Section 409 to the Act in 1958, the manufacturers of food additives (see Appendix A) became responsible for supplying to the FDA evidence of adequate testing of new food additives to ensure their safety before they could be approved for addition to foods. The sponsor of a new food additive must file with the FDA a written petition which contains adequate data to establish the safety of the food additive. The FDA then reviews the petition and if it concludes that the data are adequate *and* establish the safe use level of the food additive substance, the agency publishes an approval or regulation in the *Federal Register* which outlines the appropriate conditions of use. The level of approval can extend from very limited use in one small food category to virtually unlimited use in many food categories (Food Additive Petitions 1981). How one determines the extent of approval is described below.

The safety standard which must be met for food additives is "reasonable certainty of no harm," which is language taken from the legislative history for the food additive amendments. Just exactly what this standard means has never been formally defined by the FDA, but it may be inferred. At present, the scientific evidence must be of a specific quality and quantity [as defined by the Center for Food Safety and Applied Nutrition's Redbook (U.S. Food and Drug Administration 1982)] and there must be an adequate dose margin of safety between the no-effect dose in animals and expected levels of human consumption. In short, a no-effect dose is determined (based on animal test data); then a 10- to 100-fold safety factor is applied to this no-effect dose to define the acceptable daily intake (ADI). If the estimated daily intake (EDI) is less than the ADI, then the compound may be approved or regulated.

Color Additives (Section 706)

The regulatory procedures for the evaluation of color additives are very similar to those used for food additives. Compliance with essentially all of the premarket requirements for food additive toxicological testing is necessary.

In addition, color additives are regulated by postmarketing surveillance to assure good manufacturing practices (involves the certification of the dye to ensure that it meets established chemical specifications). The FDA maintains a laboratory which provides the chemical specifications for color additives that limit by-products and impurities resulting during chemical synthesis of the food dye.

Use of Regulatory Authority

There is a constant tension between the static language of the Act and the scientific facts and concepts used to support actions taken in implementation of the Act. On many occasions when the FDA is perceived to be making a controversial decision, it is when the language of the Act and the scientific data are not clearly congruent.

Recently, for example, as the capability of analytical chemistry has developed dramatically, it has been possible to find extremely low concentrations (parts per billion or even parts per trillion) of potentially carcinogenic substances, e.g., certain organic solvents, in the water and food we eat.

Science has provided us with the tools to detect these extremely minute amounts of these carcinogenic substances; however, the language of the Act provides the FDA little, if any, guidance in the interpretation of the significance of their presence. A rigorous interpretation of the Delaney clause of the Act would indicate that aggressive legal action should be taken against those food additives that contain even these extremely minute quantities of carcinogens. On the other hand, the concept of dose thresholds is well established for other toxicities, including neurotoxicity. For example, at very high doses, caffeine may produce definitely adverse effects (such as tachycardia). At more moderate doses, it mediates neurobehavioral effects which many people perceive as beneficial, and at very low doses, it may cause no effects at all. Thus, for potentially neurotoxic substances, there exist doses below which the toxic response will not be manifested.

In order to resolve regulatory dilemmas of this nature, two things must occur. The law must change or evolve, and science must provide us additional insight concerning the mechanisms of neurotoxicity and carcinogenicity so that the FDA can confidently justify permitting the presence of the minute quantities of neurotoxins or carcinogens in foods. This justification must be based on adequate data to demonstrate that a threshold does in fact exist for the substance's toxic or adverse effect.

One question, either implied or expressed, is often directed to the FDA following the availability of data indicating a potential toxicity problem with a substance which is presently approved for use. Why don't you just ban the compound? The answer is that the FDA must amass evidence of adverse effects and appropriately interpret its significance before any legal or regulatory actions can be taken. The quality and quantity of these adverse data must be sufficient to withstand the scientific and legal scrutiny to which it will be subjected. Agency actions taken to ban, propose tolerances, or interim list a substance must ordinarily be preceded by "notice and comment" rulemaking, a procedure during which all interested parties, including scientists, may comment or file objections. In some situations, e.g., food and color additive regulations, anyone who feels they have been adversely affected by the FDA action may also request a hearing, either before an administrative law judge or, more rarely, before a scientific board of inquiry. After final agency action, injured parties may present their arguments to courts of law [See Code of Federal Regulations, 21, Parts 10, 12–16, and 170 (1981)]. With these considerations in mind, it is clear why the FDA does not proceed with an adverse legal action against a substance unless the threat to the public health is serious, or the type and amount of scientific evidence clearly support these actions.

References

Federal Food, Drug, and Cosmetic Act, as amended (1980). U.S. Government Printing Office, Washington, DC.

Food Additive Petitions (1981). Code of Federal Regulations, 21, Part 171.

U.S. Food and Drug Administration, Bureau of Foods (1982). Toxicological Principles for the Safety Assessment of Direct Food Additives and Color Additives Used in Food.

Appendix A

Section 201(s) of the Food, Drug, and Cosmetic Act

The term "food additive" refers only to a specific set of substances defined by Section 201(s) as follows:

The term "food additive" means any substance the intended use of which results or may reasonably be expected to result, directly or indirectly, in its becoming a component or otherwise affecting the characteristics of any food (including any substance intended for use in producing, manufacturing, packing, processing, preparing, treating, packaging, transporting, or holding food; and including any source of radiation intended for any such use), if such substance is not generally recognized, among experts qualified by scientific training and experience to evaluate its safety, as having been adequately shown through scientific procedures (or, in the case of a substance used in food prior to January 1, 1958, through either scientific procedures or experience based on common use in food) to be safe under the conditions of its intended use; except that such term does not include

(1) a pesticide chemical in or on a raw agricultural commodity; or
(2) a pesticide chemical to the extent that it is intended for use or is used in the production, storage, or transporation of any raw agricultural commodity; or
(3) a color additive; or
(4) any substance used in accordance with a sanction or approval granted prior to the enactment of this paragraph pursuant to this Act, the Poultry Products Inspection Act or the Meat Inspection Act; or
(5) a new animal drug.

CHAPTER 43

Regulatory Aspects of the Use of Low-Calorie Intensive Sweeteners

Aharon Eisenberg*

The regulatory status of the major low-calorie intensive sweeteners in a large number of countries is reviewed. Guidelines developed by a governmental advisory committee on food additives are discussed. A proposed scheme for setting levels of sweeteners in foods is presented.

The past 15 years have witnessed many fluctuations in the policies of regulatory agencies regarding the use of low-calorie intensive sweeteners, both as table-top sweeteners and in foods. It would be of interest to record the developments in this field chronologically and examine critically the scientific, legal, and administrative criteria applied at different periods in various countries. However, in this chapter we will limit discussion to some of the practical problems in regulating sweeteners today.

It is axiomatic to assume that the basis for approval for sweeteners is toxicological evaluation. However, examination of the data summarized in Tables 43.1 and 43.2 shows clearly that there is no international toxicological consensus as to which sweeteners are suitable and at what levels they should be added to foods.

In fact, the changes in policy reflect and have influenced the development of toxicological methods and concepts. We should remember that when the cyclamate issue appeared in 1969, there was no agreement in the scientific community as to how toxicological testing should be carried out, and the understanding of chemical carcinogenesis was in the early stages of a process still not completed.

The publication of GLP (Good Laboratory Protocols) by the U.S. Food and Drug Administration, the U.S. Environmental Protection Agency, and the Organization for Economic Cooperation and Development in the late 1970s has greatly contributed to setting a rational policy by defining properly designed and executed studies.

*Public Health Services, Ministry of Health, Jerusalem 91000, Israel.

TABLE 43.1. Regulatory status of sweeteners in beverages.

	Number of countries where:			
Sweetener	permitted	not permitted	special permission required	decision pending
Aspartame	26	16	2	2
Saccharin	72	15	5	–
Cyclamate	28	45	4	–
Acesulfame-K	6	–	1	5

TABLE 43.2. Tolerances for sweeteners in beverages.

	Number of countries with maximum level (ppm) of:					
Sweetener	GMP[a]	< 100	101–500	501–1000	1001–5000	5001–10,000
Aspartame	4	–	4	7	7	–
Saccharin	12	19	31	3	1	–
Cyclamate	2	–	6	3	8	2
Acesulfame-K	2	–	1	1	–	2

[a]GMP—good manufacturing practice.

Israeli Guidelines for the Use of Sweeteners

The Israel Food Additives Committee has been examining this subject in order to formulate suitable principles to serve as guidelines for setting policy. This process is far from completed; however, we are of the opinion that results achieved thus far are of interest:

1. The acceptable daily intake (ADI) values adopted by the Joint Expert Committee on Food Additives of the World Health Organization/Food and Agriculture Organization of the United Nations will serve as the basis for determining permitted sweeteners and setting levels in foods.
2. Where appropriate, the use of mixtures of sweeteners will be encouraged.
3. The addition of low-calorie intensive sweeteners should be restricted to products which are both suitable and useful in planning and managing calorie-restricted diets. The basis for these decisions are guidelines set up by a group of physicians who are recognized experts in nutrition.
4. The formulation of beverages should be designed to enable consumers to drink liquids in quantities commensurate with medical recommendations in hot climates and still not exceed ADI recommendations. In practical terms, the committee set the criterion that the consumption of 2 liters of beverage should not contribute more than 50% of the ADI for a 70-kg adult. This is to take into account other sources of sweetener intake. The caloric value of the product should not exceed the level set for low-calorie beverages in Israel, namely, 15 calories per 100 ml.

TABLE 43.3. Allocation of acceptable daily intake (ADI) of sweeteners.

Sweetener	Allowable daily intake				Tablets of table-top sweetener	
	mg/kg body weight	mg/day for adult	Concn in beverage[a] (mg/100 ml)	Concn in ice cream[b] (mg/100 g)	mg/tablet[c]	No./day[d]
Aspartame	40	2800	70	280	18	62
Saccharin	2.5	175	4.4	17.5	18	3.9
Cyclamate	11	770	19.3	77	150	2.1
Acesulfame-K	9	630	15.7	63	50	12.6

[a] Based on allocation of 50% ADI to consumption of 2 l/day.
[b] Based on allocation of 10% ADI to consumption of 100 g.
[c] Equivalent sweetness: 1 teaspoon or 5 g of sucrose.
[d] Based on allocation of 40% ADI.

5. A similar exercise was carried out with certain kinds of ice creams, particularly frozen yogurts with reduced caloric content. In these cases, 10% of the ADI was allocated. Other foods found suitable will be handled similarly, and in Table 43.3 we have presented an overall scheme for the allocation of sweeteners in foods and for individual use.

There are two kinds of brakes incorporated into this scheme whose purpose is to minimize the possibility that an individual or a group in the population will exceed the ADI levels for a significant period. The first is illustrated by the levels set for beverages which allow consumers to drink quantities of liquids much in excess of normal regulatory "reasonable daily intake" figures.

The second factor is labeling. This will be discussed in the next section in detail, but it must be stressed that the success of this proposal will depend on a receptive consumer. A health education program is planned to supply interested consumers with the means to identify and calculate their sweetener consumption.

Labeling

The acceptance of the multisweetener concept poses new problems in the labeling of food products, especially beverages. Much more information has to be displayed on an already crowded label. The consumer needs two basic types of information.

The first thing that has to be clear is the nature of the sweetener or the sweeteners in the product, and the second type of information required is the amount of the sweetener in the product. In order to complete the picture, the consumer should have a reasonably accurate idea as to the recommended limits for continuous consumption by individuals. This information is supplied in the ADI recommendations referred to in this chapter.

We are aware of objections to listing ADIs on food product labels and are cognizant of the fact that this was not the intention of the experts who brought them into this world. However, some way of getting this information to consumers seeking it will have to be devised. Various suggestions have been tabled. These include concepts such as "daily beverage allowance," which expresses the quantity an adult or child could consume within ADI limits. The reciprocal concept may also be useful, namely, the %ADI for an adult or a child in a specific product, whichever is appropriate.

CHAPTER 44

Resources for Inferential Estimates of Aspartame Intake in the United States

John P. Heybach and Susan S. Allen*

This chapter examines several resources available for inferential estimation of the intake of foods and food ingredients, particularly aspartame, in the United States. The strengths and weaknesses of each of these resources are reviewed with respect to the types of interpretations and conclusions any single resource can support. Without exception, these resources indicate that although sales of aspartame as an ingredient in foods have steadily increased the actual intake of aspartame by both children and adults remains well within the Food and Drug Administration's current guidelines.

Introduction

Nutrients and ingredients, unlike foods, are not consumed directly and, therefore, their intake by the population must be estimated utilizing data based on various methods of measuring food intake.

Both government and commercial survey resources exist for deriving inferential estimates of the intake of foods, nutrients, and food ingredients by the United States population. The appropriate utilization of each resource depends upon theoretical and statistical implications which are well understood for each particular survey methodology. Emphasis here will be placed on the intake of food in categories in which aspartame is used, and the development of the various resources used to measure this intake and their validity and reliability will be described briefly. In addition, specific inferential estimates of food, nutrient, and ingredient intake based on these different methodological resources will be compared.

*The NutraSweet Company, 4711 Golf Road, Skokie, IL 60076 USA.

Traditional Methods for Inferential Estimates of Dietary Intake

Per Capita Disappearance Data

Perhaps the crudest resource available for use in estimating exposure to a food, nutrient, or ingredient is per capita disappearance data. These data are derived from production and marketing estimates provided by the United States Department of Agriculture (USDA). Although adjustments to per capita disappearance figures are made based on available information concerning imports, exports, and stock exchanges, adjustments are not made for any factors related to food loss prior to ingestion—loss due to spoilage, inedible components of food, uneaten leftovers, or use as pet food. Figure 44.1 shows per capita disappearance of aspartame in the

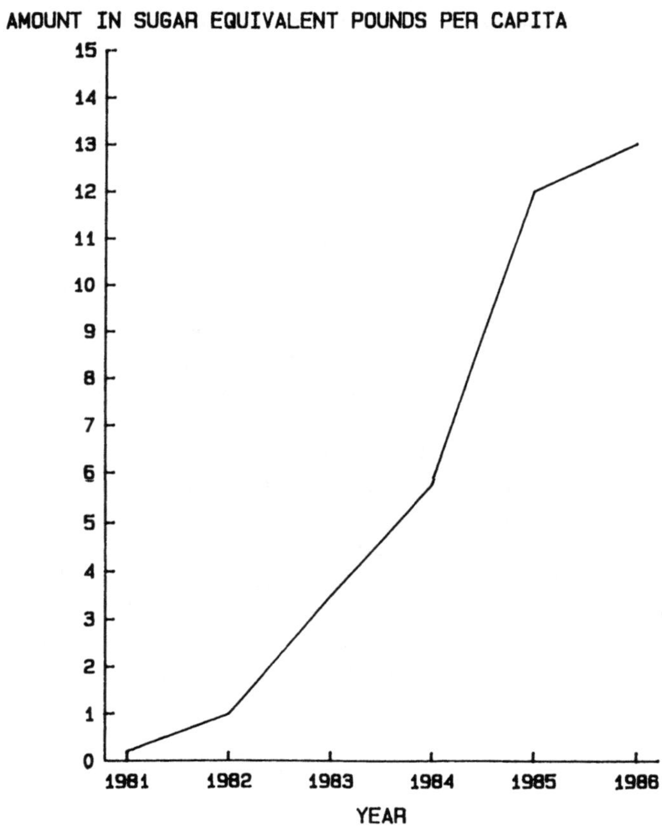

FIGURE 44.1. Per capita disappearance of aspartame in the period 1981 to 1986 in sucrose sweetness equivalent units (assuming aspartame is 200 times as sweet as sucrose). Source: USDA, ERS estimates, 1987.

TABLE 44.1. Actual versus per capita estimates of daily intake for various nutrients.

Nutrient	Daily intake: Per capita (1978)[a]	Daily intake: Actual (NFCS, 1977–78)[b]	Percentage by which per capita overestimates actual
Energy (kcal)	3330	1826	+82.4
Protein (g)	99	73.4	+34.9
Fat (g)	158	83.1	+90.1
Carbohydrate (g)	385	193.5	+99.0
Calcium (mg)	889	743	+19.7
Iron (mg)	17	12.6	+34.9
Magnesium (mg)	324	246	+31.7
Phosphorus (mg)	1491	1153	+29.3
Vitamin A (IU)	7700	5388	+42.9
Thiamin (mg)	2.08	1.26	+65.1
Riboflavin (mg)	2.34	1.72	+36.0
Niacin (mg)	25.8	18.2	+41.8
Vitamin B_{12} (g)	8.8	5.02	+75.3
Vitamin B_6 (mg)	1.95	1.4	+39.3
Vitamin C (mg)	111	82	+35.4

[a] Source: Food Consumption, Prices and Expenditures, USDA, Economic Research Service, 1964–1984.
[b] Source: Nutrient Intakes: Individuals in 48 States, Nationwide Food Consumption Survey, 1977–78.

period 1981 to 1986 in sucrose sweetness equivalent weight units. For 1986, these data suggest that the population exposure to aspartame was approximately 81 mg of aspartame per person per day (assuming aspartame is 200 times sweeter than sucrose) or a crude estimate of 1.6 mg of aspartame per kg of body weight per day (assuming a 50-kg average body weight).

Per Capita Versus Actual Intake Estimates

It is illustrative to look at the estimates of daily intake of various foods and nutrients derived from per capita data in comparison to actual food consumption estimates as measured in food intake surveys. Per capita estimates typically overestimate actual intake, often by a surprisingly large degree (see Table 44.1).

Total Diet Study

The United States Food and Drug Administration (FDA) conducts a food purchase survey, the Total Diet Study, used to estimate daily food intake. In this study, selected foods are purchased in grocery stores across the United States. These foods are analyzed for various constituents, and the results are extrapolated to a "usual diet" to assess dietary exposure to the various constituents. Data from this survey estimate that 25- to 30-year-old females consumed an average of about 54.2 g of diet soft drink per day while 2-year-olds consumed 6.8 g of soft drink per day (Pennington 1983). Although a great improvement over per capita disappearance data, these estimates are still crude and highly derivative.

Food Consumption Surveys

Food consumption surveys attempt to measure what is actually eaten by groups and individuals within groups. Various methods have been developed for the task, and their statistical implications, validity, and reliability are well understood (Beaton et al. 1979, 1983, Heybach et al. 1984, 1986, Holdsworth et al. 1984, Stuff et al. 1983).

Most commonly used to measure food intake are food intake diaries, food intake recall, and food frequency questionnaires. An example from the USDA's Nationwide Food Consumption Survey (NFCS) conducted in 1977 can be used to illustrate the intake of diet soft drinks based on these methods. The average quantity of diet soft drink consumed per day in this survey is 6 g in the 1- to 5-year age group and 47 g in adult women (data derived from USDA, Human Nutrition Information Service 1984). The USDA has recently completed analysis of an extension of the NFCS entitled The Continuing Survey of Food Intake in Individuals (CSFII), 1985. Data from this recent survey show that the amount of low-calorie soft drinks consumed by adult women has increased from 47 g/day in 1977 to 105 g/day in 1985. However, for children the numbers remain comparatively stable over these same time periods (from 6 g/day to 8 g/day). (See Table 44.2.)

Commercial food intake surveys are also conducted in the United States. A continuing food intake diary survey has been conducted on a regular basis since 1984 by MRCA Information Services, Inc. This survey has

TABLE 44.2. Comparison of estimates of mean intakes of low-calorie soft drinks from different surveys.

	Estimated mean intake (g) by:	
	Children	Women
NFCS (1977–78)[a]	6	47
NFCS, CSFII (1985)[b]	8	105
MRCA (Apr–Jun 1986)[c]	22	134

[a] Nationwide Food Consumption Survey, 1977–78. Women aged 19 to 50 years, children aged 1 to 5 years, 1 day, total population, all low-calorie soft drinks.

[b] Nationwide Food Consumption Survey, Continuing Survey of Food Intakes by Individuals, 1985. Women 19–50 years and their children 1–5 years, 1 day. Total population, all low-calorie soft drinks.

[c] Estimates derived from MRCA Information Services, Inc., April–June 1986. Women aged 13 to 14, children aged 2 to 5, 14-day average, users only, aspartame-sweetened only.

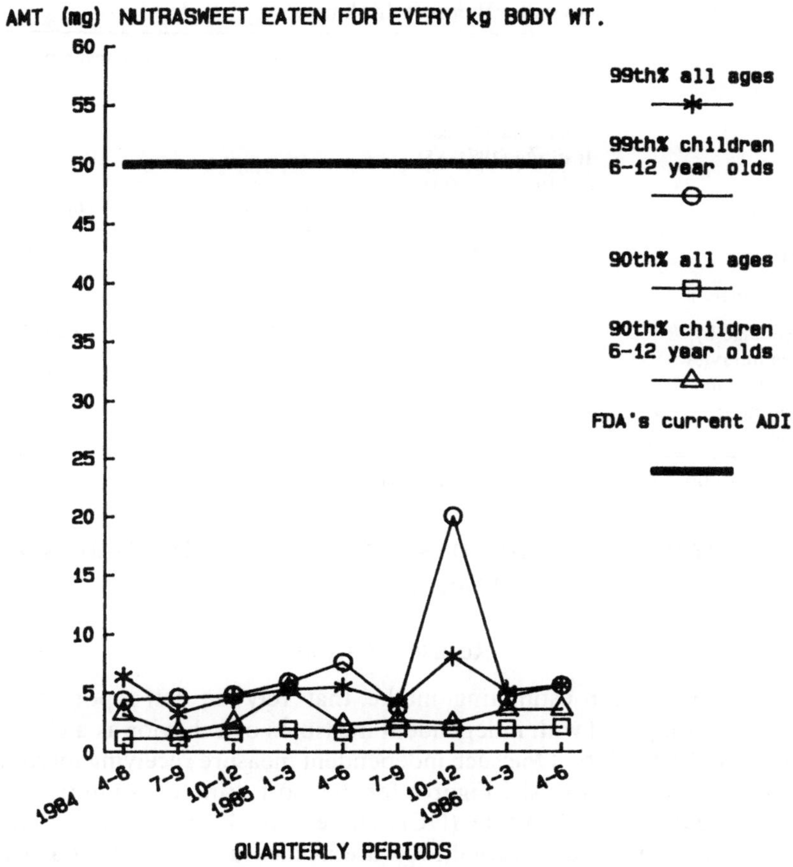

FIGURE 44.2. 90th and 99th percentile of daily aspartame consumption over 2-week periods. Source: MRCI Information Services, Inc.

served as one basis for aspartame intake estimates for federal regulatory purposes. Figure 44.2 indicates that aspartame intake as determined by MRCA is well within FDA guidelines. Recently, an in-depth analysis of food intake from the USDA's CSFII survey has yielded estimates quite similar to those of MRCA for consumption of aspartame. Over 90% of adult women in the survey who consumed aspartame had intake < 5 mg/kg. (Heybach and Smith, 1988).

As this review shows, when comparisons of estimates of intake of specific food categories are made across surveys with comparable survey instruments, good agreement can be found, suggesting a high degree of reliability. For example, when mean intake of low-calorie soft drinks from the NFCS, 1977–78, CSFII, 1985, and MRCA Information Services, Inc. Menu Census, 1986, are compared, agreement between these independent surveys is excellent.

TABLE 44.3. Refuse analysis for diet soft drink use.[a]

Instrument	National Cancer Institute (1984–85), food frequency survey (N = 54)	Nationwide food consumption survey (1981–82)	
		3-Day (N = 54)	7-Day (N = 62)
Interview	1.83	2.86	4.07
Refuse analysis	0.58	2.34	2.60
Percentage by which interview estimate differs from refuse estimate	+215.5%	+22.6%	+56.9%

[a]Source: Rathje, W. (1986). Le projet du garbage (unpublished).

Nontraditional Methods for Inferential Estimates of Dietary Intake

Refuse Analysis

The techniques for estimating intake that rely on self-report can be profitably compared with independent measures of behavior as a check on validity and reliability. One such independent measure receiving increasing attention is refuse analysis. Figures for diet soft drink consumption from the National Cancer Institute (NCI) survey, 1984–85, and the 1981–82 NFCS were compared to actual food use as measured by refuse analysis. This comparison shows a tendency for self-report to overestimate diet soft drink consumption—in the case of the NCI survey, by as much as 215% (see Table 44.3). The use of independent measures such as refuse analysis for validating self-report instruments is growing in importance in the United States.

Summary and Conclusions

There are several resources available for inferential estimates of the intake of foods and food ingredients, such as aspartame, in the United States. Each has its particular strengths and weaknesses and these need to be understood in the interpretation of data from any single resource (see Table 44.4). Finally, although sales of aspartame (as reflected in the disappearance of aspartame into the total population) have steadily increased, the actual intake of aspartame by children and adults remains well within the FDA's current guidelines.

TABLE 44.4. Appropriate use of different data resources.

Data resource for estimate	Appropriate use of estimate
Per capita	Estimate of disappearance into total population
Food purchase diary	Estimate of group exposure to a particular food or constituent for a hypothetical, defined diet
Food consumption survey	
Food frequency	Usual intake of a particular food or food category
1-day record	A reasonable estimate of a usual intake of a homogeneous *group*; a meaningless estimate of an *individual*'s usual intake
Diary or recall with more than 1-day record	With increase in the number of days on which an individual's intake is recorded, estimates of an individual's *usual* intake become more accurate (optimum approximately 11 days).
Refuse analysis	Allows for direct check of the validity of recall and diary data

References

Beaton, G.H., Milner, J., Corey, P., McGuire, V., Cousins, M., Stewart, E., de Ramos, M., Hewitt, D., Grambsch, V., Kassin, N., and Little, J.A. (1979). Sources of variance in 24-hour dietary recall data: implications for nutrition study design and interpretation. Am. J. Clin. Nutr. **32:** 2546–2559.

Beaton, G.H., Milner, J., McGuire, V., Feather, T.E., and Little, J.A. (1983). Source of variance in 24-hour dietary recall data: implications for nutrition study design and interpretation. Carbohydrate sources, vitamins and minerals. Am. J. Clin. Nutr. **37:**986–995.

Heybach, J.P., and Smith, J.L. (1988). Intake of Aspartame in 19–50 year old women from the USDA continuing survey of food intakes by individuals. Fed. Proc., in press.

Heybach, J.P., Coccodrilli, G.D., and Leveille, G.A. (1986). The contribution of the consumption of processed foods to nutrient intake status in the United States. *In* Harris, R.S., and Karmas, E. (eds.), Nutrition evaluation of food processing, Third edition, Westport, Connecticut: AVI Publishing Inc., in press.

Heybach, J.P., Vellucci, F.D., Davidson, L.L., and Coccodrilli, G.D. (1984). Comparison of selected nutrient intake estimates from two national food intake surveys. Fed. Proc. **43:**665.

Holdsworth, M.D., Davies, L., and Wilson, A. (1984). Simultaneous use of four methods of estimating food consumption. Hum. Nutr. Appl. Nutr. **38A:**132–137.

Pennington, J.A.T. (1983). Revision of the Total Diet Study food list and diets. J. Am. Diet. Assoc. **82:**166–173.

Stuff, J.E., Garza, C., O'Brian Smith, E., Nicholas, B.L., and Montandon, C.M. (1983). A comparison of dietary methods in nutritional studies. Am. J. Clin. Nutr. **37**:300–306.

United States Department of Agriculture, Human Nutrition Information Service (1984). Nationwide Food Consumption Survey, 1977–1978. Nutrient Intakes: Individuals in 48 States, Year 1977–1978. Report Number 1–2.

CHAPTER 45

Neurological, Psychiatric, and Behavioral Reactions to Aspartame in 505 Aspartame Reactors

H.J. Roberts*

Aspartame consumption was the cause or aggravating factor for a variety of unexplained and "refractory" neurological, psychiatric, and behavioral disorders in 505 aspartame reactors. The criteria included (1) prompt and gratifying relief or the disappearance of symptoms and signs following its cessation, and (2) their predictable recurrence within hours or days after rechallenge. Such patients should be observed after stopping these products—preferably before ordering costly tests and multiple consultations. Among the high-risk groups are pregnant women, lactating mothers, young children, and older persons with memory impairment.

The toxicology and licensing of aspartame, based on these observations and nationwide study, require reevaluation.

Aspartame (NutraSweet, Crystal Light, Equal) has been the apparent cause or aggravating factor of many serious neurological, psychiatric, and behavioral disorders encountered in a majority of 505 aspartame reactors. Its etiologic role was documented by (1) the prompt and gratifying improvement of symptoms and signs (except for blindness and serious brain dysfunction) after abstinence from aspartame, and (2) the predictable recurrence of complaints within hours or days after resuming such products—whether knowingly by self-testing, by the inadvertent ingestion of products not suspected of containing aspartame, or during formal rechallenge testing.

General Data

These 505 aspartame reactors included 121 patients and aspartame reactors who were personally interviewed, and 384 persons who detailed their adverse effects in a nine-page computerized survey questionnaire. The names of the latter were provided by Aspartame Victims and Their Friends (courtesy of Mrs. Shannon Roth), the Community Nutrition Institute

*Palm Beach Institute for Medical Research, Inc., 304 27th Street, West Palm Beach, FL 33407, USA.

(courtesy of Mr. Rod Leonard), and Dr. Woodrow Monte of Arizona State University.

The female preponderance was striking—375 women (74.3%) and 130 men (25.7%). The age range was from 2 to 92 years, averaging 45 years. Nearly half consumed aspartame products because of an overweight condition, real or perceived. Diabetes mellitus and hypoglycemia also were frequent reasons for abstaining from sugar.

Two-thirds experienced symptomatic improvement within two days after stopping aspartame.

Symptoms and Signs

The most significant neurological, psychiatric, and behavioral features encountered are listed in Table 45.1. Concomitant visual changes and/or auditory reactions, also listed in Table 45.1, led as well to extensive neurological evaluation.

TABLE 45.1. Symptoms reported by 505 aspartame reactors.

Symptom	Number (% of total)
Neurological	
Headaches	228 (45.1)
Dizziness, unsteadiness, or both	199 (39.4)
Confusion, memory loss, or both	144 (28.5)
Convulsions (grand mal epileptic attacks)	74 (14.7)
Petit mal attacks and "absences"	18 (3.6)
Severe drowsiness and sleepiness	83 (16.4)
Paresthesias ("pins and needles," "tingling") or numbness of the limbs	68 (13.5)
Severe slurring of speech	57 (11.3)
Severe "hyperactivity" and "restless legs"	39 (7.7)
Atypical facial pain	33 (6.5)
Severe tremors	43 (8.5)
Psychiatric and Behavioral	
Severe depression	128 (25.3)
"Extreme irritability"	113 (22.4)
"Severe anxiety attacks"	92 (18.2)
"Marked personality changes"	79 (15.6)
Recent "severe insomnia"	66 (13.1)
"Severe aggravation of phobias"	34 (6.7)
Visual and auditory	
Decreased vision and/or other eye problems (blurring, "bright flashes," tunnel vision)	121 (24.0)
Pain in one or both eyes	44 (8.7)
Blindness in one or both eyes	12 (2.4)
Tinnitus ("ringing," "buzzing")	65 (12.9)
Marked impairment of hearing	23 (4.6)
Myasthenia gravis (ptosis)	7 (1.4)

Related Observations

Reactions to Small Quantities

Some reactors evidenced severe symptoms and signs after chewing, or ingesting, especially in hot liquids, small amounts of aspartame.

Latent Period

Aspartame reactors frequently had latent periods of from several weeks to months between beginning aspartame and the onset of neuropsychiatric symptoms. Improvement occurred when aspartame was stopped. The subsequent consumption of aspartame after such "priming," however, led to recurrent "attacks" within hours or several days—a sequence reminiscent of allergic drug reactions. Some patients experienced severe reactions within minutes to an hour after their apparent *first* contact with aspartame.

Familial Ramifications

The familial incidence of aspartame reactions was 21.4%. From two to five family members were affected. Accordingly, relatives of aspartame reactors should be considered at higher risk for neuropsychiatric complications—especially when the inheritance is bilateral.

Multiple Reactions

Most aspartame reactors with neurological and psychiatric features suffered reactions in multiple body organs. These symptoms, signs, and associations have been detailed elsewhere (Roberts 1987).

Costs

The medical expenses incurred by failure to appreciate the neuropsychiatric manifestations of aspartame reactions were high. They averaged $1,719 out of pocket, and $4,647 for insurance carriers.

Clinical Caveats

Physicians must question *every* patient presenting with the foregoing neurological, psychiatric, and behavioral disorders about aspartame use. When it exists, they should recommend a brief trial of abstinence *before* ordering consultations and expensive tests, especially if invasive (e.g., carotid arteriography.) This is particularly applicable to patients with "refractory" seizures, headache, atypical facial or eye pain, depression, and the Meniere syndrome.

Younger patients who express concern about "early Alzheimer's disease," based on recent confusion and significant memory loss, should *not* be given this ominous diagnosis—or that of primary depression—until they have been observed after cessation of aspartame.

High-Risk Groups

Groups at high risk for aspartame reactions warrant special attention. They include pregnant and lactating women, young children, older persons, individuals with prior alcoholism or drug abuse, patients having iron-deficiency anemia, liver disease, renal failure, migraine, hypoglycemia, diabetes mellitus, those at risk for phenylketonuria, and patients taking drugs that could interact with phenylalanine (e.g., L-DOPA, monoamine oxidase inhibitors, and α-methyldopa).

Acknowledgments. The computer assistance of Lillian Capote, Kellie McDaniel, and George Leavitt (Management Information System, Good Samaritan Hospital) and support by Muriel Brenner, Robert Sanders, and Merrill Bank are gratefully acknowledged.

References

Roberts, H.J. (1987). Is aspartame (NutraSweet®) safe? On Call (official publication of the Palm Beach County Medical Society). January, pp. 16–20.

Chapter 46

Public Policy and Food Additives: The NutraSweet Controversy

James Wagoner*

In 1985, Senator Howard Metzenbaum introduced the Aspartame Safety Act. The legislation responded to the growing concerns being expressed in the scientific community regarding the safety of aspartame, better known under its brand name "NutraSweet." At the present time, the Senator is considering a revised version of the bill for introduction in the current Congress.

NutraSweet was approved by the Food and Drug Administration (FDA) in circumstances which can only be described as controversial. The FDA originally approved the sweetener in 1974. However, the decision was stayed after concerns were raised about health and safety problems. In March of 1976, a special FDA task force released its report on testing practices at G.D. Searle Co., the manufacturer of aspartame. That report contained the following conclusions:

> Through our efforts, we have uncovered serious deficiencies in Searle's integrity in conducting high quality animal research to accurately determine or characterize the toxic potential of its products.
>
> . . . The studies we investigated reveal a pattern of conduct which compromises the scientific integrity of studies.

Of course, one might ask what a 1976 report on testing practices at G.D. Searle could have to do with NutraSweet, a chemical sweetener approved by the FDA in 1981? The answer is simple. Over 90% of the tests submitted by G.D. Searle to the FDA in order to get aspartame approved were submitted prior to March, 1976, when the report was issued. In addition to the 25 Searle tests examined by the FDA task force, 11 were tests done on aspartame. Serious questions remain about the quality of tests used to approve this chemical sweetener.

The questions do not stop with the 1976 task force report. For in 1977, the FDA wrote to the U.S. attorney in Chicago requesting a grand jury

*Legislative Aide to Senator H.M. Metzenbaum, United States Senate, 140 Russell Senate Office Building, Washington, DC 20510, USA.

investigation of G.D. Searle Co. I quote from the letter sent by the chief counsel of the FDA, Richard Merrill:

> We request that your office convene a grand jury investigation into apparent violations of the Food, Drug and Cosmetic Act . . . and the False Reports to the Government Act, by G.D. Searle and Company and three of its responsible officers for their willful and knowing failure to make reports to the Food and Drug Administration required by the Act, and for concealing material facts and making false statements in reports of animal studies conducted to establish the safety of the drug Aldactone and the food additive Aspartame.

In 1980, the FDA established a public board of inquiry on aspartame. What did they conclude? "The Board has not been presented with proof of a reasonable certainty that Aspartame is safe for use as a food additive under its intended conditions of use."

In May, 1981, two months before the FDA Commissioner, Arthur Hayes, approved aspartame for use in dry foods, three FDA scientists informed the Commissioner that they did not believe that aspartame had been proven safe beyond a reasonable doubt. They questioned the reliability of key brain tumor tests which were submitted by G.D. Searle.

Despite all the questions raised by the chronology I have outlined, the FDA Commissioner decided to approve aspartame in July of 1981. He later approved aspartame for use in soft drinks in July, 1983.

In August, 1985, Senator Metzenbaum introduced the "Aspartame Safety Act." The bill mandates independent tests on aspartame focusing on the general effects which aspartame has on brain chemistry as well as the specific behavioral and neurological reactions allegedly experienced by some individuals—headaches, mood alterations, memory loss, et cetera.

The tests will also examine the health effects of aspartame on pregnant women and fetuses and whether aspartame consumption can lower the threshold for seizures. Another important area for investigation is how aspartame reacts to medicines, particularly MAO inhibitors which are used in the treatment of depression, L-dopa used in the treatment of Parkinson's disease, and aldomet which is used in the treatment of hypertension.

Under the bill, there would be a moratorium imposed on new uses of aspartame in food and drugs pending the completion of independent tests or for the period of one year—whichever comes sooner.

The "Aspartame Safety Act" responds to the important questions which have been raised by eminent scientists regarding aspartame's safety.

Dr. Richard Wurtman of MIT has raised questions relating to aspartame's effect on brain chemistry, Dr. William Pardridge of UCLA has expressed his concerns about fetal IQ, Dr. Elsas of Emory University has warned us about groups in the population at large. Dr. Matalon at the University of Illinois is particularly concerned about individuals who are genetically susceptible to phenylalanine—PKU carriers—and who may be a sizable risk group as far as aspartame is concerned. Nearly 5 million Americans are PKU carriers.

Two researchers in Philadelphia, Professors Gautieri and Mahalik, have done studies on mice which show that aspartame affected the vision of newborn mice whose mothers had been exposed to the chemical sweetener.

The FDA is content to have the manufacturer of aspartame, the NutraSweet company, conduct studies on these issues. How absurd. We do not need the company which is making millions of dollars on aspartame telling us it's safe.

In addition to mandating independent tests, the Aspartame Safety Act requires labeling which will inform consumers how much aspartame they are ingesting. This information is important not only for consumers who wish to regulate their intake of aspartame, but also for physicians who may be treating individuals who feel they have experienced side effects. Such side effects are likely to be dose-related, and the physician will want to know how much aspartame has been consumed. In addition, consumers have a basic right to know the makeup of the foods which they consume.

The label will also contain the maximum allowable daily intake established by the FDA. How many consumers even know that the FDA has attached such a limit to aspartame consumption? The current ADI is 50 mg per kg of body weight. It was originally 20 mg/kg. However, in 1983 the FDA decided to ignore its standard 100-fold safety factor by more than doubling the maximum allowable daily intake. Why did they decide to make an exception for aspartame? In 1983, the agency approved aspartame for soft drinks, so they decided to increase the limit knowing consumption was bound to increase. The justification the FDA used for violating its standard 100-fold safety factor was that the tests showed it was safe at the new levels of consumption. The problem is that these are the tests which have been called into question over the years.

Of course, one could agree that if you weigh 130 pounds you would have to drink 4 to 5 liters of diet soft drink to hit the limit. But if you are a child who weighs 30 pounds, you hit that limit with 3 to 4 cans of diet soft drink. That's even without the gum, pudding, breakfast cereal—all sweetened with aspartame.

Under this bill, the Secretary of Health and Human Services will be responsible for deciding how best to express the ADI on the label so consumers can understand what it means. For example, on diet soft drinks the label might read: "Maximum Allowable Daily Intake: 3 cans per 25 lbs. of body weight." There may be better ways to express this concept. The Secretary can work on that, but consumers have a right to this information, particularly given the advertising for this product which has left the impression that everyone in the population, including children, can consume as much as they want of this chemical sweetener.

Another labeling issue is whether a warning should be required concerning the use of aspartame by pregnant women and infants.

The Aspartame Safety Act requires the FDA to establish a clinical adverse reaction review committee, and to notify physicians that the agency is interested in monitoring consumer complaints regarding NutraSweet. To its credit, the agency has moved on its own initiative in this direction and has been monitoring adverse reactions for over a year. While more work needs to be done in this area, the agency has at least recognized that there needs to be an organized response to consumer complaints.

The public policy issues raised by the NutraSweet controversy will continue to be a focus for attention in the 100th Congress. With Americans today consuming over 20 billion cans of diet soft drinks, the vast majority of which are sweetened with 100 percent NutraSweet, it is critical that the concerns which have been raised about the safety of this product are resolved. In addition, the NutraSweet debate has caused many to question whether the current law regulating the approval of food additives is adequate.

As new and more chemically complex food additives appear on the market, the argument has been made that they should fall under the more extensive testing requirements which govern over-the-counter drugs.

One thing is certain. The debate surrounding NutraSweet will have a major impact on the future regulation of food additives in this country.

CHAPTER 47

The Regulatory Process and Aspartame: Why the Controversy?

James R. Phelps*

Ten years' service as a lawyer with considerable involvement in the regulatory issues that have touched aspartame probably qualifies me as a serious student of that subject. As such, it is puzzling to me that the issue of the approval of the product continues to be discussed. I must question the usefulness of continuing to debate a subject that has been so thoroughly explored and so profoundly put to rest. But the length and thoroughness of the process and the legal reviews of that process should make clear to all that this is no longer a subject worthy of attention.

The unwarranted discussion and controversy surrounding the approval process should *not* be confused with the valid discussion of scientific issues. Scientists must and will continue to discuss, test, and consider the implications of the effects of foods in the body. No one can desire limitation on the development and exchange of information that advances these scientific inquiries.

Unlike the scientific process, however, the discussion of the process by which aspartame was approved has certain end points that should, but seem not to, limit debate. My purpose in this chapter is to present the *complete* story on the approval process—including those matters seemingly ignored by critics of the approval process. It is for the reader to consider, given *all* of the information, why the approval process remains the subject of controversy.

A starting point for such a discussion seems nearly always to be the validity of the scientific work done to support aspartame's safety. Critics of the approval process frequently speak about the report of a team of Food and Drug Administration (FDA) investigators in the mid-1970s, shortly after the 1974 approvals of aspartame, which raised questions, with ugly implications, about the sponsor's research in general and about some of the aspartame studies. This is the so-called "Task Force" Report. What is

*Hyman, Phelps & McNamara, P.C., 1120 G Street, N.W., Washington, DC 20005, USA.

often overlooked are the actions taken subsequent to the report, actions which subjected the aspartame studies used to support its approval to more rigorous scrutiny than that undertaken for any other food additive.

The FDA took the task force report very seriously. Before it would even review the data supporting approval and the objections to aspartame's approval, the agency required that all legitimate doubt about the validity of the studies should be resolved. A massive review of the supporting data was, therefore, undertaken. Fifteen studies were targeted as the pivotal studies for the approval of aspartame. Three of the studies were examined in detail by FDA personnel. The remaining twelve studies were examined by an organization called Universities Associated for Research and Education in Pathology (UAREP). Neither FDA nor UAREP can fairly be characterized as an amateur organization in this kind of work. UAREP operated without any intervention by the sponsor, except insofar as the sponsor was told to provide information. FDA operated independently, as always with projects of this type.

The process of review took over two years. The UAREP report was over 1000 pages of documentation. FDA concluded that the three studies reviewed by the agency were authentic. UAREP concluded that the twelve studies it reviewed were authentic, and FDA concurred.

At no time before or since has FDA involved itself as deeply in the underlying data upon which the safety of a product has been demonstrated. And while we hear critics of the aspartame approval process talk about the "questions" raised by the process, I have never heard the plain fact that those questions were definitively answered. I have heard the task force report mentioned without even nodding acknowledgment of the UAREP and FDA reports that dispelled any legitimate doubts raised by the task force report. It is only by using this curious and illogical technique of ignoring key facts that critics of the regulatory process are able to keep their expressed concerns about that process alive.

Of course, following the UAREP and FDA reviews of the underlying data, FDA went on to hear the objections filed to the 1974 aspartame approval. To hear those objections, the parties agreed on the form of the Public Board of Inquiry (PBOI). I will not go into detail on the hearings before the PBOI and its subsequent decision. The PBOI agreed in part and disagreed in part with the objections. From there, the matter went for ultimate decision to the Commissioner, who agreed with the PBOI in its rejection of certain of the objections. The Commissioner, however, overruled the PBOI with respect to the issue of whether there was sufficient evidence to determine that tumors were not caused by aspartame in certain animal studies.

Again, critics of the aspartame approval process make much of the disagreement between the PBOI and the Commissioner. However, these criticisms again must ignore a number of facts.

First, the criticisms ignore the substantial evidence and scientific support upon which the Commissioner relied in making his decision. While people may have disagreed as to the decision at the time, *no one* has suggested that the Commissioner did not have valid scientific basis for his decision.

The facts at the time and since support that decision. What is often overlooked is that the objectors *did not appeal* the order of the Commissioner. The objectors obviously did not find the Commissioner's decision one worth appealing.

Also overlooked is the fact that the single issue on which the PBOI recommended approval for aspartame be withdrawn for further study has *never*, to my knowledge, been the subject of scientific dispute since that time. Criticisms of the approval process often have implied that the PBOI report constituted some general opinion that aspartame was not safe. In fact, the PBOI was hearing a limited set of issues, on only one of which it disagreed with the sponsor. And on *that* limited issue, no valid scientific question has been raised since. As a result, it is difficult to see how the PBOI justifies any criticism of the ultimate approval of aspartame.

As I stated, no legal challenges were filed to the 1981 approval (the so-called "dry use" approval) of aspartame. The aspartame approval *was* challenged by the Community Nutrition Institute following the 1983 carbonated beverage approval, and the results in the courts confirmed yet again the thoroughness and accuracy of the FDA review of aspartame. I will not trace in detail the history of the case before the Federal District Court, the Court of Appeals, and the Supreme Court ruling. The objectors were heard in full legal challenge, and that challenge was turned down. The most comprehensive opinion was written by Judge Mikva of the United States Court of Appeals for the District of Columbia Circuit. (Judge Mikva, of course, has a well-recognized record as a consumer advocate when a Congressman from Chicago.) This judge wrote an extensive opinion in which he made a point-by-point denial of every assertion made by the objectors.

Critics of the approval process tend not to discuss these decisions, yet the courts had before them *all* the information to that date, including the task force report.

The process of regulatory review has not ended with the approvals and the court decisions. In response to allegations of the Community Nutrition Institute about "alarming" numbers of medical complaints allegedly the result of ingestion of aspartame, the FDA and the Centers for Disease Control (CDC) undertook a monumental program to analyze data about consumer complaints. Significantly, the CDC report showed that the complaints were in general typical of the type of complaints common in the population as a whole. CDC concluded that while it could not rule out the possibility that someone could have an undefined "sensitivity" to aspartame, the complaints did not establish any cause-effect relationship to aspartame, and certainly did not present any evidence of a significant health hazard.

Again, critics of the approval process frequently cite the anecdotal report of complaints, without citing the review of those complaints, and the CDC conclusions. In short, critics of the approval process cite the existence of "questions" without noting the fact that answers already have been given.

However, regulatory review of aspartame has not ceased with the court decisions or the CDC report. Monitoring of the intake of aspartame, reviewed by the FDA, has confirmed that consumption of the aspartame is well within the allowable daily intake (ADI) set as a guideline by the FDA. This gives persons concerned with possible "abuse" of aspartame confirmation that such concerns are unwarranted.

One would believe that such questioning of the safety of aspartame in the regulatory process, with the resultant answers that the product is safe, would provide ample evidence that the regulatory process works, and works well.

Apparently, this is not the case, for people continue to cite the "questions" surrounding aspartame's approval, without acknowledging the answers received at every step in the process.

In addition to all the foregoing, the approval process has remained to this very day under scrutiny of the Congress because of the interest of Senator Metzenbaum. The Senator has listened to critics of the process and exercised his prerogative as a Congressman to initiate an investigation of that approval process by the general accounting office (GAO). The GAO spent many, many months interviewing personnel at FDA and elsewhere and looking at the records involved in the aspartame approval. GAO thoroughness is legendary, and from the outside, it appears that they left no stone unturned in the aspartame investigation.

The GAO's report has been prepared, and it appears that once again the process of approval has been confirmed as proper. It appears that the report will conclude that proper procedures were followed throughout the regulatory process, and that the agency made a thorough response to issues of study validity and safety. It appears that the GAO has no criticism of the regulatory process.

If the court actions did not do so, Senator Metzenbaum's GAO report should, once and for all, satisfy everyone that the proper, full regulatory process was followed. The actions of the GAO, UAREP, the United States Judiciary, the FDA, and the regulatory bodies of the more than 50 countries which have approved aspartame confirm the validity of the process. No other regulatory action in history has had such consistent confirmation by so many responsible reviews.

Nevertheless, legislative activity to overrule the regulatory process remains a possibility. Of course, the Aspartame Safety Act of 1985 was one such piece of legislation. In light of the frequent confirmations of aspartame's safety, such activity seems to be a continuing and useless distraction.

The taxpayers pay hundreds of millions of dollars for the FDA. The people who run that agency are given the responsibility to make decisions to implement the Federal Food, Drug, and Cosmetic Act. That is what they have done with aspartame, and the courts, GAO, and anyone else who has looked seriously at the process have given it their blessing. What else should reasonably be required?

Scientific inquiry about all foods will continue to be the proper subject of investigation and debate. But there can be no reasonable doubt that the approval process, so frequently audited and seconded, has worked properly.

Part X Summary

Conference on Dietary Phenylalanine and Brain Function, May 8–10, 1987, Washington, DC

Richard J. Wurtman*

The Conference on Dietary Phenylalanine and Brain Function has considered three sets of issues—scientific, medical, and regulatory—relating to possible effects of aspartame and of pure phenylalanine on the brain. This chapter is a personal assessment of the status of those issues.

Scientific Issues

Compelling evidence has been presented that the consumption of aspartame or phenylalanine raises the plasma phenylalanine concentration and the plasma phenylalanine ratio (Phe/LNAA) (Caballero, Filer), and thereby elevates brain phenylalanine levels (Fernstrom). Moreover, this rise in brain phenylalanine can suppress the synthesis and release of dopamine (During) and perhaps other monoamine neurotransmitters (Lou). Whether or not a decrease in monoaminergic neurotransmission actually occurs after aspartame or phenylalanine consumption depends on the species and dose being tested: in experimental rodents, relatively low doses of aspartame or phenylalanine (e.g., 200 mg/kg) cause greater increases in plasma and brain tyrosine than in phenylalanine (because the livers of such animals exhibit very high phenylalanine hydroxylase activity, and thus rapidly convert most of the ingested phenylalanine to tyrosine). Consequently, monoaminergic neurotransmission (e.g., brain norepinephrine release) can actually be enhanced. However, in rodents receiving high aspartame doses (e.g., 1000 mg/kg or greater) and, probably, in humans receiving any dose, blood and brain phenylalanine levels rise by more than those of tyrosine (Maher), and the phenylalanine suppresses dopamine synthesis (by inhibiting the enzyme tyrosine hydroxylase and by competing with tyrosine for uptake into the neuron). This species difference in the metabolism of phenylalanine is critically important to keep in mind when assessing the

*Department of Brain and Cognitive Sciences, Massachusetts Institute of Technology, Cambridge, MA 02139, USA.

relevance of particular animal studies in predicting human responses to aspartame or phenylalanine (Wurtman).

Administration of aspartame or phenylalanine has been found in some studies (Spiers, Elsas, Krause) but not all (Wolraich, Lieberman, Elsas) to affect cognitive and attentive behaviors or electroencephalographic patterns in apparently normal people [i.e., true normals and asymptomatic heterozygotes for phenylketonuria (PKU)].

Aspartame or phenylalanine can also act as a "co-epileptogen" in rodents treated with seizurogenic compounds (Maher, Garattini, Kim) or by kindling (Tilson). It is ineffective in some other animal models for human seizures (Jobe, Garattini). Tyrosine tends to be an antidote to this "co-epileptogenic" effect, as well as to the inhibition by phenylalanine of brain monoamine synthesis in animals (During) and humans (Lou).

The biochemical mechanism of phenylalanine's neurotoxicity (i.e., in PKU) may involve the brain enzyme ATP-sulfurylase, which is important in the biosynthesis of sulfatides (Hommes). At present, no information whatsoever is available concerning the threshold concentrations of phenylalanine in plasma or brain at which the amino acid begins to have cytotoxic effects or cause neuronal loss.

Medical Issues

Abundant anecdotal reports (summarized by Tollefson) describe an apparent temporal association between aspartame consumption and neurological signs and symptoms (Johns, Schomer); however, to date, no studies have been published describing well-controlled studies. One preliminary report described an association between aspartame and headache (Koehler); another failed to observe such a link between aspartame and seizures (Shaywitz). Hence, it cannot presently be determined whether aspartame has been a causative factor in these neurological syndromes. It seems critically important that such studies now be done.

Inasmuch as most of the anecdotal case reports that associate aspartame consumption with neurological sequellae describe these sequellae as occurring only after prolonged use of the sweetener, clinical studies designed to examine aspartame's role should be chronic (i.e., of at least several weeks' duration). Moreover, they should provide the sweetener according to a "worst-case" but reasonable protocol, e.g., as though it were being consumed in a pitcher of cold beverage being drunk on a hot afternoon, either without food or with foods (carbohydrates) that potentiate phenylalanine's uptake into the brain.

Regulatory Issues

At present, companies marketing products containing aspartame or other food additives are not required to file reports with the Food and Drug Administration (FDA) describing possible adverse reactions to their products. Nor, of course, do such companies (or the FDA) routinely encourage physicians to enquire about such adverse reactions (Hattan). It seems critically important that the law be changed to make such filing mandatory: as this meeting has demonstrated, once a sufficient number of reports have been received describing a particular set of responses (Tollefson), it then becomes possible to design and conduct controlled clinical studies to test the association between the additive and the reaction.

At present, laws regulating the use of new food additives do not require periodic safety updates (analogous to "Phase IV" data on new drugs) from manufacturers of these additives. Moreover, there seems to be no established mechanism for incorporating into such safety updates new information that suggests previously unsuspected ways in which an additive might be unsafe (for example, the concept of co-epileptogenicity).

At present, manufacturers of products containing new food additives are not required to list on their labels the amounts of the additive that the food (or additive package) contains. Hence it is impossible for most consumers to choose the amount of an additive that they will consume. In other countries, such quantity labeling is routinely required (Eisenberg).

Pure phenylalanine—both the L- and the D-forms—is now routinely available in health-food stores, even though its efficacy in treating the conditions underlying its sale has still to be demonstrated, and its consumption has yet to be shown safe (Young). In Canada, steps have recently been implemented to remove phenylalanine from the shelves. The legal basis underlying the sale of phenylalanine in American health-food stores is uncertain (Yingling).

It is obvious and also not surprising that this conference—the first ever, to our knowledge, on "Dietary Phenylalanine and Brain Function"—has raised many more questions from all three realms than it has answered. But it seems clear that these questions are important, and that the scientific, medical, and regulatory issues that underlie them will be with us for a long time to come.

MIX
Papier aus verantwortungsvollen Quellen
Paper from responsible sources
FSC® C105338

If you have any concerns about our products,
you can contact us on
ProductSafety@springernature.com

In case Publisher is established outside the EU,
the EU authorized representative is:
Springer Nature Customer Service Center GmbH
Europaplatz 3, 69115 Heidelberg, Germany

Printed by Libri Plureos GmbH
in Hamburg, Germany